Land for the Future

by Marion Clawson, R. Burnell Held, Charles H. Stoddard

Land for the Future

published for RESOURCES FOR THE FUTURE, INC.
by THE JOHNS HOPKINS PRESS: BALTIMORE

Resources for the Future is a nonprofit corporation for research and education in the development, conservation, and use of natural resources. It was established in 1952 with the co-operation of The Ford Foundation and its activities since then have been financed by grants from that Foundation. Part of the work of Resources for the Future is carried out by its resident staff, part supported by grants to universities and other nonprofit organizations. Unless otherwise stated, interpretations and conclusions in RFF publications are those of the authors; the organization takes responsibility for the selection of significant subjects for study, the competence of the researchers, and their freedom of inquiry.

This book is one of RFF's studies in land use and management, which are directed by Marion Clawson. R. Burnell Held, agricultural economist, and Charles H. Stoddard, forest economist, are both research associates.

Staff Editors, Henry Jarrett and Vera W. Dodds

Distributed in Great Britain by Oxford University Press, London

Printed in the United States of America
by Franklin Printing Company, Primos, Pa.

Library of Congress Catalog Card Number 60-9917

Acknowledgments

In preparing this study of land use in the United States, the authors have had unusual assistance from a number of persons with special knowledge of one or other of the major land uses. Many published reports and books and a few unpublished studies have been drawn upon freely; these are cited in the text, but we do wish to acknowledge our debt of gratitude to the many professional co-workers whose studies have helped sharpen our perspective on both the historical past and the probable future of land use in this country.

Responsibility for gathering and analyzing the data and for writing the text was shared by the authors as follows: Chapters I, II, III, VI, part of Chapter VII, and VIII—Clawson, who directed the study as a whole; Chapter IV—Held; and Chapter V and part of Chapter VII—Stoddard. During the writing of the book we had the advantage of consultation with several of our colleagues at Resources for the Future. We wish especially to acknowledge the help of Joseph L. Fisher, Harvey S. Perloff, Lowdon Wingo, Harold J. Barnett, and Henry Jarrett.

When the manuscript was in draft it was reviewed with particular thoroughness by Raleigh Barlowe of Michigan State University, by Lee M. James of the same institution, and by Julius Margolis of the University of California. Each of these men wrote a detailed commentary on the manuscript in that form.

In addition, useful comments were received from Conrad Taeuber,

Robert C. Klove, Ray Hurley, and Clarence E. Batschelet, of the Bureau of the Census; from E. H. Holmes of the Bureau of Public Roads; from E. H. Wiecking and Hugh H. Wooten, of the Agricultural Research Service; from H. R. Josephson of the Forest Service; from F. G. Renner and T. B. Plair, of the Soil Conservation Service; from Sidney S. Kennedy of the National Park Service; from H. R. Hochmuth of the Bureau of Land Management; from Rex D. Rehnberg of Colorado State University; from C. W. Loomer of the University of Wisconsin; from J. H. Blackstone of the Alabama Polytechnic Institute; from Walter J. Mead of the Santa Barbara College of the University of California; and from Louis B. Wetmore of the University of Illinois.

To all of these men, and to any others whom we may inadvertently have omitted, we gladly acknowledge our debt and extend our thanks. The book in its present form is certainly better for their criticisms and suggestions, but the authors are solely responsible for the conclusions drawn and interpretations made.

February 26, 1960

Marion Clawson
R. Burnell Held
Charles H. Stoddard

Contents

List of tables

APPENDIX TABLES

List of figures

APPENDIX FIGURES

Land for the Future

Introduction

In this book we consider the changing uses of land in the United States, historically, at present, and in the light of expectations extending to the year 2000. Our basic concern is the conflict between the demands of an expanding economy and a fixed area of land. Our aim is not so much to produce forecasts as to gain an understanding of continuing forces of transformation not new in history, but which in this young nation have taken on a pace and magnitude without precedent in any other land or time.

Our approach is three-pronged. Following these introductory pages, which explain our assumptions and the bases for our estimates, the first chapter provides background for more detailed analysis by looking at the changing patterns of land use in the United States in broad perspective and general terms. This chapter constitutes the first major part of the book.

The second part consists of a series of six chapters, each examining in detail an important use of the land. The land use first considered is urban—for cities, suburbs and the like. This is our most intensive use of the land today, one that is growing rapidly. Chapter III deals with land for recreational use. Here, largely because the pertinent data are lacking for privately owned land, we focus our attention upon the public recreation lands. Closely associated with urbanism, public recreation is another rapidly growing and intensive land use. Agriculture, as we define it, is not synonymous with farming; and so in Chapter IV we consider cropland and improved pasture, together with the necessary building areas. Forestry, which includes farm, public, and commercial forests, is treated in a separate chapter. So also is land for grazing, with its extensive area, mixed ownership, and special characteristics and problems. The last of the six detailed land use chapters deals with a number of uses which, while important as to function, are relatively minor uses of the land. Such are the reservoirs, wildlife areas, transportation routes, mining areas, and so on, which, for want of a better term we call "miscellaneous."

The analysis of each of these major land uses is based upon the total economic and social situation assumed to prevail in the United States during the time period of our study. Each of the six chapters considers broadly the total demands likely to be made for land, but each focuses on the particular land use which is its subject, examining the various aspects of that use and the problems that they raise. However, these chapters do not try to resolve conflicting demands between one use and another; it is altogether possible that the sum of the individual uses will be greater than the total area of land available to meet them. The purpose of this approach is to show what reasonably may be considered the demand for land for each major purpose, taken by itself but in light of the total economy and total demand for land.

In the third major part of this book, the concluding chapter, we reconsider and attempt to reconcile the various competing and conflicting uses; discuss the possibilities and the limitations on multiple uses of the same land to meet these many requirements; and summarize the shifts, the conflicts, and the problems which the detailed analysis has shown to be probable or, in our opinion, desirable.

In preparing this inquiry into land use in the United States, we have used data from many excellent sources. As often happens in research of this nature the definitions used are not always consistent between the various sources. There has been a tendency for each scholar or group concentrating upon a particular land use to consider it almost in isolation and to group other uses sometimes into incongruous and inconsistent categories. There is reason to suspect that certain kinds of land use are counted twice or more, and that others may be counted not at all. We decided to treat this problem —inevitable in even the best of data—by pointing out any known or suspected inadequacy as it arises in the course of the study. The six detailed chapters, therefore, make no specific attempt to reconcile differences and resolve overlapping. That task we postpone until the final chapter, where, returning to the broader perspective of Chapter I, we look at all the uses of the land together and as a whole.

ASSUMPTIONS AND PROJECTIONS

Data or judgments about the future may be presented in one of several ways. On the one hand, there is the *forecast*, a judgment usually in quantitative terms, of what is most likely to happen in the future. The forecaster weighs all the factors he thinks are relevant, and comes up with his estimate of what is most likely to happen. In many instances, he proceeds to act upon his forecast—buys wheat if he thinks wheat prices are going to rise and so on. He may or may not reveal the elements that entered into his forecast.

On another level of conjecture is the *projection*. Starting from given or assumed present positions, and proceeding on the basis of assumptions of future events, the course for the future of the matter under study is projected. There may, for instance, have been a past trend in the yield of some crop, which is extended mechanically or with given change into the future; or the trend in consumption of some raw material, as lumber or cotton, may be likewise extended into the years to come.

Population—a matter of first moment in guessing at future demands on land—may be projected upon the basis of certain assumptions as to birth rate, death rate, immigration and the like. Given the base from which the projections start, given the assumption as to the future course of certain "independent" variables, and given the relationships among the variables and the factor under study, the projection is a more or less mechanical affair. Different workers, if equally competent, would come up with identical answers.

A chief objection to the method of strict projection is that it gives in itself no measure of reasonableness; the assumption may have been most unlikely. A major difficulty has arisen in the past because projections have been accepted by uncritical readers as predictions. The Bureau of the Census has made projections of future population, for instance, based on certain assumptions as to birth rate and other factors. No one can quarrel with the projections, given the assumptions. But were the latter realistic and likely? The Bureau of the

Census has tried to avoid the connotation of forecast, yet its projections have often been used in this way.

A third approach to the future, less intuitive than a forecast and less rigid and dogmatic than projections, might be called an approximate projection, or simply an *estimate.* That is the approach and method we follow in this book. We have made projections based on certain factors, and will try to make clear the basis of our choice of factors; but in some instances the projection may be rough or rounded, rather than exact. We have chosen assumptions which we believe reasonable; yet we prefer to consider our figures for the future not as forecasts but as estimates.

Detailed and accurate forecasts of various aspects of the economy and society at future dates would be desirable for an analysis such as is attempted in this book. However, they are impossible and unnecessary. They are impossible, because no one today can *know* what the future holds, even though some reasonable assumptions may be made. But there need be no despair because precise and completely reliable forecasts of the future are impossible. The relationship between such factors as total population, per capita income, amounts of leisure, and other factors, on the one hand, and the areas of land and quantities of resources required from land, on the other, has been by no means fixed and invariable in the past, nor is it likely to be so in the future. We cannot even be sure what effect a given increase in real income per capita has had upon the demand for outdoor recreation, for example, or upon lumber consumption; and there are many reasons for believing that past relationships, even when known reasonably well, will not continue without change in the future. Consistent and reasonable estimates of future population, income, and other factors are necessary, but high accuracy and great detail are not.

Moreover, if the estimates of direction of change are right and the estimates of rate of change are not too grossly in error, the chief consequence of some errors in rate may be the date at which the anticipated situation occurs. As an example, if a population of 300 million were projected for the year 2000 and a land use analysis built upon that projection, while in fact that population came into existence in 1990 or 2020, it might well be that the land use situation would coincide with one date or the other. The date at which different situations and relationships will come into existence is important, particularly for investment decisions; yet, if the fact of the situation and the relationship can be accepted, considerable flexibility as to timing exists.

In this connection, it may be useful to distinguish between those future developments that seem most likely to be "time oriented" and those that more probably will be "population oriented." There are certain economic and other trends under way that would seem to be closely correlated with the passage of time; we judge that the development of agricultural technology and consequent changes in agricultural output from the same area of cropland is basically time oriented. Change in food habits and, consequently, in per capita consumption of certain foods, also falls in this category; so does timber growth, even assuming constant technology.

On the other hand, some trends or situations seem more probably population oriented, in that they depend more upon the numbers of people than upon any other factor; the demand for agricultural commodities seems clearly to fall in this category. In the past, when population has been increasing rather steadily, the relative effects of these two forces are not always clear, in many situations both, in fact, may operate more or less equally. However, if the rate of population growth should change, the importance of these influences might diverge. To the extent that time-oriented and population-oriented factors move at the same rate of speed, then it may not be significant which of the two is the more important; and the date at which a particular population and related situation comes into existence matters very little, because the relationship among the variables is the same whether the year is 1990 or 2010. But to the extent that time-oriented and population-oriented factors move differently, their relative importance becomes a matter of concern, and substantial divergences might develop between the various factors by a given date.

One final comment on long-range projections needs to be made. When we present estimates for a future year, these are for trend or normal situations in that year; they are not intended to fit the precise economic and other conditions that may have arisen. Thus, depression and boom years are not included, but only normals and long-term trends.

As for possible alterations to tangible factors that have definitely entered into our demands on land in the past, we assume that immigration will continue at roughly present levels. There are undoubtedly many people in the world who would come to the United States, were there no restrictions on their doing so. It is doubtful if we shall permit many more to enter in the future than have entered in the recent past; yet it is conceivable that world events might take a turn that would lead us to admit many more immigrants in the future.

There is, further, the question of effective foreign demand for our products, and our future policies of export trade. In the event of a continued impasse, a diminution of export outlets, and a general disinclination in this country to lower tariffs and accept more imports, we incline to assume that there will be little if any change in the immediate situation, short of emergency exports in another major war. On the whole, the future situation as to world trade seems generally unpredictable, apart from a probable intracontinental increase in imports and exports between the United States and Canada. And it is not improbable that the soaring rise in our own population, present and projected, will measurably offset the decline in foreign customers, and to that extent diminish our push and prospect for overseas trade.

Population Estimates

Basic to any specific estimates of the future are data on probable or expected numbers of people, including often their ages, their regions or states of residence, and other related matters. Forecasting probable future population is a hazardous undertaking. Projection, starting from stated assumptions, produces impeccable results when handled competently, but leaves unanswered the matter of the reliability of the estimates that enter the formula. Projections and forecasts, often subject to a confusion of terms, have generally failed to square with reality; and most of the best projections of the past quarter century have been seriously in error.

The difficulties seem to have been of two major kinds. First, demographers have been concerned to develop neater, logically more plausible methods of projection. This is excellent, but the interest seems to have been centered mainly upon methodology, with relatively little concern for the dependability of the data put into the formulas. And second, too much emphasis seems to have been placed upon recent trends. In the late 1920's and most of the 1930's, this was upon the falling birth rates then evident. So convinced were demographers that the experience of these years was part of a long-term and inexorable trend that they continued to use such figures long after birth rates had in fact turned upward. Then, for several years, their estimates were based upon birth rates of a few years back, so that continuously their estimates proved too low. Now, the fashion is generally to assume continued very high birth rates, but on hardly more solid evidence than the fact that we now have them.

We review this whole matter at greater length in Appendix A. The key factor in population projection is birth rate: the effect of possible variations in birth rate is much greater than the effect of any reasonable variations in any other factor, and historically the variations in birth rate have been greater and less predictable than in other factors. If the demographers and others who have projected, forecast, or estimated future population have frequently been in error, this is primarily because we (in the mass) have been unpredictable in our reproductive habits. The birth rate in the United States today is primarily a psychological and sociological phenomenon, not primarily a biological one. We could, if we wished, materially increase or materially reduce present birth rates. As students of our own behavior, we do not fully understand why we have chosen to vary our birth rate as much as, in fact, it has varied in the past thirty or forty years.

In view of the generally unsatisfactory state of population projection, we feel that high refinement in method is pretentious, and possibly misleading to those who may accept it as more accurate than it is. Our estimates are based primarily upon past trends in population and upon changes in those trends. We suspect that our simpler estimates will prove to be as accurate as the more refined. They do not, in fact, differ more widely in their results from recent, technically more perfect projections.

As everyone knows, the population of this country has grown greatly and rather regularly, from the time of white settlement to the present. But often it is not appreciated how great that rate of population growth has been, and how it has changed over the decades. The relevant statistics are in Table 1 and are shown graphically in Figure 1. For the first six decades of the nineteenth century, population grew without slackening at a rate of 32 to 36 per cent per decade. Then came three decades, from 1860 to 1890, when the rate of growth, somewhat lower but still very high and in general regular, varied from 25 to 27 per cent per decade. These nine decades were the great pioneering period of the United States, with settlement constantly pushing west; they were also the decades of great and free immigration. For the next two decades the rate of growth was again somewhat lower but rather steady, at almost exactly 21 per cent per decade. It should be noted that the length of the period of approximately constant growth was less with each decline in rate of growth.

Table 1. Population for four major regions and for United States, 1790 to 1950, and projections to 1980 and 2000

(thousands)

Year	Northeast[1]	North Central[2]	South[3]	West[4]	United States	Increase in U. S. population in preceding decade (%)
1790	1,968	—	1,961	—	3,929	
1800	2,636	51	2,622	—	5,308	35.1
1810	—	—	—	—	7,240	36.4
1820	4,360	859	4,419	—	9,638	33.1
1830	—	—	—	—	12,866	33.5
1840	6,761	3,352	6,951	—	17,069	32.7
1850	—	—	—	—	23,192	35.9
1860	10,594	9,097	11,133	619	31,443	35.6
1870	12,299	12,981	12,288	991	39,818	26.6
1880	14,507	17,364	16,517	1,768	50,156	26.0
1890	17,407	22,410	20,028	3,102	62,948	25.5
1900	21,047	26,333	24,524	4,091	75,995	20.7
1910	25,869	29,889	29,389	6,826	91,972	21.0
1920	29,662	34,020	33,126	8,903	105,711	14.9
1930	34,427	38,594	37,858	11,896	122,775	16.1
1940	35,977	40,143	41,666	13,883	131,669	7.2
1950	39,478	44,461	47,197	19,562	150,697	14.5
1980[5]	55,000	65,000	80,000	40,000	240,000	[6]18
2000[5]	70,000	80,000	100,000	60,000	310,000	[6]15

[1] Includes New England and Middle Atlantic Regions as established by the Bureau of the Census.
[2] Includes East North Central and West North Central regions.
[3] Includes South Atlantic, East South Central, and West South Central regions.
[4] Includes Mountain and Pacific regions.
[5] Projections of authors; see text for discussion of basis.
[6] Approximate only.

SOURCE: U. S. Bureau of Census reports on population. Summarized in *Statistical Abstract of the United States* for any recent year.

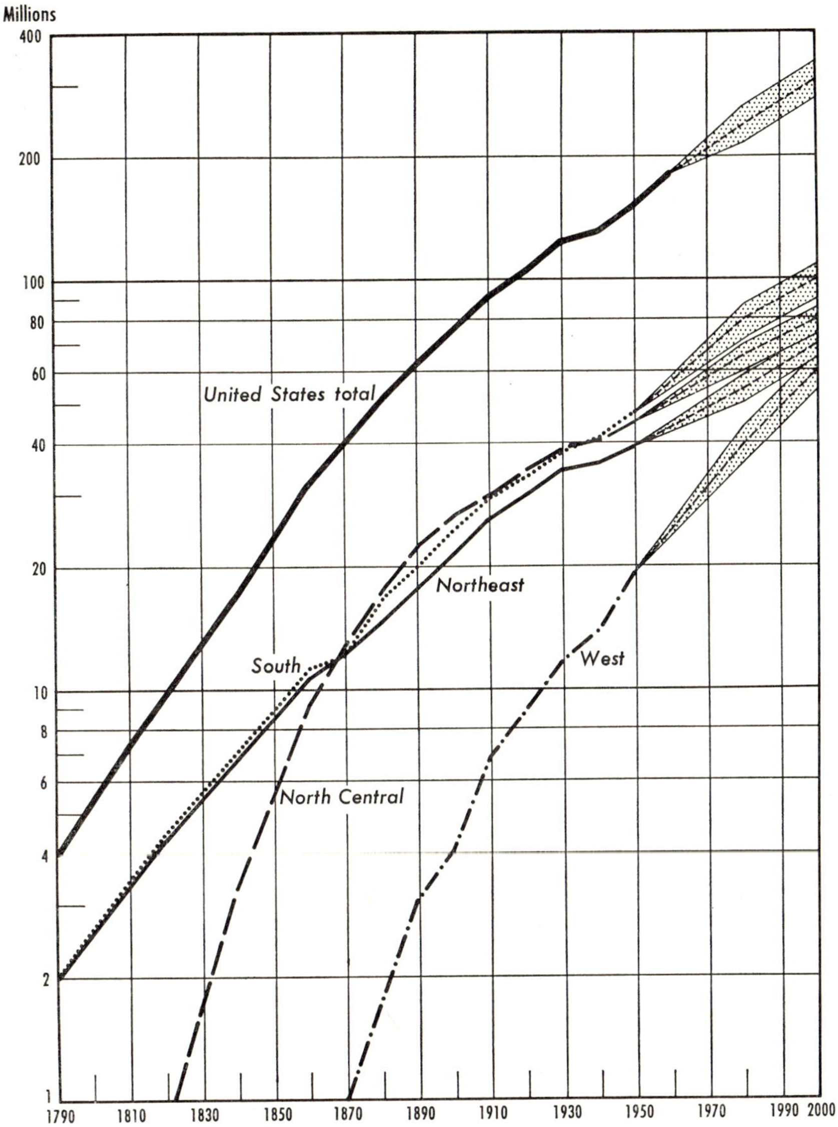

Figure 1. Population of the United States, by regions, 1790–1950, and possible growth to 2000.

The more modern period, from 1910 to 1950, shows again a diminution in rate of population growth and, for the first time, an irregularity in rate of growth—an increase after a decline. For three out of these four decades, the rate of population growth was between 14 and 16 per cent. The decade of the 1930's had a growth rate of only 7 per cent—the lowest in the history of the nation. At the time, it was easy to interpret this as but the most recent step in a continuously declining rate of population growth for the country, and the assumption of a stationary population at a relatively early date was plausible.

With the 1940's, a new rate of population increase began—a rate often hailed in popular writings as unusually high when it was, in fact, only a return to the rate that had prevailed in the two decades prior to the 1930's and was much lower than that prevailing in still earlier decades. The rate of increase of the 1940's has continued into the 1950's and shows no signs of slackening off. In late 1958 population was estimated to have reached the 175-million mark. That was a rate of increase since 1950 equal to about 18 per cent for a decade —a higher rate than has prevailed for a whole decade since 1910.

Population data for the entire United States do not tell the whole story; differences in regional growth are also important. These are shown in Table 1 and Figure 1 for the four major regions of the country. Rates of growth in the Northeast and the South have been remarkably similar since 1790. Beginning with 1880, the South has gradually pulled away from the Northeast, and by 1950 was roughly 20 per cent higher. In large part this is because the "South," as here defined, includes four states west of the Mississippi—Louisiana, Arkansas, Oklahoma, and Texas—which in earlier decades were frontier and which in recent decades have had a high rate of population growth. The North Central region, starting much later than the previous two regions, grew much faster than they did until 1870. From 1870 until 1950, the rate of growth in these three major regions was remarkably similar. The West started still later, and even to the present its rate of growth has been much higher than that of the other regions, although in 1950 its population was still less than half of that of any of the other three major regions.

In carrying regional estimates of population into the future, assumptions and forecasts as to internal migration become a matter of greater import day by day. Americans are a highly mobile people. Partly owing to the Second World War and the population migrations it stimulated or compelled through the armed services and the industrial effort, and partly because of continuing improvements in travel and communication, mobility has risen sharply in recent years. By current counts, something like 5 million people annually move from one state to another, and nearly as many move from one county to another within the same state. Over a period of years, a large proportion of the population moves from state to state, or from region to region; while some of these moves are offsetting, even for the same individuals, it is obvious that the net movement has been large and could be very great. In 1945 two out of three people in California had been born outside of the state.[1]

In the light of all such considerations, we have chosen as our estimates of the total population in 1980, 240 million; and for 2000, 310 million. These are certainly not the largest estimates that might be made with plausibility, nor are they the smallest. They contemplate for the period 1950 to 1980 a rate of net population increase larger than experienced in any decade since 1910, but about the same rate experienced thus far in the 1950's. From 1980 to 2000, the net rate of increase is assumed to be somewhat lower, roughly equal to that of the 1910-30 decades and the 1940's. It seems probable that mortality rates will fall somewhat and survival rates increase proportionately; so these estimates contemplate a modest decline in fertility rates.

In making that assumption, we are fully aware of the long-continued bias of demographers, who have almost universally argued that fertility rates will fall; and we are also aware that this bias has repeatedly been proved wrong over the past two decades. At the same time, we wonder if the current emphasis upon present or higher fertility rates may not prove to be equally in error. As our population grows, factors may well develop favoring slightly smaller families. Accordingly our estimates involve only a small decline from present fertility levels until 1980, and only a slightly greater decline thereafter. An estimate of these magnitudes for 1980 and 2000 means essentially that the experience of the 1930's is largely disregarded, but that the trend toward lower decade-rates of growth since 1790 is likewise disregarded.

To stress the uncertainties that hedge any system of population estimate, we have put a band of probable divergence in Figure 1 equal to roughly 10 per cent of the projected figure. This is surely the minimum estimate of possible error; the true figure may fall far

[1] Marion Clawson, "What It Means to Be a Californian," *California Historical Society Quarterly,* Vol. 24, No. 2 (June 1945).

outside of these limits. At the least, the estimate in Table 1 of 240 million for 1980 should be interpreted as meaning between 220 and 260 million, and that of 310 million for 2000 should be shaded between 280 and 340 million. Our rough projections are well above the highest figures in any Census projection published up to March 1958. New Census projections in process will, it appears, be higher than those made in the past; but it is not certain that their midpoint will be higher than our own estimates, nor seriously out of line with more refined projections made by specialists in this field.

Our estimates of future population for regions rest on the same approximate basis. Differences in past trends in numbers have been extended, more or less modified; and the totals by regions have been adjusted to our previous national totals. The greatest percentage increases would be in the West—over 100 per cent from 1950 to 1980, another 50 per cent to 2000, or 200 per cent increase in the fifty years. Even if these increases are achieved by the West, it would still at each period have less population than any other region. The next greatest increases are estimated for the South—70 per cent from 1950 to 1980, another 25 per cent to 2000, or 112 per cent for the fifty years. The increase in the West will be, to a major degree, from migration from other regions; the increase in the South will be from an excess of births over deaths, but modified by the amount of out-migration. Migration will play a major but differing role in these two regions, but it will also be a factor elsewhere. The increases estimated for the Northeast and the North Central regions are more nearly similar—39 to 46 per cent by 1980, and 23 to 27 per cent more by 2000.

The *number* of older persons can be estimated for 1980 and for 2000 without any assumption as to birth or fertility rates; assumptions only as to mortality or survival rates and as to immigration are needed. But the *percentage* of older people in the total population depends on the number of younger people born between now and 1980 or 2000. The age composition in any year well past 2000 will show some irregularity because of the low birth rates in the later 1920's and through the 1930's. In spite of this, many more people will fall into the older age class (65 years and older) in 1980 and in 2000 than now. In 1955 there were about 14.5 million people of 65 and over; by 1980 this is likely to rise to 24-26 million, and by 2000, to 30-35 million. At present, approximately 8½ per cent of the people are in this older age class; the proportion has risen steadily since 1900, when it was 4 per cent, due primarily to the lengthening life expectancy. If present birth rates should continue, the proportion of older people will rise somewhat by 1980 and fall slightly by 2000. If birth rates should decline in the future, the proportion of older people will rise considerably, possibly to 10 to 13 per cent of the total population. A large proportion of these older people will be retired or semi-retired, and will have at least tolerable incomes for modest living. For our purposes, a more significant fact may be that at least some of them will be comparatively footloose—able and willing to live almost anywhere in the United States where the costs of living are relatively low and the amenities of life more readily obtainable. Their numbers will probably have some effect upon the total demand for land, especially for urban and recreational purposes; they will have a much greater effect upon the regional demands for land, because of their loosened ties to their former locations.

The number of independent households in the country also will rise, perhaps more than proportionately to population increase. If relatively high incomes continue, young married couples will be less likely to double up with their parents than in the past. More of the older people will maintain separate households; there will probably be a large rise in single-person households. Older women will number more than older men, and many of them will maintain independent households. In part, this trend toward more single-person households will arise from the comparatively better income situation of older people, but in part it will stem from the better health conditions of old age.

These age and household factors are taken into account, at least in a general way, in our later analyses of the demand for land for various purposes. As suggested above, their greatest effect is likely to be in the demand for land for urban and for recreational purposes. It should also be pointed out that there is likely to be a strong trend toward ownership and maintenance, at least on a seasonal basis, of more than one living place per household. A comparatively large proportion of people in higher income brackets now own summer homes or cottages; more people will probably do so in the future, and the ownership of a winter as well as a summer vacation spot will become increasingly common. Moreover, if the work week shortens to four days or less, some proportion of the people will live in desirable but more distant locations and the worker will have a supplementary household near the place of employment. While these developments may affect only a small minority of the total population, in total they can amount to a considerable demand for land.

Income Growth and Distribution

Growth in income within a country over a period of time is affected by many factors—changes in total population, in proportion actually employed productively, in average hours worked per week, in output per hour, and perhaps by other future variants. Output per hour in turn is influenced by the rate of capital accumulation and investment, by changes in technology, and by other factors including the health and perhaps the leisure of the workers. Many social factors, such as the kind and amounts of schooling available to the average child, will also affect productivity over any considerable period of time.

An analysis of future changes in per capita income can be made on an intensive or more general basis. Partly because high precision in income estimates is not necessary to an analysis of land use, and partly because even the most refined income analysis involves some more or less arbitrary assumptions about future changes in past trends, we have chosen, as in our estimates of future population, the more general approach.

Fabricant has summarized his views on economic growth as follows:

> "The long-term rate of economic growth in the United States, measured by real national net product, has averaged about 3.5 per cent per annum. There is some evidence of retardation in rate of growth, although it is not conclusive. With respect to net national product per capita, the average rate of growth has been about 1.9 per cent per annum. Here, the signs of retardation are even fainter and may fairly be said to be negligible."[2]

If a growth rate of 1.9 per cent annually is extended from 1955 to 1980, per capita incomes in the latter year would be 55 per cent above those at present; and if extended further to 2000, per capita incomes then would be 45 per cent higher than in 1980 or 125 per cent higher than in 1955 (Table 2).

There has been some reduction in the disparity of incomes, from low to high income groups, in the past thirty years or so. The amount of change in income distribution depends in part on the particular

[2] Solomon Fabricant, "Growth and Stability in the General Economy," *Journal of Farm Economics,* Vol. 38, No. 5 (December 1956).

Table 2. Gross national product and personal income, 1950 and 1955, and estimated projections for 1980 and 2000

Item	Unit	1950	1955	1980[1]	2000[1]
Total population (July 1)	millions	151	165	240	310
		(current dollars)		(1955 dollars)	
Gross national product					
total	billion $	285	392	890	1,670
per capita	$	1,890	2,390	3,710	5,380
Personal income					
total	billion $	227	269	605	1,135
per capita	$	1,504	1,630	2,525	3,660

[1] Intensive studies also undertaken by Resources for the Future, Inc., as yet unpublished, lead to estimates slightly higher than these.

NOTE: Data for 1950 are taken from *Statistical Abstract of the United States, 1955;* for 1955 from *Survey of Current Business,* February, 1958. Projections are made from 1955 base, assuming a rise of 55% per capita to 1980 and a further rise of 45% per capita to 2000.

income definition that is used.[3] Most of the change in income distribution took place during World War II, and to a lesser extent between 1929 and 1941; since 1944 or 1946 income distribution seems to have been remarkably stable. In terms of 1950 prices, the number of families and unattached individuals with incomes below $2,000 has declined a third since 1929; those with $3,000 to $7,500 have increased almost three times; and those with over $7,500 also have increased about three times (total families and unattached individuals rose about 40 per cent). In terms of percentage of total national income, those with less than $2,000 in 1950 prices have declined from about 13 per cent to about 5 per cent of the total, while those with over $7,500 have increased their share very little or not at all. There has been some levelling of incomes, apparently primarily a "levelling-up," or an increase of the lower income groups rather than a decrease of the higher income groups.

Several factors have been responsible for or associated with this

[3] Selma F. Goldsmith, "Relation of Census Income Distribution Statistics to Other Income Data," in *An Appraisal of the 1950 Census Income Data,* Studies in Income and Wealth, Vol. 23, National Bureau of Economic Research (Princeton: Princeton University Press, 1958).

income change, among them: a mass market for many products, which in itself provides employment and relatively high purchasing power among income earners; increased bargaining power of organized labor, which has been able to channel more of the increased productivity to wages; federal, and to a lesser extent state, income taxes, which have taken a large proportion of the higher incomes.

For the purpose of this study, we assume that some further reduction in disparity in per capita net incomes will take place by 2000, but we do not make a quantitative estimate of its extent. If average per capita incomes rise approximately as we have estimated, and if there is some tendency toward a more equal distribution of incomes, then it is clear that the average incomes of those who now have the lowest incomes will rise considerably. This will, to some extent, affect the character of the demand for land and its products. A given total increase in national income will increase the demand for food, housing, and recreation to a greater extent if it is relatively larger in the low than in the high income classes.

Leisure

We consider it probable that the average citizen will have more leisure in the future than at present; estimates on this matter are presented in Chapter III, where outdoor recreation is our focus. This increased leisure will take several forms. For the worker, there may well be shorter hours per working day, or fewer working days per week, or longer paid vacations during the year. There will surely be a rise in the numbers of retired or semi-retired persons. With a large amount of leisure and with rising health standards among these older people, they will have more strength and interest to activate their leisure. The near abolition of child and youth labor, except in a few types of work, has greatly increased the leisure of this age group, but some further increases are probable. Continued extension of the average years of schooling will advance the trend in this direction. More equipment in the homes will mean more leisure for the housewife, or for the household servant, or will permit more wives to work. All these uses of increased leisure will vary according to age, income, location, and tastes of those concerned. By no means all of them will affect the use of land, but some of them will, and this seems particularly certain as to outdoor recreation.

Prices

In general, analysis in the following chapter is based upon a general price level approximating that of 1957. Within this assumed stable general price level prices of various commodities are not assumed to be stable. We recognize that there seems to be a persistent inflationary process at work in the United States today. However, this may either be brought under control, or its rate may be sufficiently slow to permit adjustments of various prices within the over-all structure to proceed fast enough to produce more or less the same relationship among prices that would exist if the general price level were stable.

Institutions

We assume that the major institutions of the United States will continue in the pattern of the past, with natural evolutionary but no radical changes occurring. We assume that private property rights in land will continue, and that shifts in land use will take place on the initiative or with the consent of the landowner, and will be carried out either by him or by those to whom he sells or leases, whether individuals or government. This does not preclude the use of zoning and other legally sanctioned public controls over private land use, for the public good. With the more intensive use of land resources that seems in prospect, somewhat more restrictive controls on individual land users may well be imposed by the public. That is part of the price of having a densely settled country. Further innovation and invention of social institutions we regard as both possible and desirable, provided that they proceed within the broad framework of accepted goals of our society.

I

The land in time and space

How can the United States best reconcile the demands that an advancing and dynamic civilization makes of land, within the framework of a still spacious but not illimitable natural endowment? How can this people in their increasing number resolve the problems that come of mounting pressures upon a given expanse of the earth? How can the all but insatiable wants and needs that we have come to regard as only our due in the American way of life be reshaped into designs better balanced for the attainment of ever higher standards?

To questions such as these there can never be final answers. Solutions will depend, primarily, on a clearer understanding of seemingly unalterable conditions and limitations ordained by nature, the dictates of climate in given locations, and by the not inexhaustible store of physical resources to be put to use. But beyond that, and even more difficult to calculate in terms of the future, is the human potential. There is no useless land. The ultimate resource is the resourcefulness of man.

The supply of land in the United States is obviously fixed in area,[1] and probably inexpansible; but the productivity of land depends not

[1] For reasons outlined on page 34, we exclude Alaska, Hawaii, and offshore possessions. Annexation of territory does not refute the basic area-productivity-demand relationships.

only on its area. Productivity depends on the inputs of intelligence, skill, and technology. Therefore, while the total area of land is fixed, the supply of land in any economic sense is not fixed. In this country we have been applying increasing capital and other inputs to agricultural and urban situations for many decades. We are rapidly developing more productive techniques for every use of land. There is every reason that we can expect to do so for many decades.

But there must be some limit to such processes. It is possible to extend almost any use of land over a greater area, by encroachment upon other uses. A considerable range of substituted uses and mingled or multiple uses is possible within certain areas. But it is manifestly impossible to expand the area in all uses of land at the same time.

Ways must be found that will lead toward a workable balance between hopes and probabilities and a reconciliation among competing demands. The chief concern of this study is to examine existing conflicts, to assess in terms of present realities and future prospects the most necessary and valid courses of change, and to suggest policies that may serve to minimize the impact of competing demand and need.

To that end, we take the broad, over-all approach, with the widest scope of concern. The land is considered as a whole. All sorts of land, and all the shifting patterns and clashing succession of uses to which land is put in an advancing civilization, are viewed as interwoven parts of the same spectacle. Land in a state of nature and land in process of cultivation or development is to be regarded ecologically and economically as one.

"Man-made changes upon the face of the Earth," said Aldo Leopold, writing in 1933, are . . . "largely the unpremeditated resultant of the impact between ecological and economic forces . . . If our system of land use happens to be self-perpetuating, we stay. If it happens to be self-destructive we move, like Abraham, to pastures new." For: "Civilization is not . . . the enslavement of a stable and constant Earth. It is a state of mutual and interdependent co-operation between human animals, other animals, plants and soils, which may be disrupted at any moment by the failure of any one of them."[2]

The validity of such an approach lies in a recognition of undying interrelationships that civilization may profoundly alter but can never

[2] Aldo Leopold, "The Conservation Ethic," *Journal of Forestry,* Vol. 31, No. 6 (October 1933).

completely ignore. The economic determinants that shift the patterns of human use and residence and shape them toward the future may be appraised in terms of accord or discord with an ectosystem or series of ectosystems of which man himself is a product and a part; the institutions and practices of a people may be evaluated in point of their possible or probable duration or survival in terms of time.

THE DIMENSIONS OF TIME

The present study centers in focus very largely on the sweep of events in the past fifty to one hundred years and on the prospects for the some forty years ahead. What have been the effects of the industrial and agricultural revolutions on our major uses of land? What next may be expected to come of the accelerating tempo of change, the increasing requirements of more people with more money and more leisure, and the shrinkage of spatial dimensions in terms of time?

Throughout the world in the past two centuries or so, and nowhere else more strikingly than in the United States, the movement of persons, of goods, and of thoughts has taken on an increasing velocity at an all but unpredictable rate. Less than two hundred years ago some of the most farsighted statesmen this nation has produced were concerned that we should ever be able to govern tolerably all of central North America as one country, simply because of the time required for travel and communication from one part to another. Prior to the telegraph, the railroad, and the steamship, travel or communication from Washington to New England was a matter of some days, and to the Pacific Coast a matter of weeks at the best. Now, well within the span of the past half-century, five days of transcontinental or transoceanic travel have been reduced by airways to as many hours; and reckonings of distance in miles-a-minute are in process of ever more rapid transition toward miles-a-second.

Some idea of what this has meant within the lifetime of middle-aged persons is charted in Figure 2 for the years 1912-50. If a like comparison of contraction in travel time were carried back on the same scale to the era before railroads, the chart would be as large as the top of a large dining table.

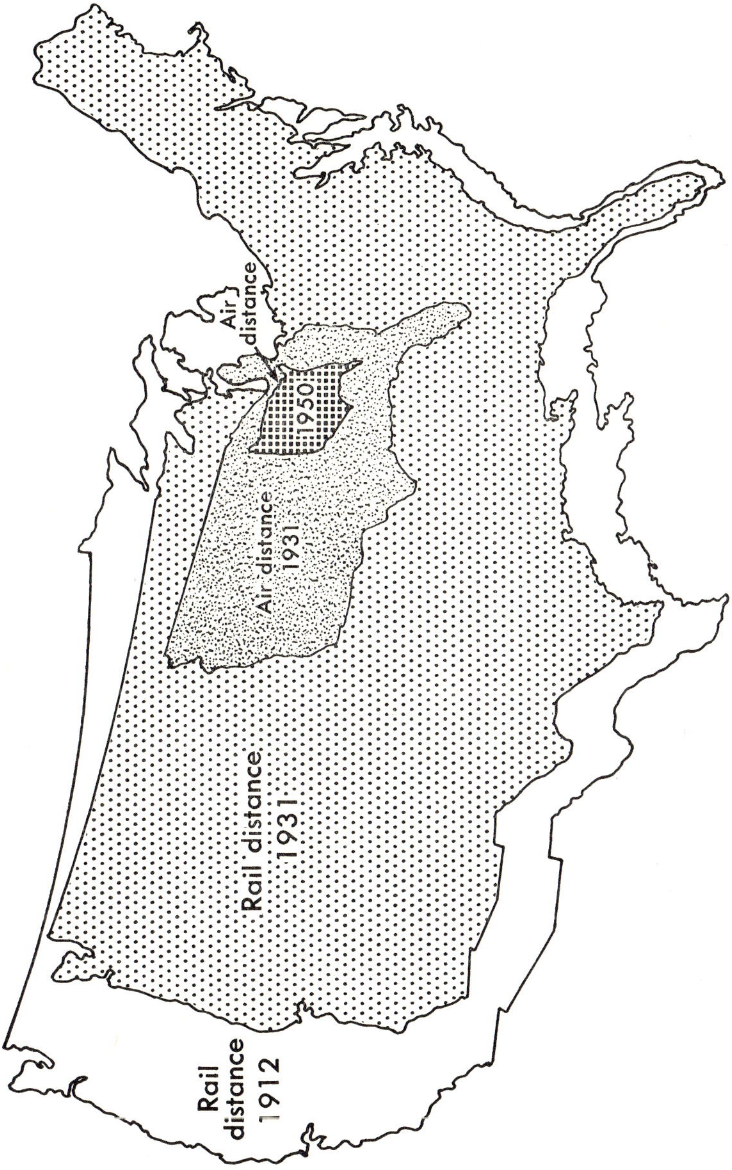

Figure 2. Contraction of the nation as shown by the increased speed of transportation, 1912–1950.

Not only mileage but areas must be considered as subject to continuing revaluation within extremely elastic frames of time. As distances shrink, areas expand in potentialities. Land may be put to all manner of uses that were impossible in an earlier day. This is not "new" land in a physical sense, yet it is new land in an economic sense.

In colonial days, a farm or small town was limited in its supply of game, forest products, and farm products to a narrow circle—wider where water transportation was available than where it was not, but to a small area even for durable natural products and to a still smaller one for perishable products. Today, nearly all physically productive land in the United States contributes to the national supply of all kinds of products; if not in the sense of direct shipment to all areas, then at least in the sense of replacement of supply that would otherwise be shipped into the local area. Wheat, beef, lumber, oil, and many other products move on an essentially national scale; and the whole milk produced for a local market takes the place, in part at least, of dairy products that would otherwise be shipped in from surplus dairy regions. As a result, the supply of land and its products has been enormously increased. Land can be and mostly is used for its most productive uses; there has been, over the past century or so, a great sorting out of land, according to its physical capabilities and uses.

Cropland and residential intensities shift and increase accordingly. So may the use of more expansive areas offering space for pleasure. Today, for instance, every citizen of the United States is a potential visitor to Yellowstone National Park, and more than a million of them go there annually. The extent to which a heightened mobility can thus, in effect, "stretch" our landed heritage is a factor especially to be figured into estimations of greater use for remaining open spaces of the West, and again for the unsettled frontiers and wilderness areas to the far Northwest, in our new and largest state, Alaska.

To review, even at a glance, the stages of shifting land use as men pressed westward across the continent gives the impression of history repeating itself with an ever-increasing velocity within shorter cycles. There are, however, many variations, conditioned by the natural endowment and traditions of locality, in the rate and extent of change. But to dwell at length on the problems and potentialities of limited localities is clearly beyond the scope of this book, which is concerned primarily with forces, trends, and situations in their broad and total aspects. National totals or averages, however, need some localization to be meaningful. If only for this reason, it is important to identify, wherever possible, situations where shifts in major uses, or major shifts in intensity and manner of use, seem probable or desirable. An additional reason lies in the attention the study may draw to geographical areas requiring more concentrated study.

To younger, more recently settled parts of the country, shorter time spans of estimation may be applied than to other sections. For like reasons, regarding changes in a given use of land the nation over, it is convenient and proper to view the shifting pattern within periods of varying duration. The differences may depend on a historical priority, the relative antiquity of the use in question, and the total store and availability of pertinent factual data; or again on a presently apparent rapidity of transition toward higher intensities of use.

So far as possible, quantitative information is presented and quantitative relationships revealed. But there are certain difficulties in applying relative measurements to some of the most striking spectacles of recent and current change. Definite data on urban and inter-urban shifts in use have, for example, barely begun to take meaningful and revealing form. The large-scale appropriation of lands for outdoor public recreation is principally a development of the past half-century, and the assembled facts on which we may base computations as to the future are likewise of recent origin, and are diffuse and scanty.

For the earliest major alterations in civilized land use, the conversion of forests and raw land into tended cropland and pasture, our period of trial and error has been somewhat longer. The terms of the accounting from colonial times to the present are the product of shifting standards and definitions. The census designations of who is or who is not a farmer today are not what they were when around nine-tenths of our population were so classified. But the over-all facts and figures on the agricultural transformation of the United States—so far as that phrase may be said today to have a separate meaning—do provide a historical and statistical framework encompassing around three centuries of past experience; and in our analysis of accomplished and prospective changes we may reasonably take as our time-span the years from the opening of the eighteenth century to the close of this twentieth century A.D.

THE FAR BACKGROUND

Written records of the ancient and medieval "New World," set down from observations and in retrospect, largely derived from Amerind myths and folklore by the first European invaders, come down to us from only around 1500 A.D. But that does not prove that the ancient and preancient Amerinds, rural and urban, were everywhere exceedingly backward or not bright. It is, indeed, never the part of wisdom to accept without question the common presumption that preliterate peoples were uninventive and incapable of progress. Archaeologists and anthropologists, constantly unearthing evidence from time out of mind in many lands, soon come to regard with wonder the preliterate accomplishments of humankind.

Viewed in the longtime perspective, the terms "Old World" and "New World" become blurred and indistinct. Distinctions customarily made between our "old" East and "new" West in appraising the western sweep of civilization across the United States tend likewise first to fade and then to take on sharper proportion in relation to the whole story of America when you remember that the earliest civilized forms of intensive land use on this continent took form, apparently, in our far Southwest.

The possibility of a preancient interchange and interdiffusion of Asian and American plants and of human stocks and cultures across the Pacific remains in dispute. Historical and geographical tags that differentiate the Old from the New World derive in the main from quite recent major shifts of population across the Atlantic to this hemisphere. That swarming of Europeans to uncrowded ground and the attendant overthrow of precedent American cultures dates back a mere four centuries to its beginnings and was brought to a virtual conclusion barely a century ago.

White men have been Americans for only a little while. Yet in these swift decades we have built here one of the most advanced and productive economic and materialistic societies that has ever existed. We like to think, and have cause to think, that it is a humanistic, livable, and durable order of life. The contemporary American tends to think of his country as the epitome of solidarity and steadfastness. And indeed, considering the cataclysms that have engulfed the world in the past two generations, and from which we have emerged not only relatively untouched but stronger, materially, there seems ground for such belief. But we will do well to remember that other cultures and nations, in their day possibly as powerful as we are today, have vanished—vanished so completely in some instances, on this as on other continents, that even the traces are hard to find.

It becomes us to examine such omens with a sense of proportion. We have reason neither to be complacent nor immediately alarmed. Our experience in our present homeland is still short. We are yet harvesting trees which were standing when our earliest forefathers first put foot on this continent. We have just arrived, and are not the first to have set in motion the processes of civilization on land quite possibly as old in its geological origins and evolutionary biological processes as any other major land mass in the world.

How old were the earliest civilizations, now obliterated or displaced? We have as yet no certain means of knowing; but there is no unquestioned evidence to show that the preancient cradlelands of American agriculture and civilization took form and initiated a dispersal of their cultures very much later than cradlelands in the Middle and Far East. Recent advances in archaeological research tend to indicate that the development and dissolution of Old and New World prehistoric societies proceeded in cycles by no means as far apart in the course of time as was previously supposed.

By such reckonings, the span of preancient, or pre-Incan, civilizations, may now be generally estimated to have come in total to 10,000 years at least. In terms of centuries, then, the relative tenure or occupancy of "Red" and "White" Americans on this soil may be roughly stated as 100 Red to 4 White.

SPACE AND PRESSURES

The means and processes by which in so short a while we have managed to provide for a resident population that now has risen above 175 millions, and to sustain millions of other people in other lands by sales or gifts from surpluses, have not been altogether admirable. But we may regard our achievements with pride and our outlook with confidence; our past mistakes and excesses are not beyond repair.

The white man has profoundly modified the natural world he

found here.[3] He has cleared forests, plowed virgin soil, and so grazed his livestock as to modify materially the grass cover on vast areas yet unplowed. He has deranged whole ecologic cultures, and completely destroyed some species. The passenger pigeons which even a century ago often darkened the sky in their flight now are extinct. The micro-climate, in and above the soil, has often been modified, with a consequent diminution of organic regeneration in the soil; and the displacement of topsoil by accelerated or man-made erosion has severely diminished soil fertility in many places.

But changes such as these give only the darker side of the picture. The land in many of its most basic characteristics remains unchanged. Our geographical position in the temperate zone and our climate have not changed. Much of the soil and a great deal of the native plant cover is unchanged or is in the reversible zone of use, overuse, and replenishment. Indeed, some of the soils are more productive now than when they were first settled.

Maintenance of the forest cover as the white man found it would in most instances have been utterly uneconomic; to have done so would have withheld from agriculture some of our most productive land, to say nothing of urban uses. Rational arguments against the conversion process must be expressed either in terms of future declines in productivity or against the needless waste that accompanied necessary or desirable shifts.

It might once have been possible to have managed the mixed-aged, mixed-species natural forests that characterized part of New England in such a way as to have preserved the original character of these forests, and yet to have harvested annually or periodically substantial volumes of forest products, for instance. But this is no longer possible, for the original forest has disappeared, and could be restored only at great cost and over long periods of time, if at all. Similarly, semi-desert ranges of southern Idaho might once have been managed to preserve the perennial native grasses, which have since been replaced almost entirely by introduced annual grasses. But today the perennials are gone from large areas, and could be re-established only with great difficulty, cost and time. This is not to argue that the best use of the resources would have been preservation of the original vegetative cover; perhaps that which has replaced it, or perhaps something still different, is better from every point of view.

[3] William L. Thomas, Jr. (ed.), *Man's Role in Changing the Face of the Earth* (Chicago: University of Chicago Press, 1956).

But it is clear that resources that have ceased to exist cannot be managed; the problem of whether they could have been or should have been so managed is now academic.

We know of no reliable estimates of trends in Indian population over long periods prior to the white settlement. Within the limitations of Stone Age technologies, however, the Americans, while sparse in number and scattered, were developing land use practices, mostly for hunting and fishing, but also for a limited tillage. In part because of natural conditions in the different regions of the continent, the technologies of the tribes and groups differed widely, and so did the forms of social organization. It is probably accurate to assume that a lean living led to raids by the more nomadic and savage tribes upon the stores of those more provident and settled in their ways, and that food shortages operated to keep population restricted, at least in some degree.

With a light and shifting population load, there existed a certain balance among various natural forces. The native vegetation represented a compromise or balance between these forces. Within the primitive forest and the prairie areas, there prevailed a certain type of ecological balance among species. Part of the environment of one species depended on the presence of another species, and among them there was both co-operation and competition for continuing life.

Into this ecologic combination must be reckoned the human factor. Even the more primitive American Indians were not as inert or innocent in accepting their environment just as they found it as has been romantically supposed. They hunted wild animals and they hunted each other. Fire to them was both a weapon and a tool. They set fires to expose their kill, to avoid ambush, or to open clearings for food crops and habitations. In this, their methods were of a kind with Stone Age practices that changed biotic cycles considerably. So, here as elsewhere throughout the world, men of the Iron Age did not enter upon landscapes completely unaltered by the hand of man. The balance of nature here was not a static one, but rather was dynamic and subject to rather rapid changes even then.

The conquering whites, whose initial technologies we now regard as simple, were better equipped to exploit natural resources far more rapidly and to an infinitely greater extent. The peak of the pioneering adventure coincided with a period in Europe when iron tools and steam power were advancing the age-old plod of hand labor, man power and horse power at an amazing rate. The new Americans imported or improvised like measures of progress and speeded them up. Here we were given great spaces over which to surge and operate, free or easy access to what seemed an illimitable expanse of undeveloped resources, and apparently insatiable markets in the old countries for the products of fresh soil. We were, moreover, a people largely indentured in various ways, to Old World capitalists and promoters; and, once the United States took form as an independent government, we remained for more than a century a debtor nation, heavily obligated to expand production enormously, and pay out.

Because white settlement here came so late in the course of world history, large populations and high population densities did not develop until the processes of technological growth and capital accumulation were well under way. We imported large amounts of both capital and technology, as well as large numbers of people; these, and our own developments of each, could be and were put to work on the natural resources so readily available. The result has been both a rapid growth of population and of income per capita. Our people have shown great energy and ingenuity, but the role of inherited natural resources, imported capital, and imported technology should not be underestimated.

It is possible—and customary among alarmed observers in this and in other countries—to draw up a damaging bill of particulars in summary of our reckless and sometimes ruinous misuse of land during the march of western settlement. The enormity of the damage done our forest resource in particular became, in this country, the subject of anxious appraisal and aroused an outcry for reform and repentance during the administration of Theodore Roosevelt and Gifford Pinchot as his Chief Forester, fifty years ago.

In these years before the First World War, Forest Service men of the Pinchot tradition cried havoc as to misuse of the public domain and predicted the imminence of a "timber famine." But at the same time, other arms and officials of the Department of Agriculture, with few exceptions, continued to exalt as "inexhaustible" the nation's farmland, and to hail the practices of private enterprise thereon as the most progressive and enlightened in the modern world.

That view, or mood, carried forward through the First World War. The pressure to produce to the hilt was beyond all precedent;

exaltation at our enlarging operations and tremendous increases of output ran high, both here and in allied lands. Those "defense" and war years coincided with mechanization in field equipment, propelled both by horseflesh and by massive tractors with internal combustion motors hauling unwieldy hitches of implements that tore up the ground roughly, but could really make time. The machines kept rolling, often beyond the bounds of safety for arable cultivation, and we added 40 million acres to our tilled domain. In the one-crop areas especially, on fresh plowland and on skinned plowland, we gave our soil a tremendous beating to provision that First World War.

Then it was, or shortly after, that our sorry record as husbandmen became a byword of reproach and agitation, at home and abroad. "Nowhere else [than in the United States]," said Gilbert Wooding Robinson of the University of North Wales, a not unsympathetic observer, writing in 1937, "has the drama of soil destruction been so swiftly enacted and on so great a stage."[4] That might justly have been taken as the final, and perhaps irreversible, verdict of history little more than twenty years ago.

The change in mood and a general reformation of our designs and practices in conservation may best be left for more detailed appraisal in subsequent chapters; but considered in sweeping retrospect, it now would seem that the wartime march of tractors beyond the dry-line and the dust storms that crossed the country afterwards served particularly to dramatize the extent and sweep of the damage and to emphasize the speed of changes for the worse. On the Plains and over most of the country there have been marked changes for the better in our field designs and methods of cultivation in the years since. For one thing, we are learning that machine methods in themselves need not hasten soil deterioration; that by redesign and provident application machine technologies which ripped off ground-cover and demolished topsoil can be made to reverse the process, to build soil and make it more fertile and responsive to cultivation, rather fast. Through a like advancement in the techniques of fertilization, genetic manipulation of new strains, and the control of weeds and pests, total yields and yields to the acre have been stepped up remarkably. Though crop acreage has remained roughly constant since 1910, total agricultural output has doubled.

[4] Gilbert Wooding Robinson, *Mother Earth, or Letters on Soil* (London: Faber and Faber, 1937).

These gains rest on a more solid foundation than in years past. The soil conservation programs are by no means fully effective as yet; but concern over soil erosion in this country within the past quarter-century, and the widespread application of contour cultivation and the restoration of cover as measures of protection, have already resulted in a greater shift from accelerated soil displacement to a relative soil stability than perhaps any other nation has accomplished in an equal period of time.

The last several decades have brought other needed changes in resource use patterns. There has been a considerable advance, nation-wide, of intensive fire control programs that hold losses to a practical minimum and so permit forest regeneration and growth; and forest replanting is proceeding on a major scale.

In all the states today fish and game commissions have become more active and experienced in improved techniques of protection and restoration; some species of wildlife now exceed in number that of an earlier time. Ways of so managing public areas as to accommodate the mounting pressures of people seeking outings, and yet to maintain the beauty and quietude of the natural scene, have been measurably advanced in recent years. And for those lands that are predominantly urban, or that may soon become so, there is a growing body of opinion that sound planning, which considers all the resources of an area, is a necessary part of long-term development, including urban renewal and highway building.

In this, as in practically every sort of resource protection and management, much of our land is being put to more intensive uses every year, but generally with greater care and skill than in years past, and with greater thought and concern for the future.

PROPORTIONS AND TERMS

The United States is richly endowed with land and its resources. Our lands lie almost wholly within the temperate zone, with a climate broadly favorable for man and his animals and plants. The population-to-area ratio remains as yet extremely favorable over our holdings as a whole. Counting in acreage of all sorts, we now have on the American mainland (i.e., the forty-eight older states) 10.9 acres per capita; and an unusually high proportion of that land is naturally arable, or fit to till.

For the world as a whole there are now about 1¼ acres of arable land per capita; for the United States, about 3 acres. We have more than the older countries of Europe, far more than the densely settled countries of Asia; and only somewhat less than the new and food exporting countries of South America and Oceania (Table 3). At present, some 15 per cent of all the cultivated land in the world lies within the borders of the United States.

Even these figures fall short of representing potentialities. We could convert more of our land into arable uses, were the population pressure severe enough. We have comparatively a high potential of good soils under forest cover, even today, after clearing much of our best forest to open it for farming. For a large nation, we have the lowest amount of wasteland per capita. Russia, China, Canada, Brazil, Argentina, Australia, and other large countries each have a very large proportion of their territory that is absolute or largely waste, as far as any real human use is concerned; we have very little.

But the per capita endowment of arable land, like that of land which can be put to other needs and purposes, can shrink quite rapidly in a world on which the burden of population is piling up fast. With us, 3 acres per person can seem a reassuring statistic. Yet a century or so ago, in the 1850's, our area of arable soil amounted to 20 acres per capita for the population of only 23 million. If the present rate of population increase continues, there will be more than 300 million Americans by the year 2000, and our arable land per capita will be down close to the present world average of 1¼ acres.

In terms of total earth-room, our 1,904 million acres—roughly 3 million square miles—still provide a spacious human habitat. We now have, by broad approximations, 6 per cent of all the land surface of the earth and 7 per cent of all its people. That figures to around 58.8 persons to the square mile for the continental United States on the average. Belgium, Holland, Japan, and the United Kingdom have, by way of comparison, upward of 500 persons to the square mile, whereas Canada, Australia, and Iceland still have population intensities averaging under 5 to the square mile.

"Population patterns," as Raleigh Barlowe has noted of Egypt in particular, "often differ as much within countries as between countries; and it is generally unwise to generalize . . . regarding popula-

Table 3. Total, arable, and forested land per capita, in world and in selected countries[1]

Country or area	Total land area per capita (acres)	Arable land: % of total land area	Arable land: Area per capita (acres)	Productive forest land: % of total land area	Productive forest land: Area per capita (acres): Total	Productive forest land: Area per capita (acres): Accessible forest	Wasteland and other: % of total land area	Wasteland and other: Area per capita (acres)
World	13.9	9.1	1.28	19.3	2.7	1.5	44.9	6.24
United States	12.8	23.5	3.1	24.3	3.4	3.0	9.2	1.18
France	3.25	38.3	1.2	20.0	.7	.7	19.3	.62
United Kingdom	1.19	30.3	.4	4.6	.07	.07	13.8	.16
China	5.2	9.35	.29	5.8	.3	.1	62.0	3.23
India	1.86	46.0	.89	[2]29.1	[2].3	[2].3	28.1	.50
Argentina	40.3	10.7	4.31	8.2	3.6	2.7	31.0	12.5
Australia	232.0	1.7	4.06	2.6	6.7	4.5	47.2	109.7

[1] Mostly postwar data, some prewar.
[2] Includes Pakistan.

SOURCE: W. S. Woytinsky and E. S. Woytinsky, *World Population and Production, Trends and Outlook* (New York: Twentieth Century Fund, 1953).

tion-land resources relationships." The average density figure for all of Egypt is 52 to the square mile, not far from the average of the continental United States. But 99 per cent of the Egyptians are concentrated on about 3 per cent of their land; and: "The Nile Valley has probably the highest density of population in the world."[5]

In the continental United States population pressures range all the way from around 88,000 to the square mile on Manhattan Island, to 748.5 in Rhode Island, to 1.5 in Nevada. A more detailed analysis of relative densities will be made in chapters to follow, especially in relation to urban concentrations and the inter-urban diffusion.

Census averages for all our land, it should here be remarked, have been based on the 1,904 million acres within the forty-eight contiguous states of the mainland. "Offshore" territory, amounting to around one-fifth of the continental area, adds another 372 million acres, separately designated in census computations. These offshore parts of the nation include such specks of recently acquired possessions as the Virgin Islands, Guam, and eight smaller inhabited islands that came under our mandate in the Second World War; but they also include the overladen island of Puerto Rico, which has attained to the status of a commonwealth; and new states of Alaska and Hawaii.

Because "offshore" situations, including those of Alaska and Hawaii, have not as yet been entered into over-all statistical averages of the census, we have followed a like practice in the tables and compilations of this book. Moreover, the economy of these other areas differs rather appreciably from that of the forty-eight states. Their land is less a substitute for other land within the older states, and their demands do not fall so clearly on the land under discussion. But there could hardly be a more striking illustration of the fallibility of using population-resource variables, derived from widely different areas, to indicate a total national situation, than is provided by the commonwealth of Puerto Rico and by the state of Alaska.

Of the first, with an area of 3,435 square miles, and a present population exceeding 600 to the square mile—despite continuing mass emigration to our mainland—Chancellor Benitez of the University of Puerto Rico, says: "Population pressure in the United States would compare with that of Puerto Rico if all the people of the world—over two billion men, women and children—landed there

[5] Raleigh Barlowe, *Land Resource Economics* (Englewood Cliffs, N. J.: Prentice-Hall, 1958), p. 51.

overnight, and, if, by the same nocturnal magic, all available mineral resources were eliminated, heavy industry disappeared, and agriculture became the main source of employment."[6]

At the other extreme, Alaska, covering 586,400 square miles of land and inland waters, with more than 365 million acres of actual land area, had as of 1950 an estimated population density averaging 0.225—around one person for each four square miles. The estimated resident population as of 1957 was 159,000; and the total of arable land from which crops have thus far been harvested is given as approximately 20,000 acres. Enter that into the reckoning of total acreage per capita for the forty-nine states, and the average rises from 10.9 to 12.9—another two acres per person. The area of arable land in Alaska increases the supply for residents of these states by barely five square feet per capita.

The forces of change are interactive and never ending. As an advancing mobility makes remote or hitherto marginal areas more readily accessible, this increases the effective land supply, but at the same time it increases the demand for that land and its resources; and it may lead to further demands and uses that did not previously exist.

In the vast transfiguration of the American landscape which has come of our successive penetration and expansion beyond new physical and technological frontiers, the fluidity of change and the fallibility of rigid predictions have been demonstrated time and again. As the great changes of the past have occurred, there was never at any given time a full adjustment to the forces then operating. Some interval always exists between the forces causing social change and the completion of that change, and by that time new forces are likely to have begun to exert their influence. The time-lags tend, however, notably to shorten as means of communication improve; and the velocity of change increases accordingly.

For some uses sheer space remains the dominant factor. On this overcrowded planet earth-room of any sort has become a resource, actual or potential. Space is a resource strictly calculable in extent, but far less calculable in terms of future use. The presently barren tundra of Alaska, and immense like spaces of the Soviet Union now in process of occupation and development, may some day come to have real usefulness, merely as usable space. We cannot be sure just

[6] Harvey S. Perloff, *Puerto Rico's Economic Future* (Chicago: University of Chicago Press, 1950), p. v.

what changes will occur to expand habitable areas. But in view of the immensity of those land areas now largely sterile, and possibly irreversible increases in human fertility, there would be a greater error in assuming no change at all.

The only common denominator among the various uses of land is area. Area itself does not take account of the divergent qualities of land—its slope, its exposure to the sun, the depth and texture of its soil, its climate, its location with respect to markets, and its other variable economic characteristics. These factors must be introduced in other ways, by details which do not neatly fall into simple groupings. Moreover, the amounts of labor and capital, and above all, of technology and managerial skill, greatly affect the human values derivable from known areas of land. But area does have the great virtue that it is quantitative and unequivocal in meaning.

Areas are measurable and subject to definition. Specifically, in the United States, "land area" is defined to embrace land temporarily or partially covered by water—marshes and swamps, river flood plains, estuaries, streams and the like less than one-eighth of a mile wide, and ponds, lakes or reservoirs covering less than 40 acres. Conversely, "inland water area" is defined to include streams and rivers wider than an eighth of a mile; ponds, lakes and reservoirs larger than 40 acres. Deeply indented embayments and sounds; other coastal waters behind or sheltered by headlands or islands less than a nautical mile offshore; and islands of less than 40 acres, are likewise figured into the inland water area.[7] By such distinctions, our continental expanse, exclusive of Alaska, comes to 1,904 million acres of "dry" land, plus 31 million acres of inland waters. And, of course, spatial dimensions of the land are measured as if it all were as flat as a map, regardless of total distances uphill and down.

Our concern here is primarily with land *use*. Land use implies a conscious decision lying behind actual use. Ideally, use of land should be studied according to the nature and degree of consciousness of the decisions underlying its management. On this basis an unused forest, or one which grew up without decisions by anyone, would not be forestry, or at least would not be conscious forestry. It would under any circumstances be difficult to measure the nature of the decision-making process underlying land use; what men think can

[7] U. S. Bureau of Land Management, *Report of the Director, Statistical Appendix 1957 (Washington: U. S. Department of the Interior,* 1957), p. 2 Table 1, footnote defining "land" and "inland water areas."

often be guessed only from what they do. Certainly data are not available on this basis, and one is forced to use the existing definitions and data, which are all based on observable characteristics, not on intent.

Use may be single, to the complete or virtual exclusion of other uses, or there may be multiple use. Among multiple uses, one may be dominant, others subordinate; or in some places two or more uses may be of approximately equal importance. In this book we emphasize the dominant uses. All land, for example, has some importance as watershed, but this we do not consider as a major use, because it is almost always subordinate to or co-ordinate with other uses. Similarly, nearly all land has some scenic value and some wildlife, but there are some areas where these are the dominant uses; and these areas we consider under the appropriate categories. Only in the last chapter is multiple use, its limitations and its possibilities considered more explicitly.

There has often been confusion between land use, as such, vegetative cover, and land ownership, whether by an individual, a corporation, or a unit of government. These are separate though interrelated ideas. Vegetative cover may vary, from forests of different kinds, to grasslands composed of different species, to deserts with their typical shrub cover. Where man has converted the land to his uses, for agriculture and for cities, vegetative cover is largely what he makes it—though every gardener knows that weeds thrive in spite of his efforts.

Land used for cropping must obviously be in farms, as units of economic activity and of decision making; the definition of "farms" is drawn so as to pick up all such land. Very little farm land is publicly owned. Included within farms, as decision-making units, is much land not used for cropping, particularly grazing land not suitable for cropping and forest land which may or may not be grazed. Because use depends upon the decisions of the landowner or controller, land ownership and tenure must be considered. To be purposefully managed for forestry, the land, whether publicly or privately owned, must obviously have trees or be capable of growing them. Moreover, certain minimum rates of growth and minimum qualities or sizes of trees are necessary if active forest management is to endure. Thus, as far as forestry is concerned, there is a close correlation between vegetative type and land use.

Livestock can find food from a wide variety of vegetative types, although there are some—the densest forests and the sheerest deserts, particularly—where there are no edible materials. Thus, the correlation between grazing use and vegetative type is not close. On the other hand, the relations between grazing and ownership or control of land may be varied and complex. The grazing lands are used under various forms of economic enterprise, combining sometimes the private ranch or farm with the public, or public and private range beyond.

Recreation may take place on land with almost any type of vegetative cover, but there seems to be a strong preference among many people for wooded areas, especially if these include small bodies of water. To a degree, therefore, there is a correlation between land use for recreation and vegetative cover. So far as ownership is concerned, we all know that it is mixed, but with public recreation places outstripping private at least in terms of area.

Vegetatively speaking, a substantial part of urban areas is the sheerest of all desert—paved streets, house tops, bare and impervious surfaces. But in most urban areas there are also trees, shrubs, lawns, and flowers. In fact, in parts of some cities there is probably a higher per-acre input of materials, capital, and labor used for growing plants than on an equal area of intensively farmed land. The large part of urban areas that is publicly owned—the streets, alleys, and sidewalks especially, but also the school yards, parks, and public buildings—is intermixed with the privately owned residential and industrial areas.

THE RECENT PAST

To some extent, the foregoing remarks can be quantified by tabular data as has been done in Table 4 for the years 1850-1954. Unfortunately, even the best data can confuse land ownership or control, vegetative cover, and land use.[8] The most complete, accurate, and

[8] These data are discussed in the following U. S. Department of Agriculture publications: H. H. Wooten, *Major Uses of Land in the United States,* Technical Bulletin 1082 (October 1953); *Basic Land Use Statistics, 1950, Supplement to Major Uses of Land in the United States,* Technical Bulletin 1082 (September 1953); H. H. Wooten and James R. Anderson, *Major Uses of Land in the United States—Summary for 1954,* Agriculture Information Bulletin No. 168 (January 1957); and Wooten and Anderson, *Agricultural Land Resources in the United States, with Special Reference to Present and Potential Cropland and Pasture,* Agriculture Information Bulletin No. 140 (June 1955).

long-continued source of data is the Census of Agriculture, taken at ten-year intervals until 1920 and at five-year intervals since. The definitions of farms and the practice in field enumeration have changed, which to some extent has affected the results; and the items enumerated also have changed somewhat over the years; yet some reasonably good data are available for 1850 to date. Data on land use in farms are obviously related to a form of land ownership and control—namely, the farm as an economic planning and management unit. Since the physical characteristics of land are easier to observe, define, and record than are the use plans and practices of its managers, data have been collected on the area of forest and woodland in farms; but we have no assurance that forestry, in any purposeful sense, has been practiced on much of it. The Census has also obtained data on areas of cropland and pasture land; these are use categories, but in practice they have been defined partially in terms of physical appearance. For the land outside of farms, estimates have had to be compiled from various sources. The forest land outside of farms is clearly vegetative cover, not forestry in the land use sense. The "other" category outside of farms, used in Table 4, is in effect a catch-all, including errors of estimate in the preceding items. It includes some of the most intensively used land, particularly that in urban use; and it also includes deserts, swamps, mountain tops, and other unused land.

Clearly, whether the categories chosen for this book or any other consistent ones are employed, no truly quantitative picture can be presented of changes in land *use*. In spite of their weaknesses, the available data reveal some significant changes in land use. The more important ones are:

1. A major expansion and geographical shifting of cropland occurred between 1850 and 1920. By clearing of forest and plowing of prairie, the total area of cropland was increased nearly fourfold in these decades. From a heavy concentration along the Atlantic seaboard at the beginning of this period, crops spread across the nation to approximately the pattern of today. Even as new lands were brought under the plow, older ones were abandoned to grow up to brush and trees. Some land once capable of farming under a subsistence agriculture could not compete as farms in a commercial agriculture.

2. Since 1920, little change in cropland in farms has taken place, either in total area or in area by localities. The total area within farms has continued to expand by some 200 million acres. This is, how-

Table 4. Land utilization in continental United States, census years, 1850-1954

(million acres)

Major land uses[1]	1850	1860	1870	1880	1890	1900	1910	1920	1925	1930	1935	1940	1945	1950	1954
1. Land in farms:															
Cropland[2]	113	163	189	188	248	319	347	402	391	413	416	399	403	409	399
Pasture[3]	—	—	—	122	144	276	284	328	331	379	410	461	529	485	526
Woodland and forest[4]	181	244	219	190	190	191	191	168	144	150	185	157	166	220	197
Other[5]	—	—	—	36	41	53	57	58	58	45	44	44	44	45	36
Total	294	407	408	536	623	839	879	956	924	987	1,055	1,061	1,142	1,159	1,158
2. Land not in farms:															
Grazing land[6]	—	—	—	883	818	625	739	661	646	578	533	504	428	400	353
Forest land[7]	—	—	—	368	344	318	162	160	203	208	184	203	186	201	238
Other[8]	—	—	—	116	118	121	123	126	130	130	131	137	149	144	155
Total	1,590	1,496	1,495	1,367	1,280	1,064	1,024	947	979	916	848	844	763	745	746
Grand total	1,884	1,903	1,903	1,903	1,903	1,903	1,903	1,903	1,903	1,903	1,903	1,905	1,905	1,904	1,904

[1] Acreages of total farmland and nonfarm land are for the calendar year indicated. Except in 1954, acreages of cropland and pasture land usually relate to the preceding year.

[2] Includes cropland harvested, crop failure, cultivated summer fallow, and cropland idle or in cover crops. Generally does not include cropland used only for pastures. However, for 1850, 1860, and 1870, it is improved land in farms, and includes some cropland used for pasture. In those years, probably included some of farm roads and farmsteads shown in later years as "other" land in farms.

[3] Includes all open or nonforested pasture in farms, including plowable pasture, and in 1945-54 includes cropland used only for pasture.

[4] Includes both pastured and nonpastured woodland.

[5] Includes farmsteads and roads and some other land. Not shown separately in 1850, 1860, 1870; probably included in those years in improved land, which is shown here as cropland for those years.

[6] Land used chiefly for grazing, but not in farms. Includes open or nonforested land, noncommercial woodland (especially pinion-juniper), shrub and brush areas, etc. Some is publicly owned in all periods.

[7] Commercial forest land primarily. Excludes forest land in parks, wildlife refuges, etc.

[8] Includes urban, industrial, and residential areas outside of farms; parks and wildlife refuges; military lands; roads, railroads, airports; deserts, swamps, rock, and other unused and waste land.

SOURCE: Hugh H. Wooten and James R. Anderson, *Major Uses of Land in the United States, Summary for 1954,* Agricultural Information Bulletin No. 168 (Washington: U. S. Department of Agriculture, Agricultural Research Service, 1957).

ever, essentially a change in land ownership and control, not a change in land use. Much land previously in public, railroad, or absentee private ownership, and generally used for grazing by farmers, with or without permission of the title holders, has in more recent decades been incorporated into farms. A great deal of this has taken place within the Great Plains.

3. The area of grazing land and of forest land outside of farms has shrunk rather steadily over the whole period. In 1955 it was only about half what it was in 1850. This is but the obverse of the increase in land within farms; in large part, it represents difference in land ownership and control, rather than in land use. But the expansion in improved land within farms over the decades surely meant a decrease in the area available for grazing or for forestry. The plowing of the Plains meant less natural grazing area, and so did the plowing of the prairies of Illinois and Iowa. The clearing of forest land in Ohio, Indiana, Illinois, Michigan, Wisconsin, Minnesota, and in many southern locations decreased materially the areas available for forestry. The total area of forested land, in the vegetative cover sense, surely decreased; the area devoted to forestry, in the management sense, has risen. Specific estimates on this latter point are presented in Chapter VIII.

4. The great period of change in major land uses, which began with the first settlement of the continent, came largely to an end about 1910 or 1920. Since then, there have been only modest changes in the area in each major land *use,* though some rather large changes in land ownership and control. Since these dates, the major changes in land use have been *within* each major use.

OWNERSHIP OF THE LAND

In certain of the more urban societies of ancient and medieval America, both land and people were regarded as possessions and chattels by the rulers of the land. Cortez and his conquistadors overcame in Mexico a compact empire, the Aztecs, obedient to regal standards, with Indians in palaces and the greater part of the population bound to the soil as serfs. The predecessor Indian empires and confederations, Incan and pre-Incan, from Peru northward, appear to have embraced concepts much like those of modern Communism, in combination with other forms of absolutism not unlike those of Egypt under the Pharaohs around the same time.

But the Indians inhabiting the broader reaches of North America and the eastern coast were singularly free, by and large, of any attachment to land either as its subjects or its owners. They had practically no sense of property. With proprietorship not only the main source of wealth in the Old World countries, but the symbol of rank and eminence, the European settlers in America were, on the contrary, not merely "land hungry," but ravenous. "It would be difficult," Alexis De Tocqueville wrote, reviewing the western thrust of *Democracy in America,* as of the 1830's, "to describe the avidity with which the American rushes forward to secure this immense booty that fortune offers." He foresaw as inevitable the all but complete "expulsion . . . of the Indian nations of America," given as they were to "that childish carelessness of the morrow which characterizes savage life . . . It is a misfortune of Indians to be brought into contact with a civilized people, who are also (it must be owned) the most grasping nation on the globe; . . . and as the limits of the earth will at last fail them, their only refuge is the grave."[9]

As the body of the continent was discovered and explored, its land was claimed by the sovereigns or by the national governments of the exploring countries—England, France, Spain, Holland, and others. In the colonial period, land was granted to lord proprietors or to land companies directly for their ownership and profit or for their colonization. These in turn disposed of much land to individuals. Land from this source passed down to the states, after independence.

About 28 per cent of the area of the United States has a title history which follows this path. Over 40 per cent of this land was later ceded by the states to the new Union. The federal government by purchase from foreign countries, as in the case of the Louisiana Purchase, or by treaty, or by war, acquired the remainder of what is now the United States. In total, slightly more than 75 per cent of our present area (excluding Alaska) at one time or another was the property of the federal government. The acquisition of Alaska by purchase from Russia for a few cents an acre in 1867 was the last acquisition of what is popularly and legally known as the public domain.

9 Alexis De Tocqueville, *Democracy in America* (Vintage edition; New York: Alfred A. Knopf, 1954), pp. 305, 355, 366.

The history of these public lands is one of the distinctive and major parts of American history, but it is a story that has often been related, and need not be examined in detail here.[10] First considered as a source of revenue, the lands were shortly looked upon as a means of encouraging settlement. Through sale at modest to low prices, through outright gift to individuals, through grants to railroad and canal companies, through grants to states for educational and other purposes, and in other ways, roughly two-thirds of the total federal holdings have been distributed to other ownerships. Following this era of disposal, and somewhat overlapping with it, was the era of reserving large acreages of forest and prime scenic land for perpetual federal ownership. These lands have in turn passed through one major phase, that of custodial management, and into another, that of intensive management.

The land disposal process was headlong, even heedless. Land was often sold or given to individuals under the impression that it was suited for farming, when later experience, and sometimes much heartache, was to prove it otherwise. Speculators were certainly a major factor in the whole process, sometimes performing a real service, other times not. Wave upon wave of settlers moved into new areas, or into the further development of areas already pioneered. It is easy now to be critical of the whole process; yet certainly it did settle our country; and, given the temper of the times, perhaps little or nothing better could have been expected.

Apart from Alaska, the present situation as to land ownership in the continental United States is roughly as follows:

[10] Some of the more recent books relating various aspects of these changing patterns of ownership are: John Ise, *The United States Forest Policy* (New Haven: Yale University Press, 1920); Benjamin H. Hibbard, *A History of Public Land Policies* (New York: The Macmillan Co., 1924); Jenks Cameron, *The Development of Governmental Forest Control in the United States* (Baltimore: Johns Hopkins Press, 1929); Walter Webb, *The Great Plains* (New York: Ginn and Co., 1931); Roy M. Robbins, *Our Landed Heritage* (Princeton: Princeton University Press, 1942); Gifford Pinchot, *Breaking New Ground* (New York: Harcourt, Brace and Co., 1947); Marion Clawson, *Uncle Sam's Acres* (New York: Dodd, Mead and Co., 1951); E. Louise Peffer, *The Closing of the Public Domain* (Stanford: Stanford University Press, 1951); Wallace Stegner, *Beyond the Hundredth Meridian* (Boston: Houghton Mifflin Co., 1954); Samuel T. Dana, *Forest and Range Policy—Its Development in the United States* (New York: McGraw-Hill Book Co., 1956); Marion Clawson and Burnell Held, *The Federal Lands—Their Use and Management* (Baltimore: Johns Hopkins Press, 1957).

	Million acres	Per cent of total
Owned by the federal government	408	21
Managed by the federal government, for the Indians	57	3
Owned by the states	80	4
Owned by counties and municipalities	17	1
Remainder—presumably owned privately, including corporations, co-operatives, trust, etc.	1,342	71
Total	1,904	100

The total area of land in the United States expanded faster for several decades than did the area in nonfederal ownership (Figure 3). By the middle of the nineteenth century the expansion of the United States was complete;[11] transfers to nonfederal ownership continued on a large scale until the end of the first quarter of the twentieth century. Part of this was the grants of land to states and to railroads; these organizations in turn, with some exceptions, disposed of all or a substantial part of their lands to private individuals. The area of land in farms in 1850 was less than half of the area nonfederally owned, and only one-seventh the area of the nation; by 1920 the area in farms was about two-thirds of the area nonfederally owned and about half of the total area. Since 1920 the area in farms has continued to rise, at nearly the past rate, during a time when net changes in federal land ownership were small and when, of course, there was no change in total land area of the United States.

One knows comparatively little about the vast interchanges in land ownership that have taken place over the decades—a subject which in any case is far afield from this study of land use. With the gap between private land ownership and area in farms, there must at all times have been a considerable acreage in "other" hands. Certainly, at an earlier time everyone lived closer to the land than now, but this by no means proves ownership of land was widespread. Today, of course, comparatively few people own nonurban land. As a people, we have grown away from the land.

One knows, also, that within agriculture there have been major changes, with farm abandonment and consolidation part of the total

[11] On the basis of the United States as it existed until late 1958; see p. 18, n 1.

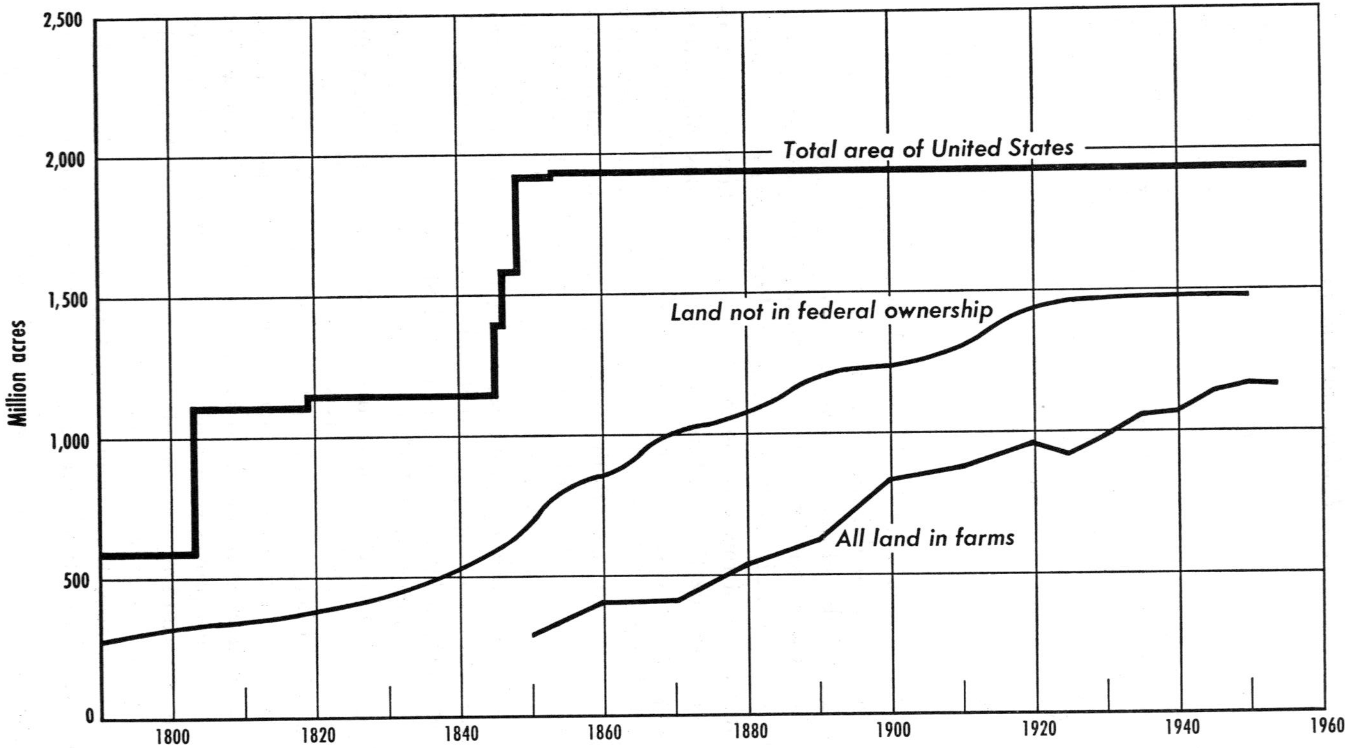

Figure 3. Total land area, land not in federal ownership, and all land in farms, 1790–1958.

picture at least from the birth of our nation. Behind the cold statistics lies many a human story of failure and migration, or of success and adjustment to the changing times.

One fact is certain: the farms and rural areas have continuously furnished a stream of migrants, mostly young and often the ablest, to the city. In the early decades, when the farm and other rural population so far outnumbered the city folk, this migration built the cities. In later decades, as city populations became the larger part of the whole, the farm-to-city migration while important has not been the dominant force in city building. City people have come to reproduce their kind fast enough not only to maintain the city but to provide its major growth. But, even to the adults of the present generation, direct connection with the soil—the rural soil—was common; and indirect connection, through the tales of parents and grandparents, was nearly universal. The generation now growing up is almost completely remote from the rural scene; so much so, that several of the larger cities have taken special steps to make it possible for young school children to see a cow.

CHANGING EXPECTATIONS

What do we Americans want of our land? No simple answer is possible; for many different people have wanted many different things at the various stages in our national evolution. Yet, we cannot well consider how land has been, is now, or may be used, without some idea of what people want from its use.

The pioneer, pushing into relatively new and unexplored territory, well ahead of the solid settlements, wanted one set of satisfactions from the land. The settlers who followed wanted something else—land to own, to produce their food, to provide some cash income, as a refuge for old age, and all this without the extreme risks of the frontier edge. The land speculator surely wanted something else; to him, land was essentially a commodity for trade; he cared little where it was, what it looked like, or what it would produce, if he could sell it profitably. The commercial logger for decades cared little or not at all for the land—he wanted only the trees then standing, for his harvesting. Often he cut what was not legally his; and often he abandoned to the public what he had owned, once he had cut it over. The early hunters—indeed, all citizens—wanted the privilege of hunting and fishing where they chose; no king's forests for them!

Through the decades relations between landlord and tenant, and between landowner and lender, have been competitive as well as cooperative. Landlord and lender alike are unpopular figures in American fiction and popular history, yet each performed real services. Tension between landlord and tenant has perhaps existed throughout, and at times has erupted into outspoken opposition and even violence. Even more has the tension between lender and borrower varied, depending very largely upon whether the general price level was going up or down. In times of prolonged price decline and depression, farmer-borrowers have had difficult times indeed.

In the early decades of our history land was nearly free, forests were more often a nuisance than not, and game was plentiful. Exploitation of these, and of other resources such as minerals, was most natural. It has been the fashion in later times to condemn the profligate and wasteful use of all resources in those earlier days; yet, we wonder, how many of us would have done differently, given the same circumstances? More, it can be argued that some measure of exploitation of natural resources—thereby to secure capital, particularly from abroad—was both good economics and sound national policy. Certainly, conservation, as it is known now, had much less economic base during the period of expansive occupation; it was more nearly a matter of faith. One can wish that the waste of our forebears had not been so extreme, their unconcern for the future so great; yet one can understand them, too.

Through the whole of the nineteenth century appropriation of the public land into private ownership was one major objective. While in retrospect one tends to overestimate the value of land then on the frontier but now in intensive farming, the fact remains that perhaps never before or again will such excellent land, forests, mineral deposits, and other resources be available on terms such that nearly everyone could get at least a slice of the pie. The role of free land in American history has been stressed by many writers, and it is unnecessary to repeat it here. Indeed, it may well be that many have overestimated that role, large as it was. But there are no longer public resources available in large amounts and virtually free; and the popular attitude today is strongly against "give-aways" of resources.

The past generation has produced major changes in attitudes on conservation—soil, forest, range alike. The conservation movement, as usually described, goes back farther; but for a long time it was chiefly the possession of dedicated leaders, not the practice of the landowner and land user. Whereas a generation ago most farmers knew little about soil conservation and were not seriously concerned with it, today they typically know, care, and do something about it. Whereas a generation ago nearly all forest owners were concerned with profitable liquidation of their stands, today an increasing percentage is committed to sustained yield operation. Whereas a generation ago the typical rancher knew little about the plants his animals grazed on, today his son both knows how to manage his range for maximum sustained output, and does it. This same generation has seen also the rise of parks—national, state, and local—from a few exceptional cases into large systems with widespread public use and public support. This is not to suggest that all is yet perfect even under today's conditions, much less than for those of tomorrow. But major changes in attitude have occurred, and these are increasingly being translated into action.

For the future, one can only guess, or surmise. But there seem rather clear signs that people generally will be more conscious of land and other resource problems, more anxious to insure satisfactory conditions for their children, more willing to invest now for future gains, than has been true in the past in this country. The trend seems clearly in this direction; in addition, as population grows such attitude and actions will be well-nigh forced upon us. Regardless of how we now judge the exploitation of a century or less ago, we cannot repeat it under the conditions of the future.

An expanding economy and a dynamic society more selective and more productive in every use of land are the postulates of this book. In agricultural land use, particularly, recent history and the present state of technology provide every reason for expecting further increases in output with fewer people farming, and higher average yields from each acre, each dollar of capital input, and each man. We are not certain how it will be achieved in every particular; but we conceive this to be an all but certain extension of present trends and forces.

Estimates of change extending up to forty years ahead, such as we propose to make, may reasonably be considered long range, with all the problems inherent in projections of this character. Two broadly different types of influences will reshape events and circumstances in the years ahead.

On the one hand, presently evident trends and forces may continue, with appropriate modification, to influence land use in the future. "Evident" trends and forces will be more clearly so in the future, looking backward, than they are in the present, looking forward; and "appropriate modification" is a matter upon which there may well be differences of opinion. Nevertheless, the observable trends and forces now operating often can be measured. The rationale behind these trends and forces may suggest that the future will be different from the past, or it may not. A projection of past trends and forces, with whatever modifications the investigator thinks appropriate, merely makes explicit for the future what should be implicit in his analysis. The problem becomes greater with the time-span of projection; and the importance of modifying extrapolations of past trends also increases.

In a different category are the wholly new forces that may be anticipated, as these may affect the future use of land. The dividing line between a modification of a past trend and a wholly new force may not always be sharp, but a few examples from the recent past may help. In 1850 electricity was a wholly new force; few scientists and no economists envisaged anything remotely approaching it, to say nothing of trying to measure its economic and social impact. Flight by heavier than air machines is in the same category; while flight had for long been a dream of man, in 1900 it apparently was no closer to fulfillment than in earlier centuries. Similarly, during the next forty years at least four major innovations or changes might drastically affect land use in the United States (and in the rest of the world):

1. *Weather modification,* on a large scale and to a significant degree. To the extent that this comes about, and to the extent that the modification of weather is substantial, the use of land might be greatly changed, not only as a direct result, but indirectly, as the result of changes in other areas.

2. *Major germ plasm changes,* in forest trees, cultivated agricultural crops, and other plants or animals. Through atomic radiation, further genetics research, or in other ways, markedly new strains or varieties might be developed that would affect land use to a major degree. Not all these changes might be intended or desirable—some might come from excessive amounts of radioactive materials in the atmosphere.

3. *Fusion energy* might be developed on a massive scale and at much lower costs than presently conventional sources of energy.

4. *Interplanetary economics* offers untold possibilities for the dreamer or the comic-strip artist. We assume that some form of interplanetary travel is within the realm of probability before the year 2000; but we anticipate no significant changes in economic processes or in the natural order of earlier processes, as a consequence.

Undoubtedly, a longer list could be produced which forty years or more from now might rate a higher score for accuracy. But in this book unforeseen and perhaps unforeseeable innovations of the future are merely hailed in passing; their ultimate importance, and indeed the direction of their effect, is beyond the reach of sensible quantitative estimate at this time.

Perhaps the most striking aspect of current speculation as to the future economy and society of the United States is a prevailing note of confidence. The economy is expanding by leaps and bounds; population is increasing; per capita incomes are rising rather rapidly and steadily; the average citizen has more leisure than he has ever had before; and large sections of the people are mobile to a degree unmatched in the world's history. There is a buoyant expectation, apparent in most professional forecasts and in the public mood alike, that the future will bring more and more of all this—more people, better living, an ever-increasing fulfillment of the American dream. This buoyancy is the more remarkable now that man has found a ready means to his destruction. As a people, we are not willing to introduce devastating war into the formula for estimating our individual futures. In this book, also, no major wars are assumed for the period under study.

This boundless confidence, definitely in keeping with our traditions and achievements as a people, is in sharp contrast to the spell of doubt and uncertainty that dominated our thinking around twenty-five years ago. Then, in the midst of the Great Depression, some of our best scholarly opinion foretold an early stationary population, economic maturity with its consequent slower economic growth rates, and similar gloomy prospects. People in general were confused and discouraged by continued massive unemployment. Yet it was in this period, with its Civilian Conservation Corps, Works Progress Administration, Public Works Administration, and other public programs to provide employment, that so much was done for the benefit of our natural resources.

State and national parks and forests were expanded and improved to provide outdoor recreation on a scale then not clearly probable for the near future. And it was from that time on that emergency programs of farm relief and soil conservation redesigns put us so rapidly along the way to a sweeping reformation of practices that can and must, it now appears, release much good land for purposes other than the primary production of food and fabric. The days of pioneering have not ended. But the too simple and sometimes brutal concepts of pioneering that prevailed in our initial march of occupation had to be, and have been, enlarged.

II

Urban uses of land

The focus of this chapter is on land used or withdrawn from other uses for urban purposes, within cities and between cities. The widespread interdispersal of rural and urban patterns of settlement and resettlement is quite new in this country and still to be sharply defined. We shall especially consider the sprawl of residential, industrial, and business demands and requirements beyond city limits and suburban fringes over much of the land, and also the changing land uses within city limits.

Whether they take form in compact centers or are more widely diffused, the economic and social effects of urban dominance are of pre-eminent importance in any study of land use. The idea of a rural people and the values of a rural society are yet strongly held in the United States; but despite immensely productive agriculture and vast expanses of forests and grazing lands, we are now a people of preponderantly urban occupations and interests. In 1950, 64 per cent of the population lived in cities or towns with 2,500 or more people; they possessed 74 per cent of the country's total wealth;[1] they found 68 per cent of all jobs (Figure 4).[2]

[1] Calculated by the authors, taking estimates of total wealth in the United States, as reported in recent volumes of *Statistical Abstract of the United States,* and estimating the proportion of the real wealth located within (not owned within) cities.

[2] Gunnar Alexandersson, *The Industrial Structure of American Cities* (Lincoln: University of Nebraska Press, 1956).

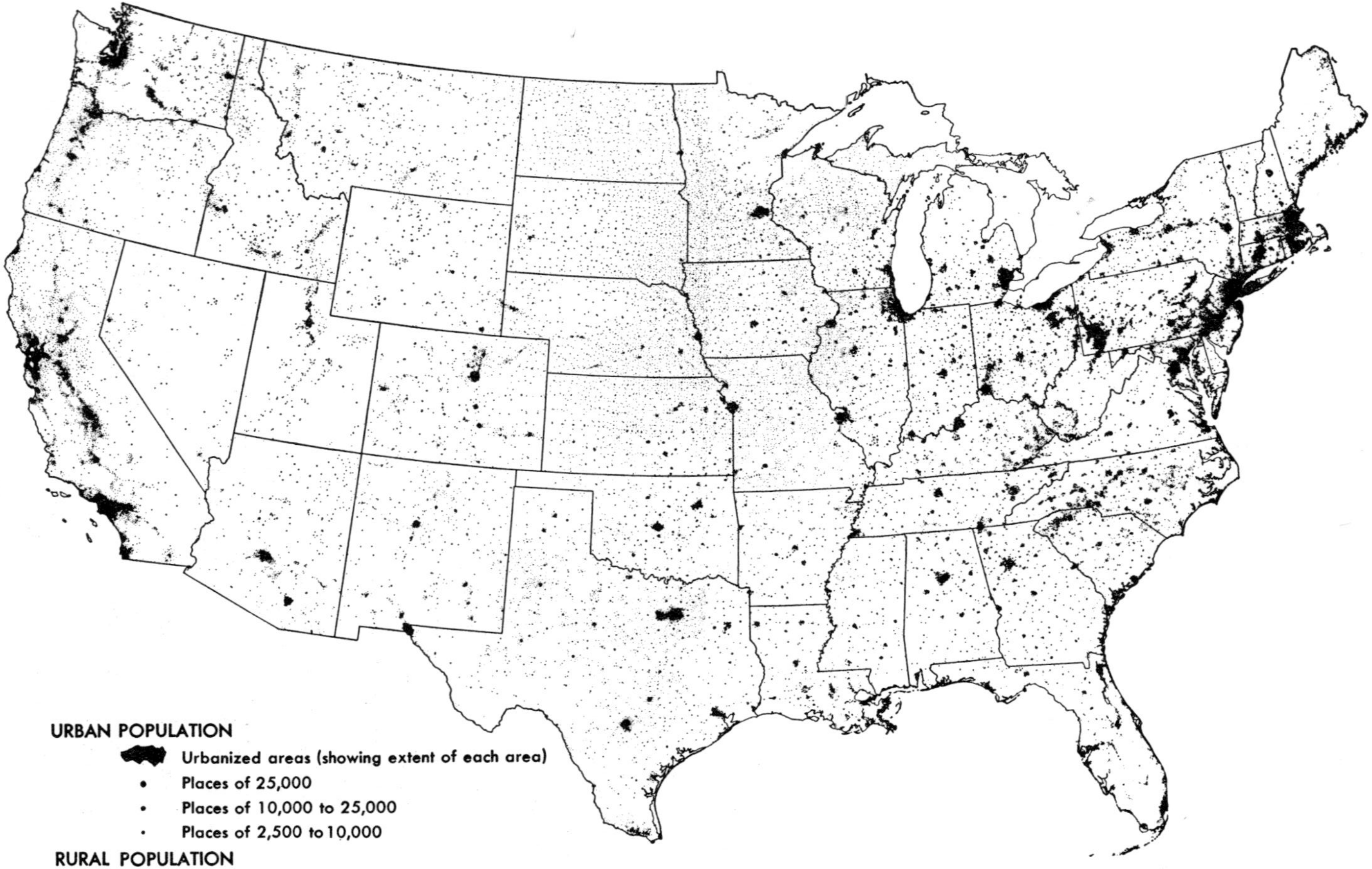
URBAN POPULATION
Urbanized areas (showing extent of each area)
Places of 25,000
Places of 10,000 to 25,000
Places of 2,500 to 10,000
RURAL POPULATION
Places of 1,000 to 2,500

The total area of land given over to urban pursuits or withheld for them may seem inconsiderable—less than 1 per cent of all the land in 1950. But the total impact of urban demands and standards within a competitive economy is reason enough for considering urban uses first. The lowest values of bare urban land are usually some $2,000 per acre, and the values of urban land can amount to $40,000 or more per acre; in contrast, rural land values per acre rarely exceed $1,000 and more commonly are less than $400. Similarly, rentals for urban uses are usually far greater than those for agricultural, forestry, or other major land uses. Urban land is, as an average, perhaps two hundred times as densely populated as rural land, and is valued at perhaps a hundred times as much. These rough comparisons suggest why urban uses of the land generally override other uses. Based on private profit calculations, land will be shifted from other to urban uses whenever returns from the latter are larger, and this will occur whenever there is any considerable demand for the land for urban use.

So, all in all, the one land use trend on the contemporary American scene, of which we may be reasonably sure, is the continuing growth of cities in various forms, and the urban use of land. The chief concern of this study is with the amount and kinds of land the process requires or pre-empts. As to this, in intra-urban and inter-urban situations alike, much will depend on efficient use of the areas given over to, or withheld for, urban aims and purposes. In both spheres, sound planning and vigorous translation of plans into action can lower costs and increase benefits.

ECONOMIC AND SOCIAL ROLE OF CITIES

Urbanism, in the sense of a large proportion of a total population living in cities, as a way of life and of producing a living, is a comparatively recent phenomenon in world history—a matter of the past two centuries or less. It required an agricultural as well as an industrial revolution to establish a large part, often far more than half, of

---

Figure 4, opposite. Population distribution, urban and rural, 1950. (Map from *National Atlas of the United States,* U. S. Bureau of the Census.)

the total population within cities. Until agriculture's productivity rose so that a small segment of the people could produce enough food for everyone, comparatively few people were able to live in cities and engage in nonagricultural lines of production. An industrial revolution also was necessary, to provide a source of productive employment and a means of earning money with which to purchase the food raised by the minority.

Today, however, many countries are primarily urban in character. This is true not only of some of the smaller European countries, which import a large part of their food supply, but also of the United States, a large country with a large and highly developed agriculture. It is, in fact, only because our agriculture is highly developed that the country can be so urban and yet so largely self-sufficient in food supply. Urbanism is much more than a matter of the location of residences of a majority of the population; it involves also many cultural, social, and political aspects, as well as those of an economic kind.

"Though 'cities' have existed in one form or another close on 7,000 years," writes Eric E. Lampard, "the past two centuries, 1750-1950, have witnessed an unprecedented urbanization of people and economic activity in areas affected by the industrial revolution. During this brief moment of history, many towns and villages in Europe and North America have ceased to be mere regional markets for craftsmen and cultivators; they have become vibrant centers for almost all the manufacturing, servicing, and distributive functions developed in an expanding economy. Only food raising and certain extractive processes remained tied to the countryside. The coincident growth of cities, population, and nonagricultural employments seems to have been a characteristic feature of all economically advancing societies."[3]

And Amos H. Hawley has said:

> "One of the most significant social consequences of the industrialization of society is the extraordinary increase in the size and in the organizational complexity of the aggregates in which men live. It has long been the practice to conceive this phenomenon in terms of the city, and an enormous literature, scientific and non-scientific, has accumulated about the city concept. But the processes of cen-

3 Eric E. Lampard, "History of Cities in the Economically Advanced Areas," *Economic Development and Cultural Change,* Vol. 3 (January 1955).

tralization and expansion unleashed by the Industrial Revolution have refused to yield to the confines of corporation limits. Every advance in the efficiency of transportation and communication has further extended the radius of convenient daily movement, thereby diffusing the urban mode of life over a greatly enlarged area. The result has been the emergence of an entirely new type of urban unit. Indeed, the city is the creature of the nineteenth century; its successor in the twentieth century is the metropolitan community. This new urban unit is an extensive community composed of numerous territorially specialized parts the functions of which are correlated and integrated through the agency of a central city.

"Although the development of the metropolitan community has progressed rapidly since the turn of the century, its maturation is far from complete even yet. Manifold changes are still at work in the spatial distribution of population, industry, retailing and services of all kinds and, what is more important, in the total structure of relationships in the city and its tributary area. Because these movements have such radical and hence vital implications for the administration, financing, and functioning of collective life there is urgent need for sound knowledge of the metropolitan community."[4]

In the United States, the growth of cities and the rise of economic output have been highly correlated for a hundred years or more (Figure 5). Total national income, in constant prices, rose about ten times from 1850 to World War I, and has more than doubled since then. Almost exactly paralleling the increase in the earlier period was the growth of urban population.

Cities have been classified into certain primary types, or combinations of types. Harris and Ullman, for example, list three major sources of support for cities: as central places performing comprehensive services for a surrounding area; as transport cities performing break-of-bulk and allied services along transport routes, and as specialized function cities.[5] In a primarily agricultural area, for instance, many types of services can be performed more efficiently if grouped

[4] Amos H. Hawley in Foreword to Donald J. Bogue, *The Structure of the Metropolitan Community—A Study of Dominance and Subdominance* (Ann Arbor: University of Michigan Press, 1949).

[5] Chauncy D. Harris and Edward L. Ullman, "The Nature of Cities," *The Annals of the American Academy of Political and Social Science,* Vol. 242 (November 1945).

at one point, where volume is larger and where customers can get the various services they need. In any system of transportation from one region to another, there necessarily must occur some points where transportation methods or volumes of goods change. Some writers have designated these as "nodes"; a change from water to land trans-

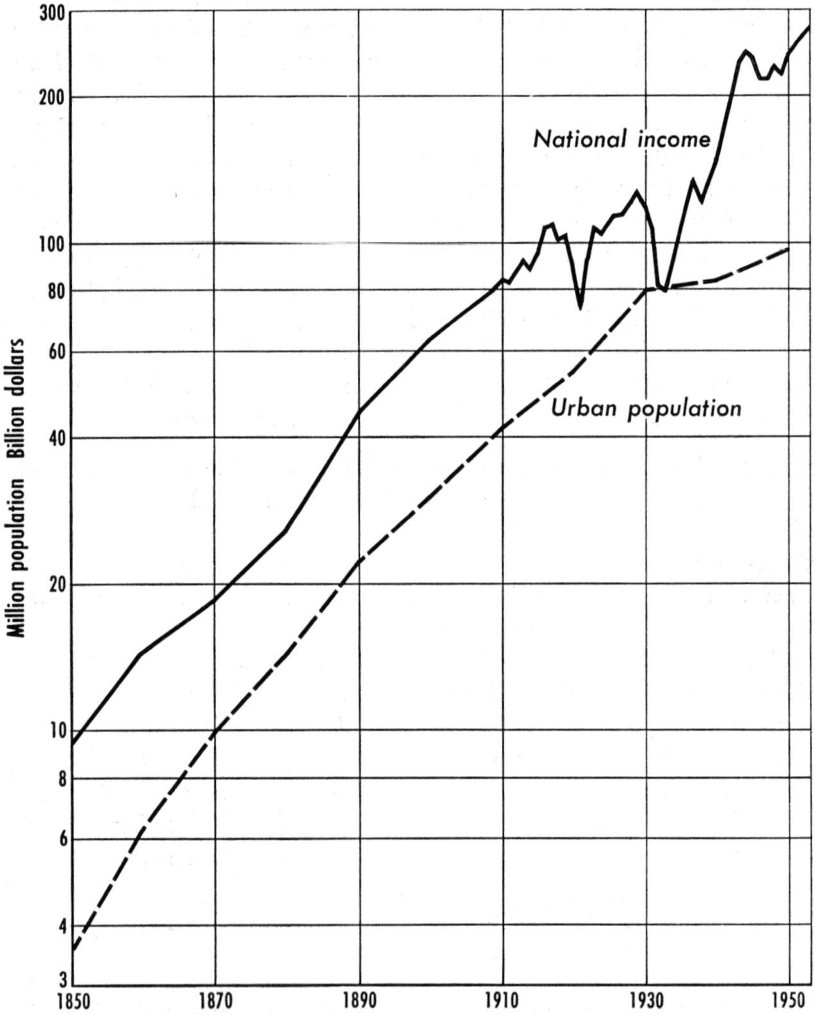

Figure 5. Growth in national income (in 1950 dollars) and in urban population, 1850–1950.

portation is a common example. A town or city almost invariably grows up at such a point. Some cities develop specialized functions, such as mining or manufacturing, which singly or together are based upon various factors of resources, markets, or transport routes.

Isard has shown how the location of cities is related to resources and to transport, how specialized functions develop in each, and how the role and the size of cities change over time. As new cities develop, they may take away some functions from the earlier cities, and thus eclipse them.[6] The existence of a labor force, originally attracted by one set of factors, may in turn serve as a force toward the attraction of other economic activities. The presence of an economic activity which attracts male workers may in turn provide a supply of labor from women and children which makes possible other economic activity in the city. Isard stresses that in time a region comes to have a hierarchy of strategic nodal sites, each occupied by a city with varying functions and perhaps of differing size.

Alexandersson groups economic activities generally found in cities into "city forming" and "city serving."[7] Wherever people are found in any number, schools, retail trade, bakeries, many kinds of personal services, certain types of printing and other communications, and many other activities, will rise to serve their needs. In contrast, certain other types of economic activities are found only in limited locations, and there they serve as city forming. He includes not only mining and forest processing and transportation-nodal functions, but also such types of manufacture as automobiles, which are found in only a few locations. Economies of scale and the relation of volume of output in these specialized areas to total market demand may be critical for some of these activities. A city may grow so large that some activities, such as wholesaling, which would normally be city serving, grow to be city forming also, because of service to other cities. Our larger cities owe their existence to a complex combination of forces, and a substantial part of their total economic activity is for their own use.

Corresponding to the many economic functions of a typical larger American city will be many variations of land use within the city. In terms of land area, the largest use is for residences, including the streets, public areas such as schools, and retail trade areas closely

[6] Walter Isard, *Location and Space-Economy* (New York: Technology Press of Massachusetts Institute of Technology and John Wiley and Sons, 1956).

[7] Alexandersson, *op. cit.*

associated with residential land use. But major transportation in all its forms will require some land;[8] and so will industry.[9]

RESOURCE REQUIREMENTS OF CITIES

The significant resource requirements of cities relate to their economic hinterlands more than to their actual sites. If a city has a large (in an economic more than a physical sense) hinterland, within which it is dominant as compared with other cities, it will be a large and important city. This hinterland may be based upon land resources as in agriculture, or upon mineral, fuel, power, or other resources; part of it may have important transportation-nodal functions; or there may be various combinations of these characteristics. It may well contain, in turn, smaller cities with their own hinterlands and subsidiary cities or towns. This is the hierarchy of strategic nodal sites to which Isard refers.

Within the hinterland there exist economic, political, and social relationships constantly shifting one with the other, yet, usually as a system, remarkably stable as the whole urban cluster develops through the years. Important as are these relationships to the use of urban land, their detailed examination would lead far afield from the limits of this study. Demographers, geographers, and regional economists have in recent years engaged in their study, believing that in the formation and reformation of their patterns clues may be had to the growth of some cities and the decline of others. Some of their findings, sufficient at least to prompt interested readers to refer to their works, are summarized in Appendix B.

The actual site qualities required or preferred for urban use are secondary to these economic considerations. Moderate slopes, good drainage, both surface and within the soil, and good soil qualities are, of course, helpful; construction costs are less under these circumstances, and transportation easier. But several American cities have developed where these conditions were not present. San Francisco,

[8] In this study, transportation outside of cities is treated separately as a land use, in Chapter VII.

[9] There are also certain city-related land uses, such as recreation outside of the city, or municipal watersheds, which use land. We consider these in other chapters.

for instance, is largely built upon rugged terrain; its fine harbor, strategic nodal position with reference to a large hinterland, and the absence of better land sites have led to the use of land that otherwise would not have been chosen for urban purposes. Good harbors, with deep water leading to the ocean and with firm banks for building, are helpful for cities much of whose economic base is transportation-nodal; but, if the other requirements for a transportation-nodal site are present and if there is no competing site of better physical character, harbors can and will be developed under relatively unfavorable circumstances. Los Angeles is a good example.

When a city expands, it is likely to choose first the more nearly level or gently sloping, better drained, and often more fertile soils. Soil fertility is important for residential use, but topsoil can be hauled in. For other urban uses, soil fertility is unimportant; but the good internal soil drainage that is important for many other urban land uses is likely to be highly correlated with good soil fertility. For residential use, gently rolling topography is likely to produce more variety, especially of outlook and view, and hence be desirable; steeper areas also may be desirable, but generally are more costly on which to build. The use of "good" land—meaning land productive for agriculture—for urban purposes, when a major portion of it will be covered with impervious concrete, asphalt, or roofing, is often decried; but in view of the greater economic returns to be had from urban use shifts of this kind are inevitable unless they are restrained by public action.

Urban uses of land are dependent primarily upon the *site* quality of the land, not upon its own physical characteristics, and still less upon its vegetative and ecological characteristics. Cities will expand into forested and nonforested areas with almost equal ease. In this regard, urban land use is at the opposite extreme from forestry and grazing (discussed in Chapters V and VI), where natural vegetation and ecological relationships are essential elements.

By and large, land is shifted from rural to urban uses, and from one urban use to a more intensive one, in response to economic pressures. As the total economic activity in a city's economic hinterland grows, so does the total economic activity within the city—assuming that some other city does not capture all the growth. More economic activity eventually requires more land. As the total land area of the city increases, the use of particular parcels and districts shifts. The largest and most readily observable shift is in residential use—the expansion of residential use onto previously rural lands, the intensification of residential use by construction of large apartment houses, and the doubling up of people in the older and now crowded residential districts. At the same time, some formerly residential areas shift to business and trade or to manufacturing and transportation. Some of these latter shifts take place in the heart of urban areas but, increasingly in recent years, some occur at urban margins and even in the hearts of open country. The new residential areas are commonly characterized by suburban shopping centers, where adequate parking space and the requirements of a new style of architecture combine to require relatively large areas of land.

PAST AND PRESENT USE OF LAND BY CITIES

How much land have people in urban areas used, and how much will they use or "need" in the future? The area of land for residential purposes depends partly upon the number of people in cities, but also upon the density of land use—the number of people to the square mile. The area used for transportation, industry, and other purposes depends to a large extent upon the kinds of economic activity in a city. The relationship between total land area and population may be no less important than the changes in numbers of people, in estimating future land use needs.[10] Underlying these various factors are basic questions to do with land use classification and concepts, which have developed over the years.

Basic Urban Land Use Concepts

In order to examine the trends of urban land use, it is necessary first to decide what an urban area is. The Bureau of the Census publishes some data for four different kinds of urban-like areas.[11] These kinds of areas are not mutually exclusive, but rather overlap greatly; and the amount of information about each also varies substantially.

[10] For a comprehensive discussion of this subject, including a listing and analysis of other recent writings, see F. Stuart Chapin, Jr., *Urban Land Use Planning* (New York: Harper and Brothers, 1957).

[11] These classifications are treated in greater detail in Appendix C.

The kinds of areas and the major characteristics of the data are as follows:

1. The "standard metropolitan area" (SMA) is based on a group of counties containing at least one city of 50,000 or more inhabitants. By far the greater part of the total area of all SMA's is nonurban in character. Data on this basis have been tabulated only after World War II.

2. The "urbanized area" is very similar to the SMA in that it, too, is centered upon at least one city of 50,000 or more inhabitants, but spreading only to include the urban fringe around each city. While including less nonurban land than the SMA, the urbanized area also excludes smaller independent cities. Data on this basis are available only beginning with 1950.

3. The "incorporated city" of 2,500 inhabitants and above is more a political than an economic entity. While cities include some nonurban land and exclude some urban land, they more nearly coincide with the urban land use category than do any other areas for which data are available. Moreover, these data have the great virtues of applying to small as well as to large cities, and of being available (in part) from 1900 to 1950.

4. The "urban place" includes all incorporated or unincorporated places of 2,500 inhabitants or more. Unfortunately for our purpose, the data about urban places is limited to population, and does not include area of land.

Each of these four classifications has its special values for particular purposes. For those classifications including substantial areas of land other than urban, the possibility of double counting of the land area arises. Land outside of cities might be included under urbanized areas or SMA's, which might also be considered under agricultural land, or forested land, or even grazing land. The length of the historical record is also important for our purposes. For these reasons, the incorporated city is used as the primary basis for detailed analysis of urban land use throughout this study, but for purposes of comparison the implications of data based on the other Census classifications are also considered.

It will be readily seen that some of the apparent lack of agreement among land use data can grow out of the lack of a clear set of concepts concerning urban land. Broadly, two concepts are to be distinguished: (1) actual *use* of land for urban purposes, and (2) land *withdrawn* from other uses because of potential urban use or because of proximity to urban areas. In this study, land is considered as "used" for urban purposes if it has upon it residences, trade or business places, industries, transportation, public places of various kinds, recreational areas, and so forth. Use may well change from time to time. Some land, however, cannot by any reasonable definition be said to be "used" for urban purposes; it is idle. The total of the idle and the urban used land is, for lack of a better term, termed here "withdrawn" land. As far as other land uses are concerned, the total withdrawn area is used by the city, since it is the city which has taken it away from other use.

Refinements and definitions might be introduced into each concept, but the basic difference remains. There is the approach of the city planner, who is concerned with the "need" for land for various purposes in the city he is studying, and there is the approach of the general land economist, who is trying to consider the use of land for many purposes, and knows that the land the city takes out of other uses—whether actually used yet or not—is highly important. Each concept is accurate and useful; the difficulty comes when it is confused with another. As will be seen, there is an important difference in the area of land involved, depending upon which concept is used.

The matter is further complicated by the unit of land area and of population that is being studied. Broadly, there are two alternatives: (1) the city as a legal or political unit, and (2) the city as an economic unit. The former may include some land not included in the latter, if large vacant areas within the city borders are actually used for other purposes. But in modern times in the United States the city as an economic unit is often the larger of the two. The economic city spreads beyond the boundaries of the political city to take in urbanized areas that lie in satellite cities or that are unincorporated.

The nature of this data problem is quickly seen in the schematic table (Table 5). In one direction, a differentiation is made between (1) land actually used for urban purposes, and (2) land withdrawn by the urban area from other land uses; in the other direction, a distinction is made between (A) land within cities as political units, and (B) land within cities as economic units. In practice, almost no data on actual urban land use conform completely to any of the four possibilities shown in the table. The older Census and other data for cities mostly relate to cities as political units. Not all of their area is actually used for urban purposes. While it may be presumed that all land within such cities is withdrawn from other uses, there is no direct evidence that this is correct; and it is highly unlikely that all withdrawn land is included. Data of this kind conform most nearly, but not exactly, to A2 in Table 5. Some Census data on urbanized areas and standard metropolitan areas attempt to relate the economic city to the use of land, but, the results are not fully satisfactory, and in no case do the data distinguish between land used and land withdrawn.

Table 5. Scheme for analysis of data on urban land use

	A. Land within cities as political units	B. Land within cities as economic units
1. Land *actually used* for urban purposes—occupied by residences, other buildings, schools, streets, railroads, parks, etc.—includes industrial and commercial as well as residential use—includes public as well as private use.		
2. Land *withdrawn* by urban area *from other uses* —includes vacant areas within urban area, or outside of it if rather clearly withheld from other uses, in addition to foregoing direct use areas.		

Although available data do not fit precisely this conceptual framework, a clear idea of it, or of some alternative framework, is necessary for meaningful discussion.

In the case of the land "needed" for urban uses, some land for playgrounds, local parks, and similar uses is normally included. This land may be considered as used for urban purposes, or for recreational purposes. Most statistics on urban land include such areas. The remainder of the chapter includes such park and recreation areas as parts of citics; but in Chapter III, where recreational use of land is discussed, the same areas will again be examined in their character of recreational lands. In Chapter VIII, when considering the balance among various land uses, an attempt will be made to exclude duplications of this kind.

Maximum Acceptable Urban Land Density

To provide a basis for later comparison with actual urban land density, some estimate of maximum acceptable urban land density is desirable. In general, the higher the density of urban land use, the lower are many costs for facilities such as water supply, sewage disposal, and transportation. However, there is a limit above which higher densities of urban land use are not acceptable. And, for those who can afford them, lower densities are often desirable.

Ludlow has made some careful calculations on this matter.[12] He starts with standards of acceptability for number of rooms per person, area of floor space per person, and residential lot sizes. From these he calculates maximum allowable density of purely residential land use, with no allowance for transportation or other services. To this he adds an allowance for such services within the local community; and then, finally, an allowance of area for the central business district, the larger social services, and the industrial and commercial areas needed in the employment of the workers. His results are summarized in Table 6.

Table 6. Form of residential housing

Density	Detached houses	Row houses	3-story apartments	13-story apartments
Persons per acre of:				
net dwelling area	25	65	160	350
residential neighborhood area	19	42	72	112
total developed urban area	14	23	29	35
Persons per square mile of total developed urban area	8,950	14,700	18,575	22,400

The concern here is chiefly with the last row of figures—the number of persons per square mile for the whole urban complex, with maximum acceptable densities for different patterns of residential

12 William H. Ludlow, "Urban Densities and Their Costs: An Exploration into the Economics of Population Densities and Urban Patterns," Part II in Coleman Woodbury (ed.), *Urban Redevelopment: Problems and Practices* (Chicago: University of Chicago Press, 1953), pp. 101-220.

living. In practice, of course, few cities or other urban areas for which statistics are available will consist of one of these types of residential living alone. In practice, also, some persons will choose and will be able to afford more room, or a lower density of land use. However, it is also true that in practice many people live in more crowded conditions than have been assumed here as the maximum acceptable. Actual densities are very much lower than these maximum acceptable ones, mainly, we believe, because much of the land within cities is idle or inefficiently used.

Estimates of maximum allowable densities for different types of housing can be shown in a generalized scheme of relationship between population density and distance from city center (Figure 6).

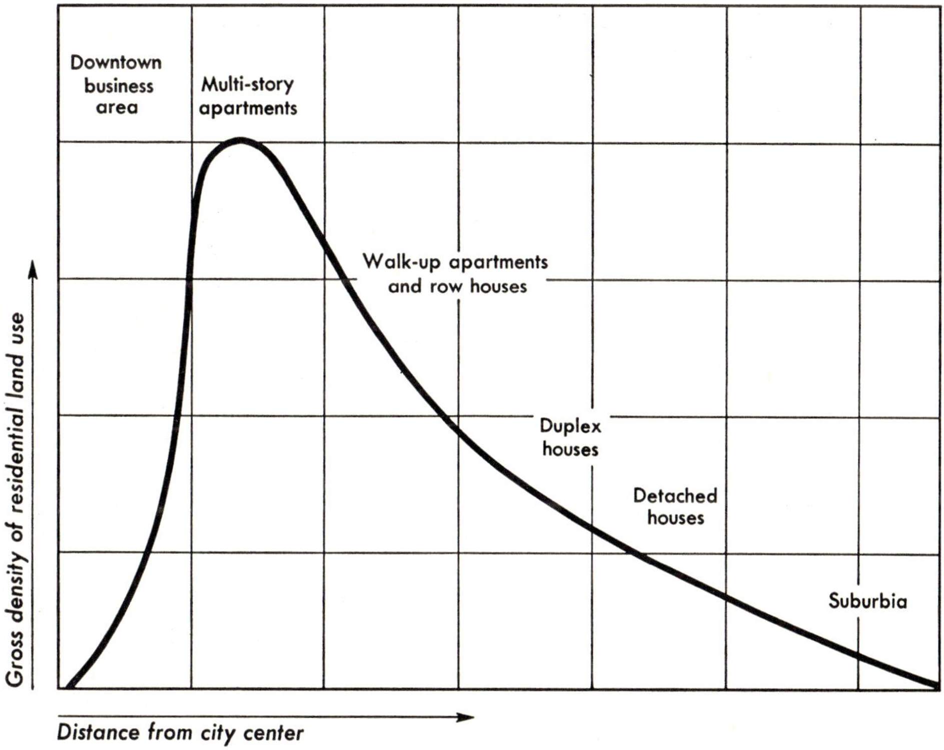

Figure 6. Schematic relationship of residential density to distance from city center.

The pattern is never as regular as the diagram suggests. Major transportation areas (such as railroad yards) and industrial areas would result in low residential density, around which density might rise somewhat as around the business district. In any particular town the distances between the different types of housing would vary. Moreover, the directional pattern of zones of different densities might differ greatly in one city from that in another. Along a main street, for instance, business development might stretch for many blocks or miles, while on the cross-streets the transition to detached houses might occur quickly. Transformation of previously detached single-family housing into multi-family occupancy, as often occurs when an old neighborhood degenerates into a slum, might well make a different pattern of density. The development of a suburban shopping center would also change the geographical pattern. The figure should be regarded as a generalized picture of the residential density patterns in different parts of cities.

Land Used Compared with Land Withdrawn

As a first step in measuring how much land the city withdraws from other land uses, consideration must be given to the need for some vacant land within the city available for building when needed. Ludlow's calculations are based upon complete occupancy of the land, but clearly this is impossible in a growing city, no matter how frugally and wisely land use is planned and carried out. Some vacant lots are always necessary; under the best planning and management their area is small and their time of vacancy short. In practice, however, a great deal of the land left vacant within cities has been withdrawn for a particular purpose which has not yet come about; meantime its immediate use may contribute little to the city's economy, and often it has no productive purpose at all.

Bartholomew and associates have prepared detailed plans for a considerable number of cities of varying sizes, types, and locations.[13] The cities selected were necessarily those which were sufficiently interested in city planning to spend money for preparation of the Bartholomew plans; thus, they do not constitute a statistically random sample. Nevertheless, the cities show relationships that may have con-

[13] Harland Bartholomew, *Land Uses in American Cities* (Cambridge: Harvard University Press, 1955).

siderable validity. The plans were prepared over a period of several years, and hence the data have the advantage of applying to a longer time-span than a single year; and most of them were made after 1940. The data relate to cities as political, not economic, units. The cities studies were grouped into central and satellite cities. Some of the satellite cities were residential and others industrial in character. In a few instances data are shown for the total urbanized area of which the central city was the core. Land uses within the cities were classified. All land used for residential, industrial, transportation, public, and recreation uses was considered "developed"; all other land was "vacant." Developed land, in this terminology, is equivalent to the "used" land of our classification; but the total of developed and vacant land may be either more or less than our category of "withdrawn" land, although perhaps it averages close to it.

Among the fifty-three central cities studied, there was great variation in relationship between total area within city boundaries and the developed area (Figure 7 and Appendix D, Table D-1). In the case of nine of the fifty-three cities, or 17 per cent of those studied, less than half of the total area was developed, or the total area was more than twice the developed area. The most extreme case was Corpus Christi, Texas, where the developed area was little more than 10 per cent of the total area. On the average, 56 per cent of the total area was developed, or the total area was 1.79 times the developed area. These averages are affected considerably by the one extreme case, but in most cities there was a considerable area of vacant land.

Among the thirty-three satellite cities there was even greater variation in the relationship between developed and total area (Figure 8 and Appendix D, Table D-2). Of the thirty-three cities, nine or 27 per cent had more than twice as large a total area as developed area. On the average, 64 per cent of the total area was developed, or the total area was 1.57 times the developed area. The absence of an extreme case among this group of cities affects the averages to the advantage of developed land.

One measure of the withdrawal of land from other uses may be the area of land actually plotted and subdivided. Once land has been plotted and subdivided, and streets laid out—even if not improved—the land is unlikely to be used for agriculture, forestry, or other use. It may be "ripening" for eventual or hoped-for urban use. Vacant land may be productive in the sense that ready availability of sites can facilitate development for a variety of uses. However, the area

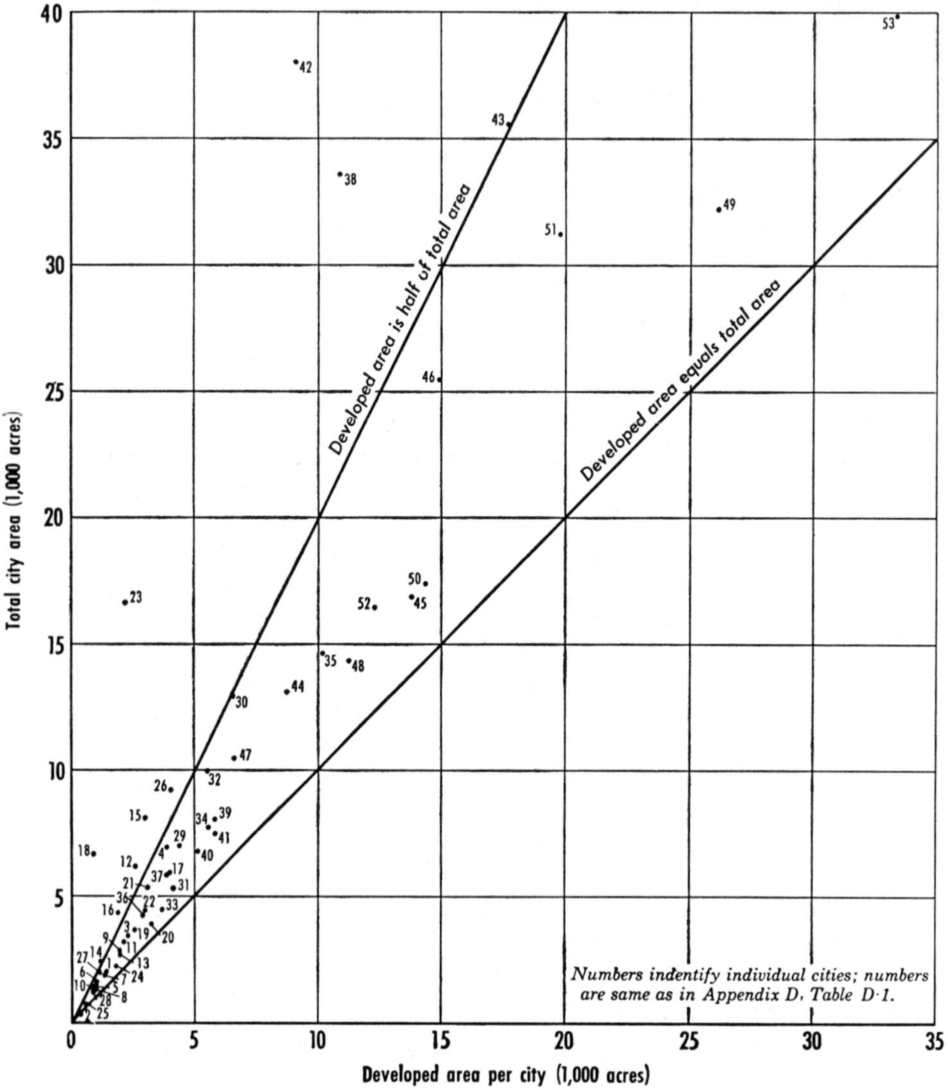

Figure 7. Relation of total and developed areas per city, 53 selected central cities.

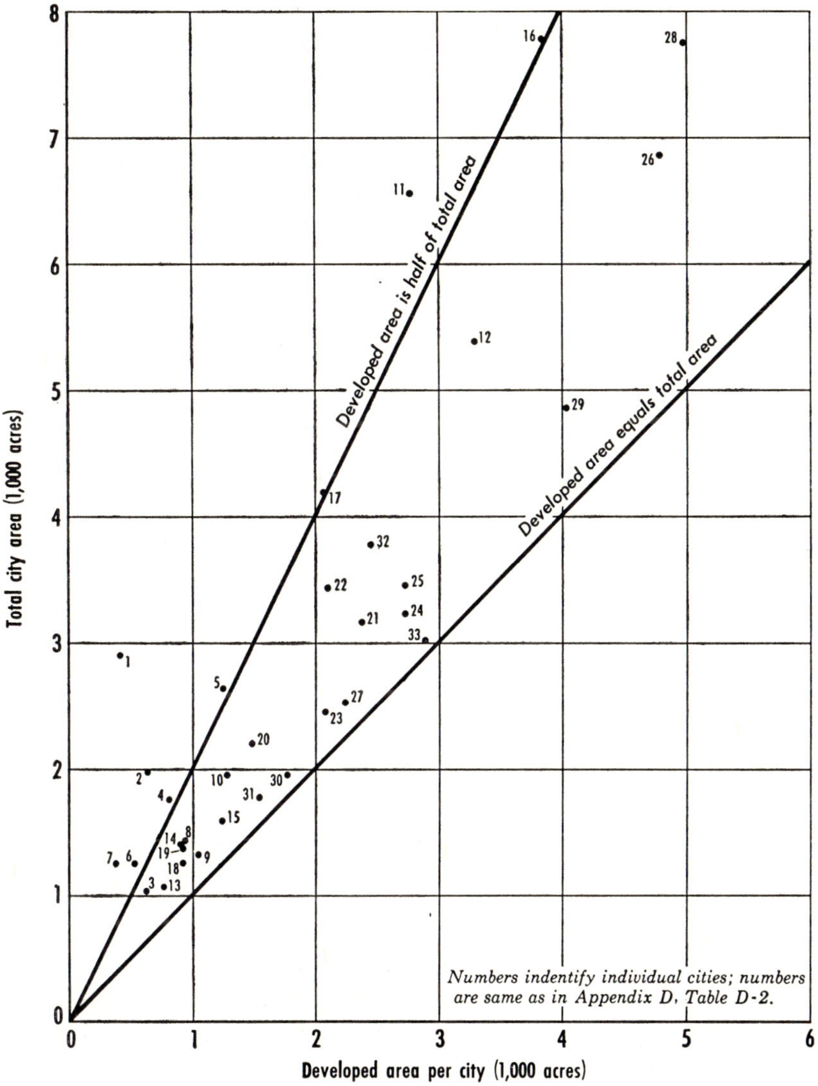

Figure 8. Relation of total and developed areas per city, 33 selected satellite cities.

required for this purpose is not large, and tracts ideally should not be held out of use for long periods of years. In practice, the areas plotted and subdivided are often far larger than necessary and they develop very slowly, and sometimes not at all, into active urban use. When an area is subdivided and the plan filed with the appropriate local agency, under most state laws this dedicates to public ownership the streets and alleys and sometimes other public areas as well. Thus, even if one man owned all the lots he still would not own the entire area. This is obviously a further handicap to use of the land for agriculture, forestry, or grazing.

For many decades prior to World War II, land around the larger towns and cities had been subdivided in anticipation of future settlement, far in advance of actual need. In general, the process was subdivision by a "developer" or speculator who sold, or tried to sell, the lots to individual buyers, each of whom would arrange to build as he chose and could finance.

This process has been studied in some detail for New York State.[14] Large-scale speculation in urban subdivision began in that state in the first half of the nineteenth century, and received a tremendous boost when the Erie Canal was completed. The typical process was for a speculator to buy a farm or other rural land, often on credit, then to subdivide and sell lots as he could, often extending credit to purchasers and in turn borrowing from banks or others to finance his own operations. As the lots were sold and some building occurred, a demand for public services such as water, sewerage, and power arose. As the latter was met, it often meant heavy bonded indebtedness for the municipality, which in turn meant higher taxes to the areas affected. The combination of higher taxes and speculative prices drove the land out of agriculture—in fact, destroyed its value for that purpose by raising the costs well above any possible income. Because many of the lots were unsold or not built upon, partly because the area subdivided was so far in excess of that actually needed, the lands often became tax delinquent, especially with the first economic depression and pricking of the speculative bubble. This in turn put a much heavier burden of taxes on the developed properties, effectively shifting much of the burden of speculation to those who had not engaged in it. With mounting tax delinquency and private debt, much of the land accumulated charges greater than its value. Properties

[14] Phillip H. Cornick, *Premature Subdivision and Its Consequences* (New York: Columbia University, Institute of Public Administration, 1938).

were foreclosed upon, debt and tax charges accumulated; and the processes of clearing title have been so costly and slow that no one, including the municipality in question, could afford to obtain clear title to the land. It thus became worthless and unused.

Table 7. Percentages of all taxable parcels vacant in 1935-36, for selected areas in New York

Area	Percentage of taxable properties in vacant lands: Whole area	Central part of area	Outlying districts
City of Buffalo	14	2	23
Erie County, all towns	74	—	—
suburban towns	—	—	79
intermediate towns	—	—	42
outlying towns	—	—	18
City of Rochester	19	3	29
Monroe County, all towns	—	—	54
suburban towns	—	—	63
outlying towns	—	—	32
City of Syracuse	25	4	[1]12 [2]38
Westchester County	59	—	—
4 cities	48	—	—
24 villages	—	48	—
unincorporated area	—	—	78
City of Yonkers	60	13	[1]36 [2]70
City of New York	22	7	35

[1] Intermediate wards.
[2] Outlying wards.

SOURCE: Based on data in Phillip H. Cornick, *Premature Subdivision and Its Consequences* (New York: Columbia University, Institute of Public Administration, 1938).

Cornick makes it clear that substantial areas were in this status in 1935-36 (Table 7). For the larger cities, from 14 per cent to as high as 60 per cent of all parcels of land were vacant or unimproved, and for some counties the ratio was even higher. The proportion of vacant lands was relatively low near the center of the city, but rose rapidly around the edges of the central city and in the adjacent suburbs, where speculative activity in relation to actual building had been the greatest. It then usually declined at greater distances, which had been too remote even for the most over-optimistic speculators. A large part of this subdivided but vacant land had been in that status for several decades when Cornick made his study. And a major part of the tax delinquency at the time of the study—when all tax delinquency was relatively high, due to the Great Depression—was from the vacant lots.

An earlier study shows essentially the same general picture for a Michigan city.[15] The study covered the period 1909 to 1931, inclusive. During those years, the vacant lots in the Grand Rapids metropolitan area ran consistently from 39 to 46 per cent of all subdivided lots. The authors point out that lot subdivision is highly speculative, varying greatly from year to year. While building also varies from year to year, during the period studied building was much more regular than lot subdivision. The number of lots available at the beginning of this period was almost as great as was used during the whole period; and the numbers of vacant lots at the end of the period would have, by the most generous calculations, sufficed for seventeen years of building. Indeed, at a more probable rate of building, they would have sufficed for thirty-five years. But, in spite of the relatively large numbers of vacant lots, subdivision continued in new areas. The situation differed greatly in different parts of the metropolitan area; in the central part of the city, by the time this study was made, vacant lots were only 10 per cent or less of all lots—a degree of vacancy that probably was healthy. On the other hand, in the more outlying areas from 60 to 80 per cent of all subdivided lots were vacant. The authors present no direct analysis of the length of time various lots had remained unoccupied, but it seems clear that many had been vacant for years. In this case, of course, accumulated taxes, interest, and other charges would mount considerably.

Chicago is another city studied in some detail, which has experienced subdivision far beyond need.[16] Some of the lots subdivided well before 1870 are still vacant in spite of the great increase in population the city has since experienced. Some of the lots subdivided at later periods are also vacant; one boom period for subdivision was be-

[15] Ernest M. Fisher and Raymond F. Smith, *Land Subdividing and the Rate of Utilization* (Ann Arbor: University of Michigan, School of Business Administration, Bureau of Business Research, 1932).

[16] Herman G. Berkman, "Decentralization and Blighted Vacant Land," *Land Economics,* Vol. 32, No. 3 (August 1956). See also references cited therein.

tween 1919 and 1929, but in 1950 a majority of these lots still were vacant. In the meantime, they had incurred large costs to the municipality for streets, sewers, water, and other services. Many of the vacant lots were tax delinquent, some for many years; and often it was impossible to locate their "owners." In many cases the original size of lot was too small to meet modern building needs, so that lot consolidation was necessary but extremely difficult. As a result of all these factors, much of the vacant land is worse than worthless—the costs of assembling it from the many and scattered "owners" into units usable today, and of paying off the accumulated charges, are greater than its value. The land lies vacant, and settlement proceeds around and beyond it. In an effort to deal with this problem, legislation has been enacted to provide simpler methods for consolidating and clearing title, and for direct writing off of some of the accumulated charges, partly by the state of Illinois and partly by the federal government. The constitutionality of this legislation has been upheld, but it is still too early to judge of its effectiveness.

Similar situations exist in or around nearly all American cities; the cases cited here have been selected merely because detailed studies of them have been made.

So far as we know, no comparable studies have been published for areas subdivided since World War II. But in two unpublished studies of Michigan areas somewhat similar and yet somewhat different results were found.[17] Where subdivision followed the old pattern of a speculative subdivider selling lots to individuals who then sought financing to build houses (as in most of the southwestern Michigan area studied), the results were similar to those in prewar subdivisions. A large proportion of the lots remained undeveloped; the author of this study concludes that 41 per cent of the subdivisions were premature. Most of the subdivisions established immediately after the war were poorly planned, hurriedly undertaken in hopes of cashing in upon the expected boom in housing and the clear need for additional building areas, and thus their proportion of use was comparatively low. Later subdivisions were better planned and had a higher rate of development, but vacant lots were still plentiful.

[17] John E. Hostetler, "Subdivision Trends in Southwestern Michigan," Master's thesis, Michigan State University, 1957; Louis A. Vargha, "Independence Township: A Township in Transition, A Study of Suburbanization in a Selected Portion of the Rural-Urban Fringe of Oakland County, Michigan," Master's thesis, Michigan State University, 1958.

In the area studied northwest of Pontiac and Detroit, the early subdivision followed the pattern already noted—vastly more lots established than were used, mounting charges against the land, much tax delinquency, and consequent seriously fouled up titles. Subdivision since the war on the whole has been different. It has generally been made by housing constructors, who acquired blocks of land, subdivided it into lots, had or obtained capital, constructed houses, and sought to sell the house and lot as a package. This type of subdivision has had its problems; not all the lots have been used promptly, but the rate of development has been much faster, and the amount of land in idle lots much lower, than in the older type of subdivision. However, it is also true that larger tracts of land, as yet unsubdivided, lie between the subdivided tracts, and that there is a tendency for them to become idle because of the mounting tax charges and mounting land values which make them unattractive for agriculture.

In many states, subdividers are now required to post bonds or collateral to ensure street and other public developments. This is a powerful incentive to subdivide only when market prospects are reasonably good, and should operate to reduce premature and excessive subdivision.

The full effects of any changes in subdivision practices since World War II are not yet evident. The high level of economic activity since the war has perhaps induced many people to pay taxes on vacant lots that would have gone delinquent during the depression. There have been more public controls over land subdivision in recent years than earlier. Perhaps more important than either of these factors, methods of land subdivision, dwelling construction, and financing have all changed. Single lots are no longer commonly sold to individual buyers, who then arrange for building and financing as best they can. Instead, the usual method now is subdivision of large tracts, construction by one builder who sells the completed houses to individual buyers, and financing under some sort of group arrangement, often underwritten by the federal government in one way or another. Vacant land within and adjacent to cities is no longer likely to be in the form of single vacant lots, or small groups of them, but of larger tracts not yet subdivided and developed. The unit of land vacancy is thus much larger than formerly; but there is no evidence that the proportion of vacant land is lower.

Many students of urban development—including those noted—decry "premature" subdivision of rural land into lots. It is true that much land is effectively taken out of agricultural or other production in this way, that large costs are incurred, and that substantial loss occurs to society and to some individuals. But the question may well be raised: is subdivision really "premature" from a private profit viewpoint, assuming no restrictions over it? Subdivision of rural land in order to facilitate its development for urban purposes offers excellent prospects for profit—prospects that sometimes are realized. The difference between the value of land, based on its earning power for agriculture or other rural use, and its value even for modest housing, is often considerable. In urban uses of most kinds, the cost of raw unimproved land is often a very small part of the cost of the land provided with public services of all kinds plus the cost of the building itself. The raw land price can often be boosted considerably, proportionately speaking, without making the total cost prohibitive. The prospects for profit from subdivision may lie in actual development of the land, or only in its sale to gullible "suckers." If the subdivider misjudges the market and loses money, then the subdivision was "premature" from his viewpoint.

If there are prospects of future gain through subdivision, they are promptly discounted into higher present land values. If it is the consensus of land speculators that a lot will be worth $500 twenty years from now, and if the effective interest rate is 6 per cent, then that lot will come to have a present value of $156 (ignoring taxes and other costs). Present prices of rural land will be bid up, until the expected gains in value are fully absorbed in carrying costs, including taxes, interest on investment, etc. The full possibilities of gain will be absorbed in the interest on the asked price. In some situations the consensus of future values is too low, and the smart present buyer will reap a gain; in other cases, the present consensus is too high, and the present buyer will lose. There seems good reason to believe that the latter situation has prevailed more commonly than the former. But, in the absence of controls over the rate of subdivision and in the absence of means of capturing any "unearned increment" in land values, the (imperfectly) competitive market for land will lead to a large degree of "premature" land subdivision with consequent large areas of idle land around the edges of built-up areas.

In considering the relationship between area *used* for urban purposes and area *withdrawn* because of urban use, at least four major categories of land use arrangements seem evident: (1) land actually used, in the sense that Bartholomew uses the term—not only residential and other building use, but streets, railways, industrial and commercial establishments, public buildings and grounds, recreational areas and other uses within a generally urbanized area; (2) vacant lots, usually within city boundaries; (3) larger vacant tracts, either inside or outside of city boundaries, but engulfed by or intermingled among developed tracts in such a way that because of size of tract, high general real estate taxes, special taxes for water supply or sewage disposal or other purposes, or because of speculative land values, uses other than urban are largely or wholly precluded; and (4) land outside of presently developed urbanized areas which nevertheless is subject to some of the foregoing influences to a degree that makes its productive use for other purposes difficult.

To our knowledge, no well-defined statistics of vacant land have been collected, even for sample areas, showing uses in and around cities according to the foregoing classes. There would be considerable difficulty in obtaining such data. Specific standards would have to be developed to supplement the general definitions. There would be a problem in defining the boundaries of the areas to be surveyed; the legal boundaries of cities are obviously inadequate, but no other easily determinable alternative exists. The boundaries of the areas to be studied would shift outward as the cities grew. In spite of the difficulties, data are urgently needed. Although urban use of land is perhaps the most important major use of land in the United States today—certainly it is the use that will increase the most over the future—in any over-all and quantitative sense we know least about it.

Even if quantitative information is seriously deficient, however, it seems certain that there is a great deal of vacant land in and adjacent to urbanized areas. Within city boundaries it is probably not less than one-third, on the average, and in many cities it runs to over half the total area. There is a general belief that the situation is worse in the new suburbs.

"Because of the leapfrog nature of urban growth, even within the limits of most big cities there is to this day a surprising amount of empty land. But it is scattered; a vacant lot here, a dump there—no one parcel big enough to be of much use. And it is with this same kind of sprawl that we are ruining the whole metropolitan area of the future. In the townships just beyond today's suburbia there is little planning, and the development is being left almost entirely in the hands of the speculative builder. Understandably, he follows the line of least resistance and in his wake is left a hit-or-miss pattern of development.

"Aesthetically, the result is a mess. It takes remarkably little blight to color a whole area; let the reader travel along a stretch of road he is fond of, and he will notice how a small portion of open land has given amenity to the area. But it takes only a few badly designed developments or billboards or hot-dog stands to ruin it, and though only a little bit of the land is used, the place will *look* filled up.

"Sprawl is bad aesthetics; it is bad economics. Five acres are being made to do the work of one, and do it very poorly. This is bad for the farmers, it is bad for the communities, it is bad for industry, it is bad for utilities, it is bad for the railroads, it is bad for the recreation groups, it is bad even for the developers."[18]

Present Use of Land by Cities

In considering present "use" of land by cities, it is essential to bear in mind our earlier distinction between land actually used by cities and land withdrawn by cities from other use. In neither case does the available information fit precisely, but some data more nearly fit one category than the other, and some of the apparent differences in data arise from differences in concepts. Data are available from two sources: (1) the Bartholomew studies of particular cities, mentioned earlier (page 66); and (2) Census data for urbanized areas, standard metropolitan areas, and cities.

Bartholomew and associates have shown the area of developed land within the cities they studied, according to the kind of use made. It will be recalled that their term "developed area" applies to approximately the same kind of land use as our term "used area." There is a marked difference in total area used per 100 of the total population and an even greater difference in the area for some purposes, according to the size of the city (Table 8). The smallest central cities studied were less than 50,000 in total population; they required nearly 10 acres of developed land per 100 population, or a population density

[18] William H. Whyte, Jr., "Urban Sprawl," *Fortune* (January 1958).

Table 8. Urban developed area for various purposes, per 100 population, in cities and urban areas of various sizes and types[1]

Type of city or urban area	Size of area (in total population) (1,000)	Number of areas	Acres of occupied area per 100 persons, for: Residential use	Commercial use	Industrial use	Railroads	Parks and playgrounds	Public and semipublic property	Streets	Total
Central cities	up to 50	28	3.94	.31	.57	.50	.51	1.32	2.82	9.97
	50-100	13	2.98	.21	.38	.39	.52	.87	2.66	8.01
	100-250	7	3.33	.23	.47	.43	.46	.90	2.21	8.03
	250 and over	5	2.02	.21	.43	.22	.43	.48	1.25	5.04
	all	53	2.73	.23	.45	.33	.46	.75	1.94	6.89
Satellite cities	up to 5	7	7.42	.82	1.77	1.22	.81	6.17	8.79	27.00
	5-10	6	8.03	.31	.53	.82	1.51	3.11	7.11	21.43
	10-25	10	6.77	.28	.21	.40	.62	1.69	3.27	13.24
	25 and over	10	2.32	.18	.78	.34	.20	.40	1.55	5.77
	all	33	3.65	.22	.69	.40	.38	.95	2.40	8.69
Urban areas	all	11	4.16	.39	.84	.92	.68	3.75	4.10	14.84

[1] Data are taken from tables in Appendix to Harland Bartholomew, *Land Uses in American Cities* (Cambridge: Harvard University Press, 1955). Data are for various dates, mostly after 1940.

of about 6,400 persons per square mile of developed area. For the central cities of 250,000 and over, the land required per hundred persons was only slightly more than half as great, or the density per square mile was nearly twice as high. Satellite cities were much smaller, on the average, and so land used per hundred persons was much higher, or densities much lower; but for cities of the same size, the differences may have been due entirely to the smallness of the sample.

The great saving in land used, in the larger as compared with the smaller cities, was in area required for streets, for public and semipublic property, and to some extent for residential use. In the very small towns, more land was used per hundred persons for streets than the land requirement for all purposes in the larger towns. Considerable savings in land use for residential purposes occurred for the smaller towns, but the savings among the larger cities were more modest. The area used for commercial, industrial, and park and playground use showed no consistent relationship between city size and area per hundred persons. Some savings were made in area for railroad use, but in terms of the total area involved the saving was not large.

Among the cities studied by Bartholomew and associates, some relationship was shown to exist between size of city and the percentage of its total area developed (Table 9). Developed area was a larger proportion of total area for the larger than for the small cities. But the relationship was not close; the variations are probably due to the limited number of cities studied in each case. It should also be noted that population density, on a developed area basis, reached Ludlow's estimated maximum acceptable figure for detached houses only in central cities of 250,000 and over, and only in satellite cities of 25,000 and over. For smaller cities, densities were below this maximum acceptable level. Moreover, all cities, but especially the larger ones, surely contained apartment houses, duplexes, and other multiple-unit structures. For these, maximum acceptable densities are higher. For no group of cities did density reach the allowable figure for row houses.

A rather consistent relation is shown between size of city, measured on a logarithmic scale, and average density of population, whether the latter be measured on total area or on developed area (Figure 9). A regularity of relationship of urban population data with the logarithm of number of persons in the city has been noted

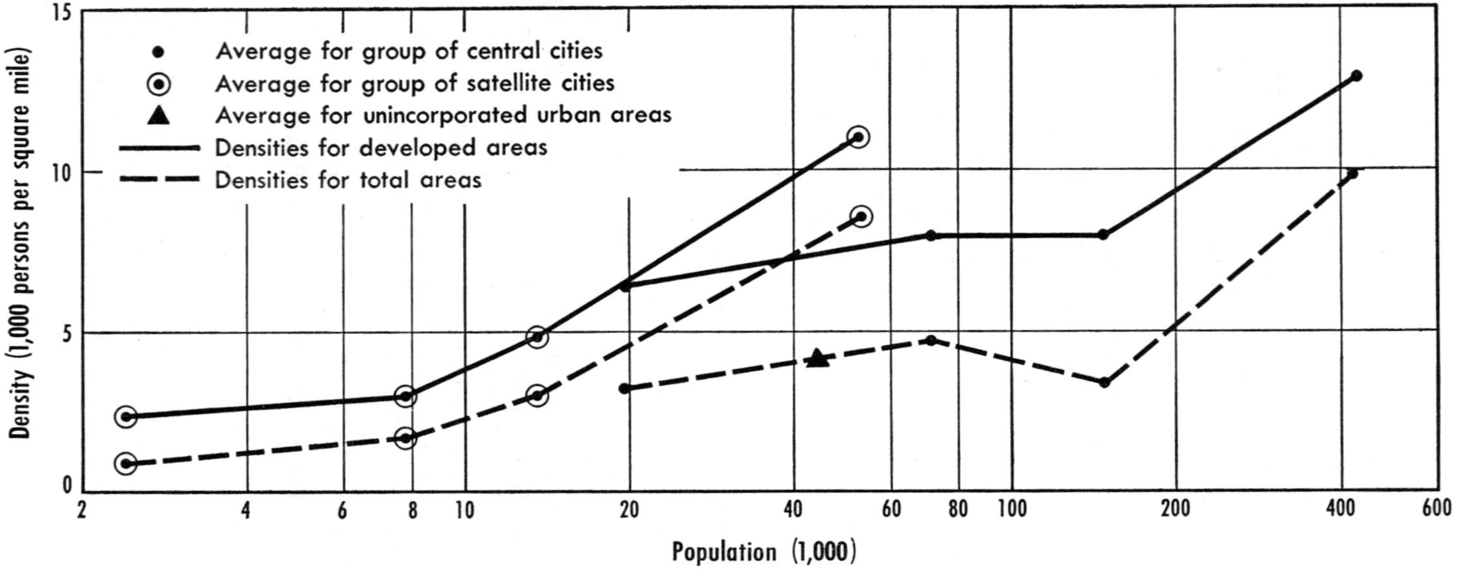

Figure 9. Average population density for groups of cities.

Table 9. Population densities of sample cities and urban areas, of various types and sizes[1]

Type of city or urban area	Size of area in total population (1,000)	Number of areas	Population at time of study (1,000): Total, all areas	Population at time of study (1,000): Average per city or area	Total area (acres): Total	Total area (acres): Developed	% of total area that is developed	Population density (persons per square mile) on basis of[2]: Total area	Population density (persons per square mile) on basis of[2]: Developed area
Central cities . .	up to 50	28	551	19.7	109,499	54,935	50	3,220	6,420
	50-100	13	932	71.7	127,141	74,729	59	4,690	7,970
	100-250	7	1,030	147.1	196,725	82,630	42	3,350	7,975
	250 and over	5	2,105	421.0	136,904	106,172	78	9,850	12,700
	all	53	4,618	87.2	570,269	318,467	56	5,180	9,290
Satellite cities .	up to 5	7	17	2.4	12,782	4,621	36	852	2,350
	5-10	6	47	7.8	17,647	10,098	57	1,705	2,990
	10-25	10	135	13.5	28,800	17,922	62	3,000	4,820
	25 and over	10	524	52.4	39,185	30,221	77	8,560	11,110
	all	33	724	21.9	98,413	62,861	64	4,710	7,370
Urban areas[3] . .	all	11	475	43.2	[4]	70,440	[4]	[4]	4,340

[1] Data are taken from tables in Appendix to Harland Bartholomew, *Land Uses in American Cities* (Cambridge: Harvard University Press, 1955). Data are for various dates, mostly after 1940.

[2] Slide rule division.

[3] The central city of these urban areas is included above in the data for central cities.

[4] Since these include unincorporated areas, no data on total area are available.

by other students.[19] As the city grows in population, its use of land and its basic social characteristics seem to be a function more of its relative than of its absolute size. As far as use of land is concerned, this may be due to the fact that the area of land within the city increases as the square of the distance from the center, so that the use of land is more simply related to distance than to area. But the explanation for this relationship among social data is not clear.

The lines on Figure 9 look somewhat erratic, probably because of the small number of cities studied. The considerable divergences among individual cities are shown in Appendix D, Tables D-1, D-2, and D-3. While each shows great variation, the satellite and central cities show no consistent differences for cities of the same size. The unincorporated urban areas, fringe areas around the central cities, do show a consistently lower density than either kind of city. For the whole group of areas, an increase of about 72 per cent in population meant an increase in average density of about a thousand persons per square mile of developed area.[20] Even when only developed areas are considered, population density is affected by factors other than total population of the city, although the latter does have a strong effect.

The other source of data on the present land use of cities is the 1950 census for urbanized areas, incorporated cities, and standard metropolitan areas. These data are each subject to the drawbacks of classification already noted (see pages 360–63), and in addition each type of data has its special limitations. In many respects, so far as land use is concerned, the urbanized area most nearly conforms to the economic city. It includes the incorporated city or cities and the contiguous or nearly contiguous clearly urban use area. With few exceptions, all the land inside these larger incorporated cities would be withdrawn from other uses; it is only rarely that such land would be managed for purposes other than urban. On the other hand, it is

19 Otis Dudley Duncan and Albert J. Reiss, Jr., *Social Characteristics of Urban and Rural Communities, 1950* (New York: John Wiley and Sons, 1956). Duncan and Reiss use the logarithmic relationship through choice of class intervals of city size, the upper limit of each interval being roughly twice its lower limit; and when data are plotted, each class interval is at an equal arithmetic distance from the other. We employ the logarithmic relationship more directly, but the effect is the same.

20 The formula is: $Y = 4{,}310 \log X - 11{,}800$, where Y is persons per square mile of developed area and X is total persons in the city or area of measurement. The relationship has a gross correlation of about .73.

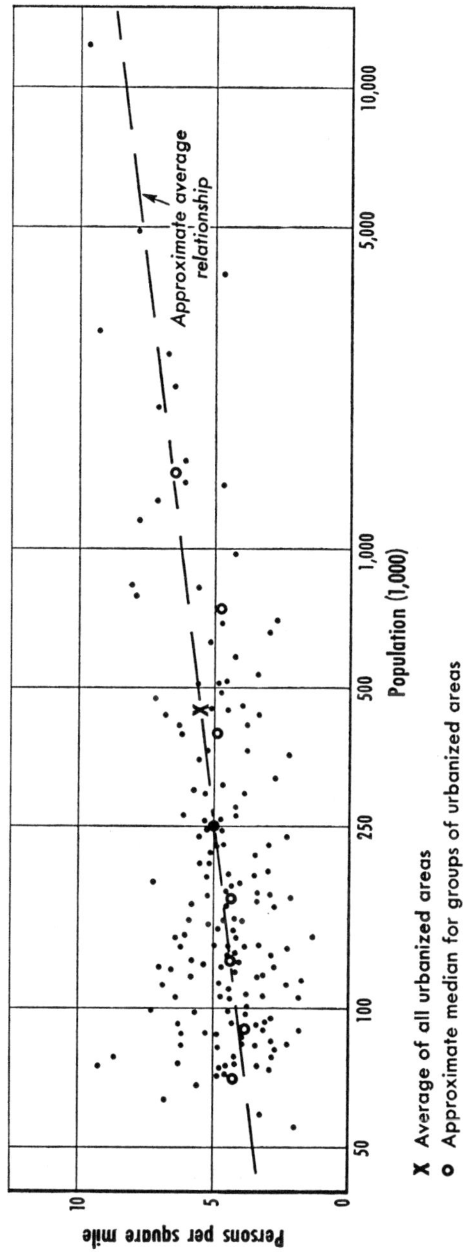

Figure 10. Population density in relation to total population, all (157) urbanized areas, 1950.

also unlikely that all this land is actually *used* for urban purposes. With few exceptions, the land outside of the incorporated city or cities is both withdrawn and used. However, it is most unlikely that all the withdrawn area outside of city boundaries would be included. It is likely that some additional surrounding area is being withheld from other uses, in the expectation or hope that it will be subdivided and converted to urban use before long.

For the 157 urbanized areas reported in the 1950 census, there was a relationship between size of population and average density (Figure 10). As in the case of the other similar relationships, the density is related to the logarithm of the population of the area.[21] The relationship is not close, the simple correlation coefficient being in the neighborhood of .42. It is obvious that many factors other than size of city affect the density of settlement, especially when measured on this gross or nearly withdrawn basis. The proportion of idle land within city boundaries is surely one such factor; the type of economic activity in the city, its age, its relationship to other larger cities, the region of the country, and other factors are probably involved.

For the purpose of this study a major difficulty with the data for urbanized areas is that they relate only to urban groupings that include one or more cities of 50,000 population, when by far the greater part of the total area of land used for urban purposes is in cities and urban groupings smaller than this. Another difficulty is that they exist only for 1950.

If the basis of analysis is all incorporated cities in the United States for which area information was available in 1950, a rather close relationship is disclosed between average population per city and average density (Table 10, Figure 11, and Appendix D, Table D-4). Of the nearly 4,000 incorporated cities of over 2,500 persons in the United States, about 85 per cent had a population of less than 25,000. This proportion included only about 28 per cent of the total population of incorporated cities, but it also included over half of the total area in incorporated cities of 2,500 or more population. The largest cities, those with 250,000 or more population, represent only about 1 per cent of the total number of cities over 2,500; they have 40 per cent of the total population, and less than 19 per cent of the land.

[21] The formula is: $Y = 2{,}150 \log X - 6{,}700$ where the symbols have the same meaning as previously. However, density here is more nearly on a withdrawn than on a used area basis, thus perhaps accounting for the lower density and the lower correlation.

Table 10. Numbers, population, area, and average density, by size of incorporated cities, 1950

Cities with a population of	Number of incorporated cities		Total population (1,000)		Total area reported (sq. mi.)	Average population of cities reporting area[1]	Average area of cities reporting area[1] (sq. mi.)	Average persons per sq. mile, cities reporting area[1]
	Total	Reporting area	All cities in size group	In cities reporting areas				
2,500-5,000	1,549	1,419	5,502	5,029	2,967	3,540	2.09	1,695
5,000-10,000	1,089	975	7,535	6,742	3,093	6,910	3.18	2,180
10,000-25,000	753	678	11,445	10,400	3,853	15,320	5.68	2,695
25,000-50,000	249	249	8,711	8,711	2,439	35,000	9.81	3,575
50,000-100,000	126	126	8,941	8,941	1,762	70,900	14.02	5,075
100,000-250,000	65	65	9,484	9,484	1,577	145,900	24.26	6,015
250,000-500,000	23	23	8,242	8,242	1,343	358,350	58.39	6,135
500,000-1,000,000	13	13	9,189	9,189	1,000	706,840	76.92	9,189
1,000,000 or more	5	5	17,405	17,405	1,241	3,481,000	248.20	14,015
All cities	3,872	3,553	86,454	84,143	19,275	23,650	5.41	4,360

[1] Slide rule division.

SOURCE: See Appendix D, Table D-4.

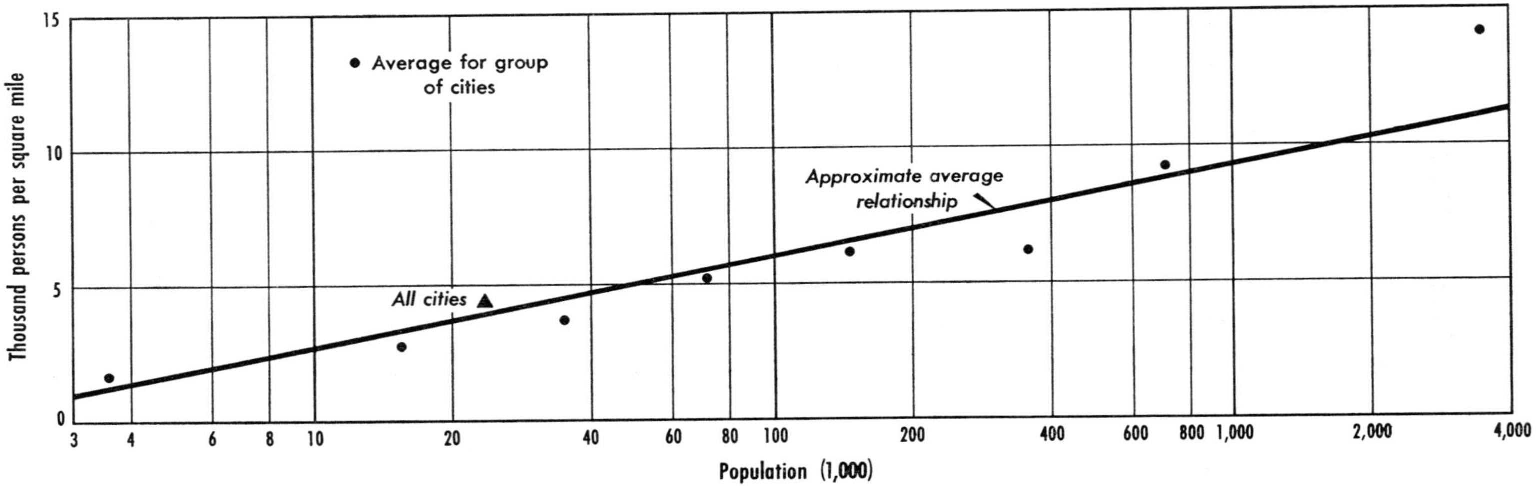

Figure 11. Average population densities for groups of incorporated cities, by size class, 1950. (See Table 10.)

The relationship on Figure 11 also is based on the logarithms of population. [22] A doubling of city size on the average adds 1,000 persons to the density per square mile.

The data in Table 10 seem clearly to indicate that as the city grows larger, the density of its central portion rises faster than its average density. Thus, as the average city increases from the 2,500-5,000 population range into the 5,000-10,000 range, its gross area increases from 2.09 to 3.18 square miles while its total population rises from 3,540 to 6,910 persons. This would indicate a density in the added area of over 3,000 persons per square mile, which is higher than the average density of cities of either size class. This relationship existed for all size classes of cities in 1950, as Table 11 shows.

Table 11. Density relationships in cities of increasing size, 1950

Size of city	Average density for entire city	Density, added population divided by added area
2,500-5,000	1,695	—
5,000-10,000	2,180	3,090
10,000-25,000	2,695	3,370
25,000-50,000	3,575	4,765
50,000-100,000	5,075	8,520
100,000-250,000	6,015	7,325
250,000-500,000	6,135	6,230
500,000-1,000,000	9,189	18,850
1,000,000 and over	14,015	16,200

But the most casual knowledge of American cities is sufficient to indicate that population density around the edges of the city is lower than that nearer the center. What happens, as cities grow, is that the density rises rapidly in the central portion, and also rises but is proportionately lower, in the next adjoining outer areas.

The general relationship between city population and density also holds fairly well within regions (Table 12). In every region the larger cities have much higher densities than the small ones, and in nearly

[22] The formula is: $Y = 3{,}295 \log X - 10{,}500$, when the symbols have the same meaning as in the earlier formula.

every region there is an orderly increase in population density as city size increases. There is some variation in average density of cities of a given size, when the regions are compared. The very low densities in the smaller New England cities is noticeable; this is due to the fact that some such cities are essentially rural New England "towns," with much farming and forest land within their borders. In some cases, variations in density among the larger cities are due to the small numbers of cities in a group. Average densities of all cities in a region are nearly meaningless, because they are so markedly affected by the size of the cities. Densities are above average in cities of every size group in the Middle Atlantic region; otherwise, there is no clearly consistent variation by region.

It would be interesting to analyze the effect of type of city on average density or on the relation between size and density. Do small cities lying within SMA's have density patterns different from those of independent cities of the same size? Are the densities of heavily industrialized cities different from those of lightly industrialized cities of the same size? There is evidence that cities where growth has been most rapid since 1920—the cities that have grown up in the automobile age—have lower average densities than older cities of the same size.[23] With all the limitations of the Census data, a more extended study than is possible in a study of land use in general, might produce some useful insights into urban uses of the land.

The great differences in average density that are evident among cities, may be examined directly from the Census data for cities of 25,000 and over.[24] Here, the most common density in 1950 was 4,000 to 5,000 persons per square mile (Figure 12 and Appendix D, Table D-5). This was especially marked for cities with population of 25,000 to 50,000, but was less pronounced and less regular for the larger cities where however the density, commonly, was higher. Perhaps more striking than these tendencies was the strongly skewed nature of the frequency distribution. The density was obviously limited on the lower side, to a zero density, and in practice to one somewhat

[23] The authors are indebted to T. Lowdon Wingo, Jr., of Resources for the Future, for this information, taken from some unpublished studies.

[24] The Census reports do not give data on density per city for cities smaller than 25,000. It could be calculated, of course, from the data on total population and area, but the general relationships shown for the larger cities are clear enough to justify drawing an assumption of rather similar relations for the smaller.

above this, for these are—after all—cities. On the side of high densities, there was no such obvious limit, and in fact some cities had densities five times or more the average or modal density.

Bogue estimates that in 1949 the standard metropolitan areas of the United States included over 128 million acres, or nearly 7 per cent of

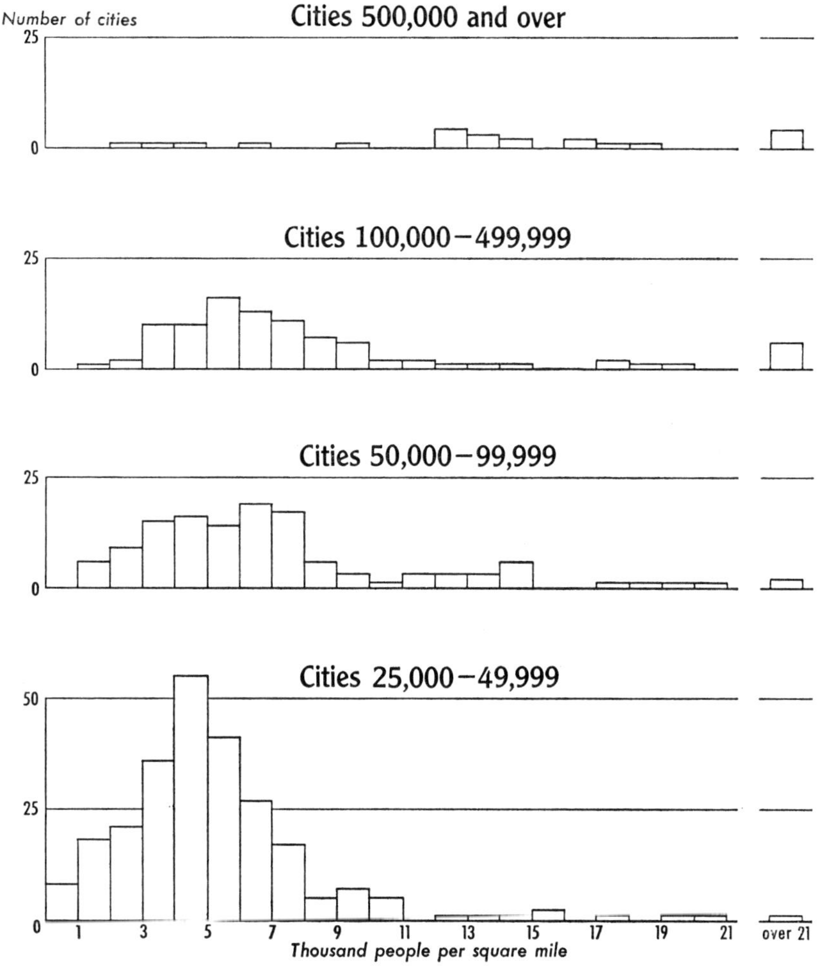

Figure 12. Number of cities with specified population density, 1950.

Table 12. Average population density (persons per square mile) in incorporated cities[1] of different sizes and in various geographical regions, 1950

Cities with a population of	New England	Middle Atlantic	East North Central	West North Central	South Atlantic	East South Central	West South Central	Mountain	Pacific	United States
2,500-5,000	341	2,001	1,533	1,845	1,615	1,770	1,758	1,782	2,065	1,695
5,000-10,000	449	3,000	2,265	2,110	2,120	2,065	2,285	2,260	2,208	2,180
10,000-25,000	793	4,275	3,275	2,925	2,910	3,310	3,275	3,280	3,150	2,695
25,000-50,000	1,593	4,380	4,800	3,650	4,085	4,400	3,540	3,460	3,775	3,575
50,000-100,000	3,970	9,880	6,130	3,520	4,840	4,150	3,615	2,865	5,635	5,075
100,000-250,000	7,290	8,780	6,300	3,615	6,840	5,240	4,965	4,060	4,840	6,015
250,000-500,000	—	14,650	7,430	5,850	8,940	5,220	4,075	6,210	5,615	6,135
500,000-1,000,000	16,700	13,520	10,280	12,000	12,520	—	3,250	—	17,250	9,189
1,000,000 or more	—	22,520	15,730	—	—	—	—	—	4,375	14,015
All cities	2,295	7,740	5,190	3,585	3,985	3,070	3,200	3,105	4,255	4,360

[1] Based on those cities reporting area. Slide rule division.

SOURCE: See Appendix D, Table D-4.

the total land area of the nation.[25] At that time, almost half of the land within the SMA's was in farms, and a fourth of it was cropland; moreover, as nearly as can be estimated, well under one-fourth of the nonagricultural land was actually used for urban purposes or withdrawn by the city from other uses. The SMA's are simply *not urban* so far as land use is concerned (Figure 13). The use of a unit of statistical summarization which is sufficiently inclusive to gather in all the people who are economically and socially integrated into the city life has much to commend it; but in the process, a great deal of nonurban land has been included.

The results can sometimes be bewildering. For example, New York County, New York, has a land area of 22 square miles and a population close to 2 million persons, or a density of nearly 90,000 per square mile; and Kings and Bronx counties each have densities more than a third as large; but the New York-Northeastern New Jersey metropolitan area includes nearly 4,000 square miles with an average density of 3,278 persons per square mile—a density equal to that of the average city of 25,000 persons. The city of Chicago has a total area of 207.5 square miles and a total population of over 3½ million, or an average density of 17,450 persons per square mile; but the Chicago metropolitan area which includes five counties in Illinois and one in Indiana, has a total area of 3,617 square miles, a total population of nearly 5½ million, and an average density of 1,519 persons per square mile—less than the average small city of 2,500-5,000 population.

The population density for all SMA's is 407 persons per square mile; this compares with the average density for whole states as follows: Connecticut, 410; New York, 309; New Jersey, 643; and Rhode Island, 749. The average for the SMA's is pulled down greatly by such extreme cases as San Bernardino, California; the inclusion of all of San Bernardino county, most of which is part of the Mohave Desert, brings in over 20,000 square miles, or nearly 10 per cent of the area of all SMA's in the nation, with an average density of but 14 persons per square mile.

Under these circumstances, it is not surprising that the distribution of population densities for SMA's is markedly different from that for

[25] Donald J. Bogue, *Metropolitan Growth and the Conversion of Land to Nonagricultural Uses,* published jointly by Scripps Foundation for Research in Population Problems and Population Research and Training Center, University of Chicago (Oxford, Ohio: Miami University, 1956). The Census volume, *County and City Data Book 1952,* shows a slightly larger total area.

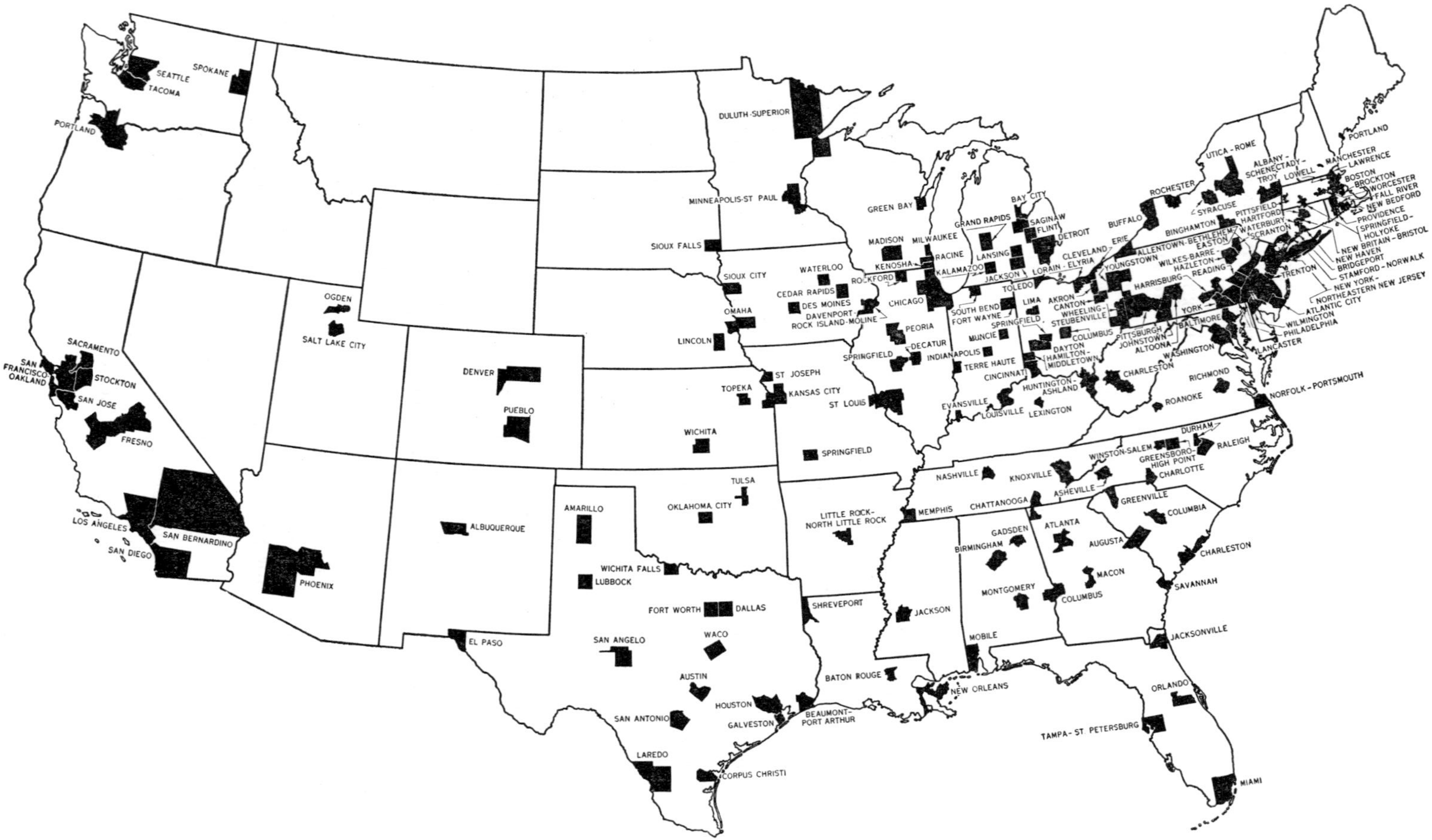

Figure 13. Standard metropolitan areas of the United States, 1950. (Map from U. S. Bureau of the Census.)

cities (Figure 14 and Appendix D, Table D-6). The most common density for SMA's in 1950 was from 100 to 200 persons per square mile; that for the larger cities was 4,000 to 5,000. Densities in over 85 per cent of all SMA's averaged less than 1,000 persons per square mile, but they were as low as this in only 2 per cent of the cities of

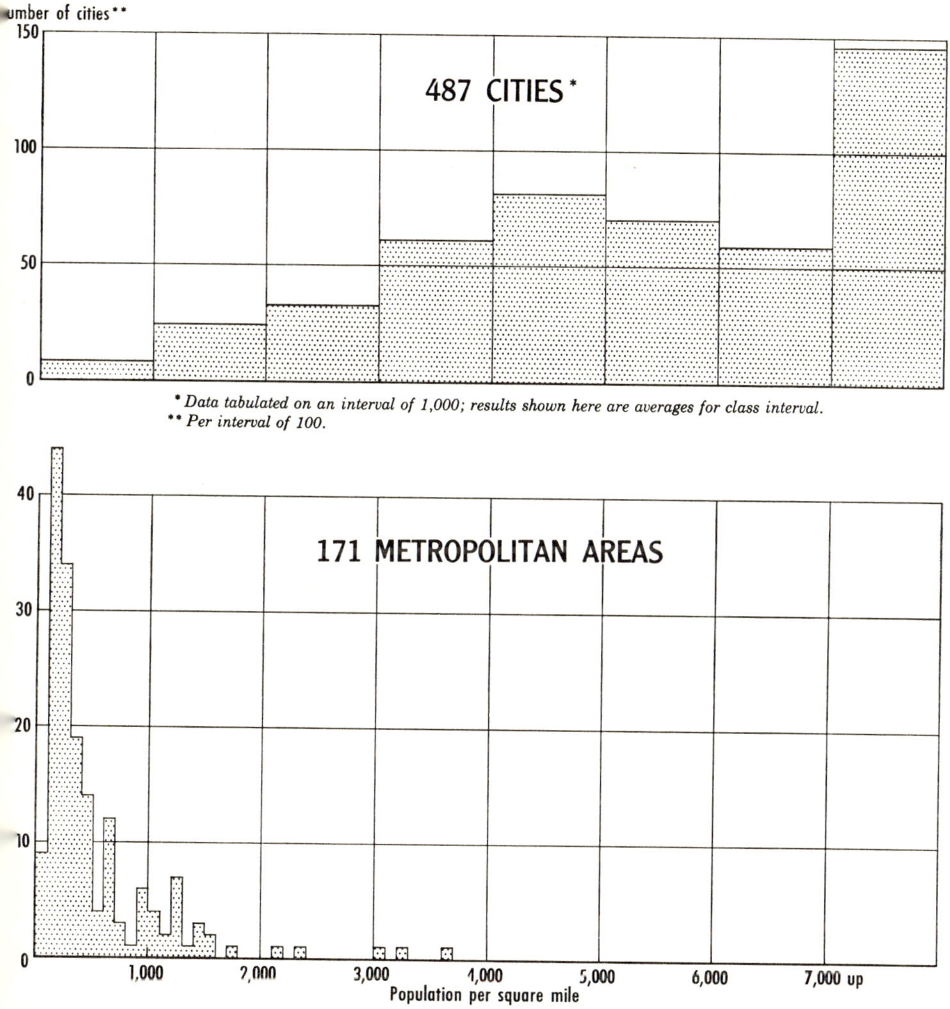

Figure 14. Number of cities of 25,000 population and over, and of metropolitan areas, by population density, 1950.

25,000 and over. In only 3 per cent of the SMA's were densities higher than 2,000 to the square mile, but in over 93 per cent of the cities of 25,000 and over had densities averaged more than this. From a statistical viewpoint, the data on densities for SMA's and those for cities are simply not from the same universe.

Summary. It is now possible to summarize this discussion on present use of land for urban purposes.

1. The available data are ill-suited to any economic analysis of urban land use. A few sample data relate rather closely to land actually used; but these are mostly for political, not economic, cities, and the sample is small. Census data conform neither to the concept of land actually used for urban purposes nor to the concept of land withdrawn by the city from any other land use, although they are nearer the latter than the former. Census data are available for incorporated cities as legal entities, but these are sometimes artificial in terms of economic activity, and are in any case affected by the accidents of history. Some data are available for larger urban groupings as economic entities (urbanized areas). The SMA's include all the population and economic activity centered around the larger cities; but to do so they must also include vast areas of nonurban land uses, thus distorting the whole land use picture.

2. The average density of land use, and hence the area required for any given population, depends to a great extent upon the size of the city. Small cities are more lavish than large cities in their use of land; this is natural, and from their point of view, logical. But there are great variations in this relationship between size of city and land use. We have expressed this relationship by the formula $Y = a \log X - b$, when Y is the number of persons per square mile, X is the total population of the city, and a and b are constants for each city. The following results were obtained:

Bartholomew sample cities $Y = 4{,}310 \log X - 11{,}800$ $(r = .73)$
Census, all urbanized areas $Y = 2{,}150 \log X - 6{,}700$ $(r = .42)$
Census, all incorporated cities $Y = 3{,}295 \log X - 10{,}500$

The Bartholomew data are for developed area; a similar relationship, but at lower densities, would exist for total area. The Census data are for all land within city boundaries—certainly not all used, probably nearly all withdrawn, but not all the latter. While the relationships are not close for individual cities, yet the total effect is great.

3. Within all incorporated cities of 2,500 population and over, cities of less than 25,000 contain at least half of the land. These small cities and other still smaller towns, both incorporated and unincorporated, use perhaps as much as three-fourths of all land withdrawn from other uses. Some of these small towns and cities are satellites to larger cities, but many are independent. These are the cities whose large use of land would seem to offer the greatest opportunity for intensive study. On the other hand, both the incentive to more intensive land use and the social and governmental mechanisms whereby it might be brought about operate against more frugal use of land in small cities.

4. Although the data are inadequate, there seems good reason to believe that there is much idle land within city boundaries; studies of sample cities would indicate that the idle land averages one-third or more of the total. Some idle land available for expansion or new uses is desirable, but it appears that the amount now idle is far in excess of a reasonable amount, and that great economies could be achieved in its better use. On the other hand, much of the land is ill-suited to use, primarily because of the size of holdings into which it has been divided. For some of it, titles are fouled up and thus its value is further reduced.

5. All incorporated cities of 2,500 and over report 13 million acres within their borders. An estimate of total area within urban places is perhaps ½ to 1 million acres larger; this takes account of the urban population outside of incorporated cities. There are nearly 14,000 "places" of less than 2,500 persons, that are classed by the Census as rural; they average less than 800 persons each. The total area of land in them is unknown but probably is in the rough magnitude of 10 million acres. However, it is more than likely that the area is included in the statistics of surrounding land uses—forestry, grazing, agriculture, etc. All of these figures represent more than area used, but somewhat less than area withdrawn from other uses. Bearing in mind the data's limitations and imperfectly defined relationships, it is still possible to make a very rough estimate of the area occupied by all urban population as that term is defined by Census:

withdrawn area: 17 million acres
used area: 11 million acres

Past Changes in Land Use by Cities

Cities' use of land in the past is even more difficult to assess than their use today. Three sets of data are available: (1) data on area in incorporated cities of different sizes, from 1900 to date, by census periods; (2) data from a special study by Roderick D. McKenzie[26] and from the Bureau of the Census, showing area and population for identical large cities from 1850 to the present; and (3) some calculations on changes in land use within SMA's between 1930 and 1950. Each set has its limitations.

The area and population of cities above the 25,000 or 30,000 population level is shown in the census for 1900 and subsequent periods (Table 13). These cities included from 65 per cent to 70 per cent of the total urban population at each census period; because urban population was increasing relative to total population, their share of total population increased rather regularly from 27 per cent in 1900 to 40 per cent in 1940. However, judging by the results of studying the city of 1950, one would expect that their share of the total area of all incorporated cities was much less, probably less than half. These tabulations include all incorporated cities within the specified size classes at each census period. As cities grew in size, they "graduated" from one size class to another, and new cities were being added at each period, especially in the smallest size class. There was, therefore, a changing grouping of cities between census periods. The one exception to this, from 1900 to 1910, is explained in a footnote to the table.

When these historical data on density for cities of different sizes are combined with the data for 1950 on a single chart, a remarkable uniformity in relationships is evident (Figure 15). With few exceptions, each group of larger cities showed a higher average density than the next group of smaller cities. Some irregularity was shown in this pattern of progression toward higher density with larger size, where the largest cities are concerned. These groups contained fewer cities, so that when a city such as New Orleans with its large area and consequent low density, shifts from one size class to another, the average density of each group of cities is materially affected.

The relationships in Table 13 and Figure 15 are indicative of general tendencies, but are not closely applicable to individual cities. So

26 R. D. McKenzie, *The Metropolitan Community* (New York: McGraw-Hill Book Co., 1933), Appendix Table IX, pp. 336-39.

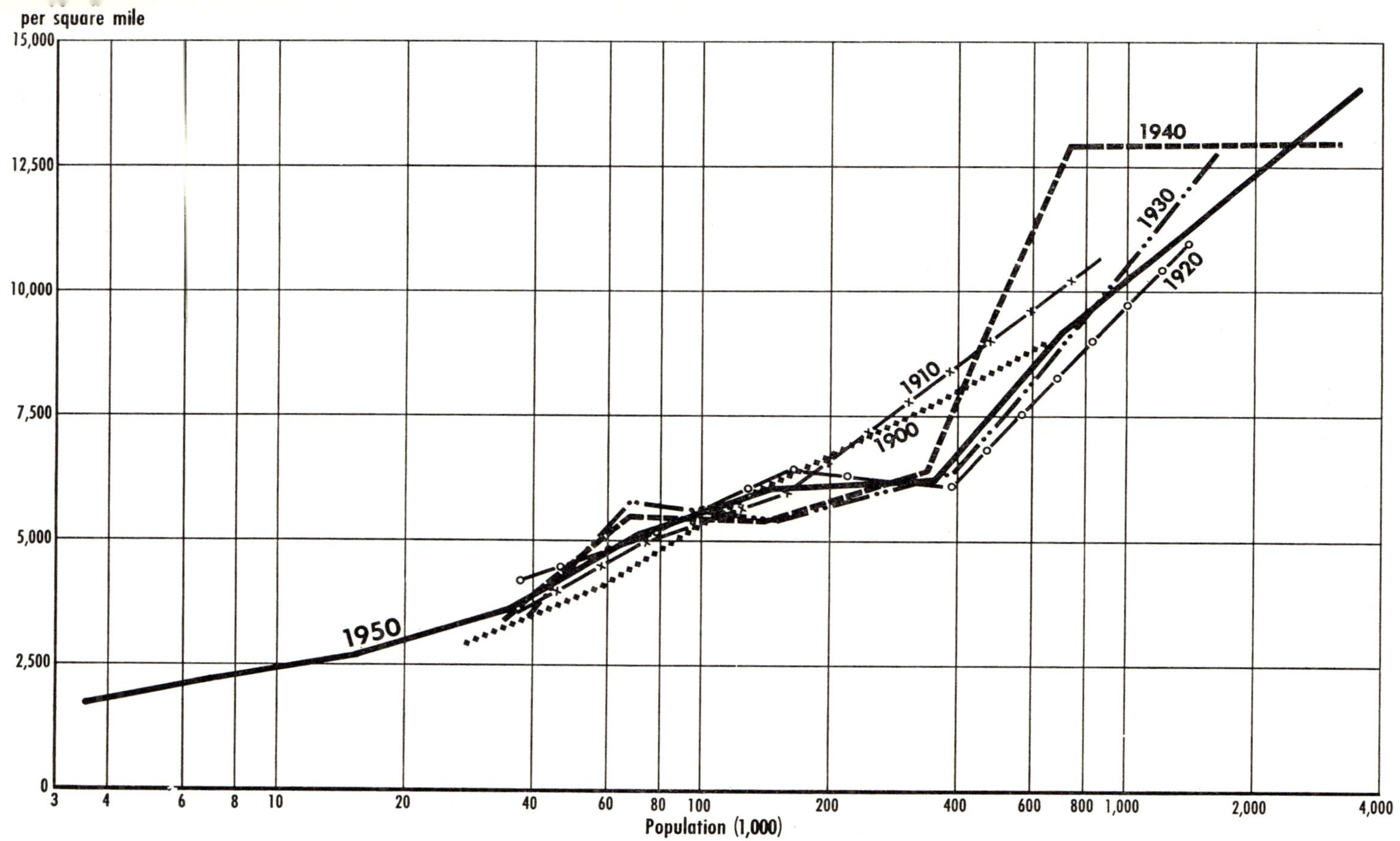

Figure 15. Population density in relation to total population for incorporated cities, by census periods, 1900–1950.

Table 13. Number of cities, population, area, and density, all cities of 30,000 and over, by census periods, 1900-1940

Size of city (population)	Number of cities	Population[1] Total for group	Population[1] Average per city[2]	Land area (acres) Total for group	Land area (acres) Average per city[2]	Population density (persons per square mile)[2]
1900[3]						
30,000-50,000	75	2,088,100	27,900	[4]465,200	6,200	2,870
50,000-100,000	59	2,919,628	49,500	[4]458,100	7,770	4,065
100,000-300,000	32	3,582,355	112,000	407,282	12,750	5,635
300,000 or over	18	11,617,020	645,000	[4]841,800	46,750	8,835
Total	184	20,207,103	109,800	[4]2,172,382	11,820	5,955
1910						
30,000-50,000	75	2,835,354	37,800	[4]507,300	6,770	3,575
50,000-100,000	59	4,178,915	70,800	[4]543,000	9,210	4,920
100,000-300,000	32	5,108,237	159,600	548,927	17,150	5,955
300,000 or over	18	15,193,901	843,000	[4]914,800	50,750	10,620
Total	184	27,316,407	148,500	[4]2,514,027	13,660	6,955
1920						
30,000-50,000	106	4,059,220	38,300	621,802	5,865	4,175
50,000-100,000	78	5,404,767	69,300	694,994	8,910	4,970
100,000-300,000	48	7,818,985	163,200	787,837	16,410	6,350
300,000-500,000	9	3,471,830	385,759	363,662	40,407	6,115
500,000 and over	12	16,576,823	1,381,402	910,037	75,836	10,930
Total	253	37,331,625	147,600	3,378,332	13,350	7,070

1930						
30,000-50,000	121	4,681,506	38,700	856,320	7,070	3,495
50,000-100,000	94	6,295,607	67,000	698,992	7,440	5,765
100,000-300,000	69	10,911,515	158,000	1,271,432	18,420	5,490
300,000-500,000	12	4,693,000	391,000	471,533	39,350	6,370
500,000 and over	13	20,849,400	1,605,000	1,046,560	80,500	12,780
Total	310	47,431,028	153,000	4,344,837	14,010	6,830
1940						
25,000-50,000	200	6,973,000	34,865	1,310,000	6,550	3,403
50,000-100,000	105	7,230,000	68,800	829,000	7,890	5,581
100,000-250,000	55	7,793,000	141,500	926,500	16,850	5,381
250,000-500,000	23	7,828,000	340,500	778,000	33,800	6,432
500,000-1,000,000	9	6,457,000	717,444	319,500	35,500	12,919
1,000,000 and over	5	15,911,000	3,182,200	780,000	156,000	13,051
Total	397	52,190,000	131,400	4,943,000	12,430	6,753

[1] As of June 1, 1900; April 15, 1910; July 1, 1920; and July 1, 1930. Data for 1940 rounded to nearest thousand.

[2] Slide rule division.

[3] These are cities falling in this class in 1910; population is 1900 reported population; area is 1910 area minus area added since 1900 (this assumes no areas eliminated in this decade).

[4] All cities reported area of land and water; most cities reported land and water areas separately; but for those cities reporting only total area, area of land was estimated on basis of relation between land and water in cities reporting each separately.

SOURCE: 1900 and 1910, *Financial Statistics of Cities Having a Population of Over 30,000: 1910*; 1920, *Financial Statistics of Cities Having a Population of Over 30,000: 1921*; 1930, *Financial Statistics of Cities Having a Population of Over 30,000: 1930*; and 1940, *City Finances, 1942*, all by the U. S. Bureau of the Census. Area data for 1940 calculated by slide rule from data on square miles.

far as these are concerned, it seems probable that correlation between their density and size is low. Many factors can affect the boundaries of a city—general governmental policy in the particular state and city as to annexation of territory, presence or absence of adjoining cities, historical accident, attitudes of leading citizens—these and more, besides the strictly economic factors.

Within the size range for which data are available in the census years before 1950, the 1950 data for incorporated cities seem generally representative.[27] There is no clear evidence of a shift in the size-density relationship between census periods. It is assumed, in analysis presented later in this chapter, that the 1950 relationship is in fact typical or representative for the whole range of city sizes in the past, and that it will be generally representative for the future as well.

McKenzie, in his statistical series for selected cities, gives the area at census dates from 1850 to 1930 for all cities of 100,000 or more in 1930 (Appendix D, Table D-7).[28] Census data on areas for these same cities are available for 1940 and 1950, and population data are available for all census years. A summary of these data according to two sizes of cities, based upon their 1950 population, is given in Table 14 and Figure 16. The cities included in each group were the same throughout, although they changed in size, so that each group represents a constant or fixed number of cities. By 1950, one of these cities had fallen slightly below 100,000 in population; in addition, fifteen other cities had over 100,000 but cannot be included here because data are lacking for the earlier years.

The ninety-one cities of 100,000 or more population in 1930, for which McKenzie assembled data on area, include roughly half of the total urban population of the United States—somewhat more than this in earlier years and as much as half even as late as 1930, and somewhat less than half in the last two census periods. However, the area of land included within them was only about one-fourth of that in all incorporated cities of 2,500 and over in 1950, and an even smaller percentage of the land in all towns and cities.

It is apparent that these cities have grown both in population and in total area over the years, and at a somewhat similar rate. Part of their population growth was the result of annexing territory in which people were then residing, but part was due to population growth of

[27] Census data are published for cities of all sizes above 2,500 in 1940, but are summarized only for cities above 25,000.

[28] *Ibid.*

Table 14. Area, population, and density of selected cities, by two size groups of cities, 1850–1950[1]

Region	Item[2]	1950	1940	1930	1920	1910	1900	1890[3]	1880[3]	1870[3]	1860[3]	1850[3]
I. Cities 500,000 and over in 1950												
NE (5)	Area in cities	583.7	563.8	564.7	553.4	547.8	534.1	276.8	275.3	240.6	206	—
	Population	12,022	11,405	10,905	9,287	7,945	5,966	4,497	—	—	—	—
	Density	20,600	20,250	19,350	16,800	14,500	11,170	16,250	—	—	—	—
NC (7)	Area in cities	662.0	648.3	654.2	547.3	464.7	427.8	381.3	193.8	120.0	71.2	—
	Population	8,906	8,249	8,160	6,505	4,938	3,756	2,685	—	—	—	—
	Density	13,450	12,700	12,450	11,900	10,650	8,775	7,050	—	—	—	—
S (4)	Area in cities	499.5	412.3	406.7	373.5	302.1	295.3	295.3	278.4	290.0	244.4	—
	Population	2,918	2,402	2,043	1,697	1,307	1,120	893	—	—	—	—
	Density	5,850	5,825	5,025	4,550	4,325	3,800	3,025	—	—	—	—
W (2)	Area in cities	495.5	492.9	483.8	406.3	142.9	85.5	71.4	71.4	71.4	71.4	—
	Population	2,745	2,139	1,872	1,084	736	445	349	—	—	—	—
	Density	5,550	4,350	3,870	2,670	5,150	5,200	4,900	—	—	—	—
All (18)	Area in cities	2,240.7	2,117.3	2,109.4	1,880.5	1,457.5	1,342.7	1,024.8	818.9	722	593	333
	Population	26,591	24,195	22,980	18,573	14,926	11,287	8,424	—	—	—	—
	Density	11,850	11,450	10,900	9,875	10,250	8,400	8,225	—	—	—	—
II. Cities 100,000 to 500,000 in 1950												
NE (27)	Area in cities	470.9	463.2	468.0	455.7	409.1	390.6	371	359	315	287	267
	Population	4,417	4,320	4,253	4,053	3,382	2,538	1,943	—	—	—	—
	Density	9,375	9,340	9,075	8,900	8,250	6,500	5,240	—	—	—	—
NC (20)	Area in cities	716.2	671.1	657.9	559.9	487.8	405.8	282	142	114	83	49
	Population	4,338	3,852	3,741	3,072	2,191	1,549	1,188	—	—	—	—
	Density	6,050	5,740	5,700	5,500	4,500	3,810	4,210	—	—	—	—
S (17)	Area in cities	882.0	561.8	560.7	340.7	275.3	158.2	113.5	95	92.5	81	75
	Population	4,370	3,407	3,068	2,088	1,487	906	697	—	—	—	—
	Density	4,960	6,075	5,475	6,140	5,400	5,725	6,150	—	—	—	—
W (9)	Area in cities	532.0	509.5	509.5	469.0	437.4	303.9	185.6	157.5	161.5	146.5	125
	Population	2,716	2,045	1,893	1,494	1,146	521	363	—	—	—	—
	Density	5,110	4,020	3,720	3,180	2,620	1,720	1,955	—	—	—	—
All (73)	Area in cities	2,541.1	2,205.6	2,196.1	1,825.3	1,609.6	1,258.5	952.1	753.5	683.0	597.5	516
	Population	15,481	13,624	12,955	10,707	8,206	5,514	4,191	—	—	—	—
	Density	6,225	6,175	5,900	5,875	5,100	4,375	4,400	—	—	—	—

[1] See Appendix D, Table D-7 for sources and for names of cities in each group.
[2] Area in square miles, population in thousands of persons, densities in persons per square mile. Numbers in parentheses following regional abbreviation are numbers of cities in group.
[3] Area data for these years for a few cities were unavailable; see Appendix D, Table D-7. These totals were estimated on assumption that area of cities in existence changed proportionately to changes in area for cities where data were available, and that area of cities not in existence was zero.

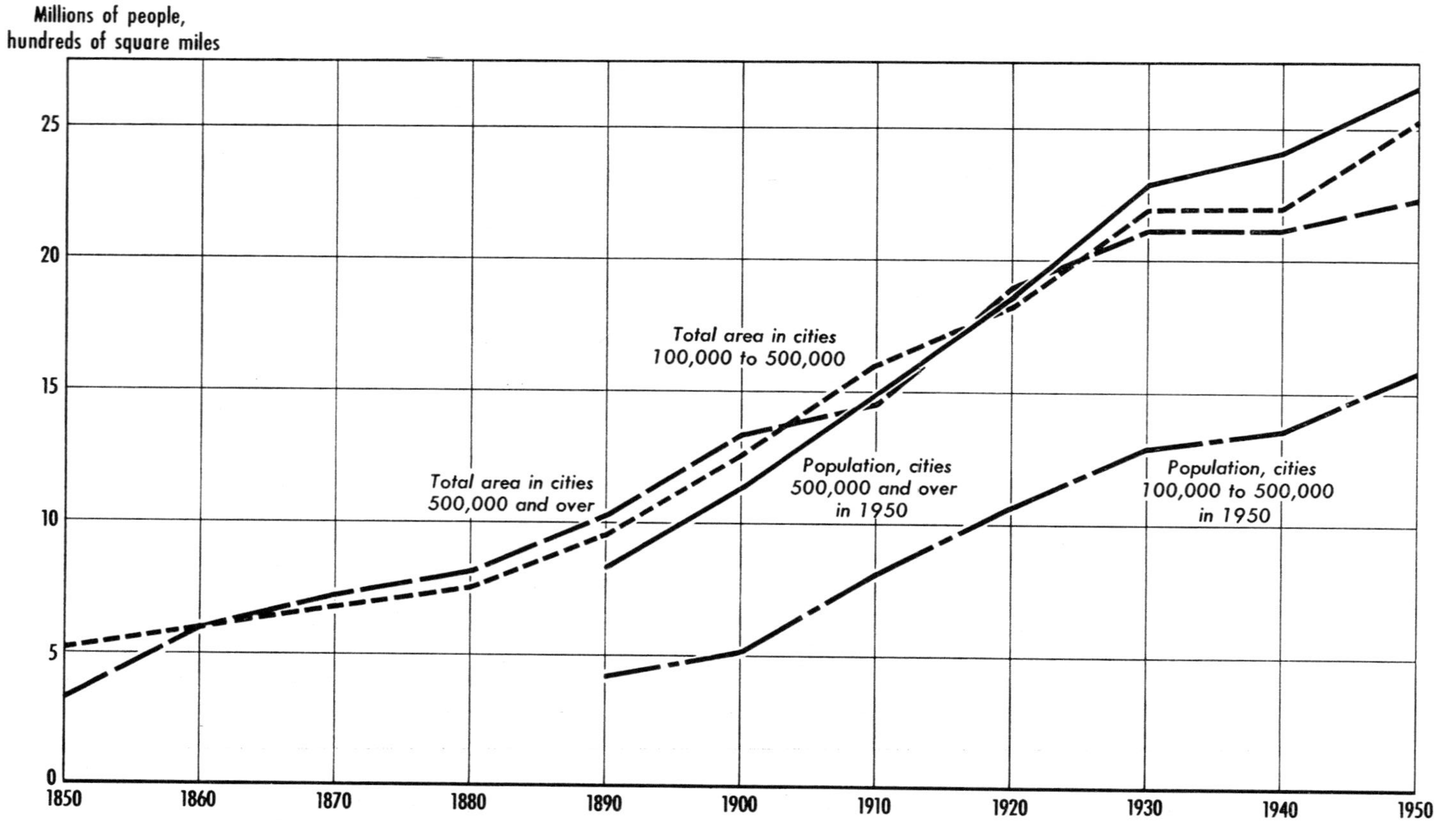

Figure 16. Total population, 1890–1950, and total area within cities, 1850–1950, for two groups of selected cities.

the city after annexation. The cities of half a million or more persons in 1950 have increased in total population from 8½ million in 1890 to 26½ million in 1950, a threefold increase; at the same time, their total area increased from slightly more than 1,000 square miles to over 2,200 square miles. The smaller cities increased in total population from slightly over 4 million in 1890 to nearly 16 million in 1950, or nearly a fourfold increase; at the same time, their total area increased from less than 1,000 square miles to over 2,500 square miles. The increase in both population and area was remarkably close to being regular from decade to decade. For each group of cities, therefore, some increase in average density took place, since population rose faster than area. However, average size of city also rose, and presumably this alone would lead to a higher average density.

This nearly proportional expansion of city area with growth in population is contrary to a common impression that population growth takes place largely outside of a city's boundaries. The data make it clear that the larger cities of the nation definitely have expanded their boundaries as their population increased. An examination of the data for individual cities shows that growth in area was sporadic—long periods of no change, often followed by a major change in boundaries and in area. The process of area growth is most irregular for any one city. In the future, growth in city boundaries for the largest cities will tend to be more difficult than in the past, because many of the larger cities are now ringed with other incorporated cities, whose absorption into the central city would be most difficult.

The data also reveal great differences in population density at any given time and throughout the whole period for cities in different regions of the country. Average density has always been highest in the Northeast. Many of the large cities in the region have added very little to their area in the past fifty years; thus, despite a relatively slow population growth within the cities, average density has risen considerably. In the North Central region, where population growth has been more rapid, city area has also expanded, and at a greater rate; hence, the rise in average density has not been as great as in the Northeast. For the largest cities in the South and West, increase in area has been almost as fast as increase in population; consequently only a slight change in average density has taken place. While the smaller cities show the same general difference in average density between regions, in each case the extent of the difference is less.

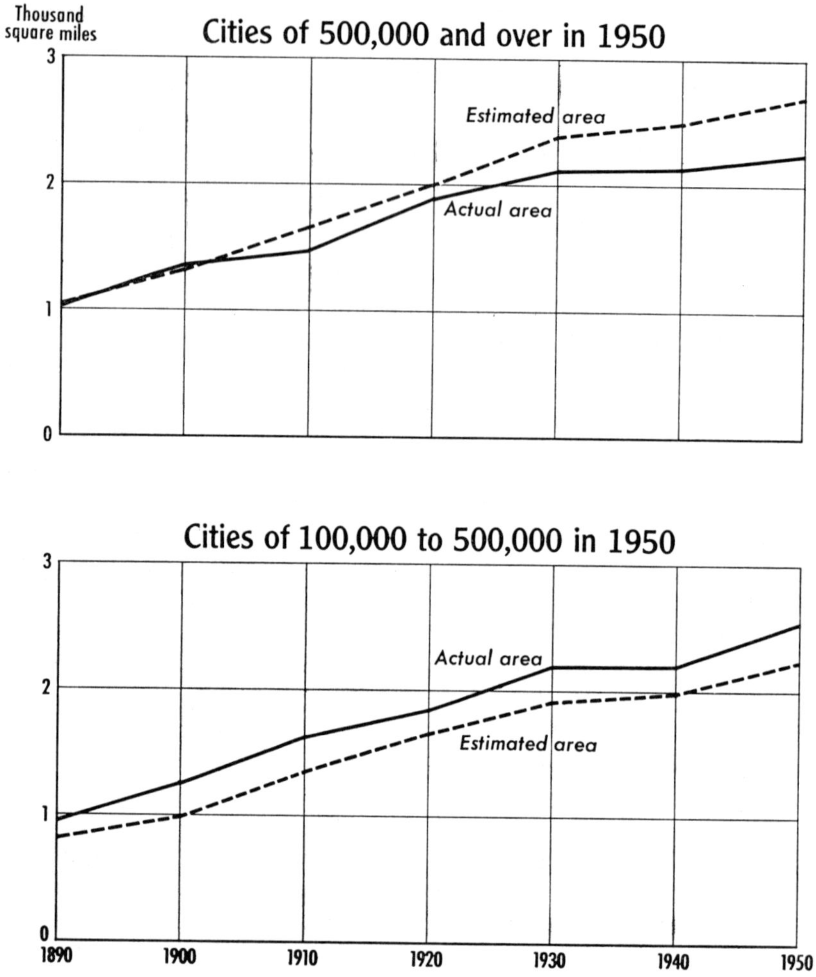

Figure 17. Actual area and area estimated on the basis of city size, for two groups of larger cities, 1890–1950.

The area in each of the groups of larger cities can be estimated for earlier periods, on the basis of the relationship between city size and city density presented for 1950.[29] The estimated and the actual area virtually coincide for the largest cities in 1890 and 1900, but in later

[29] See p. 87, n 22.

decades the actual area did not expand as rapidly as the estimated area (Figure 17). This means that city size was about 20 per cent smaller and population density correspondingly greater in recent decades than the 1950 average relationship would indicate. For the smaller cities, the estimated area at all dates was somewhat below the actual area, but by a rather constant amount. These calculations are further evidence that these cities of 100,000 and more in 1930 have, over several decades, expanded in area almost as fast as the average relationship between city size and city area would lead one to expect.

Table 15. Average density and apparent density of added areas, identical cities, 1890–1950

	Cities over 500,000 in 1930		Cities 100,000-500,000 in 1930	
Census period	Average density	Population increase divided by area added, preceding decade	Average density	Population increase divided by area added, preceding decade
1890	8,225	—	4,400	—
1900	8,400	9,010	4,375	4,310
1910	10,250	31,650	5,100	7,660
1920	9,875	8,610	5,875	11,660
1930	10,900	19,250	5,900	6,060
1940	11,450	153,500	6,175	70,500
1950	11,850	19,440	6,225	6,600

Calculated from Table 14.

Since, in general, the average density of population was rising from decade to decade for each group of cities, it is obvious that the number of persons added each decade, when divided by the area added during the same period, produces an apparent density higher than the average at either period. This is, in fact, the case (Table 15). With only one exception for each size group, the apparent density of the added area was greater than the average density for that size group of cities at both the beginning and end of the decade. In this respect, the changes here measured over time for the identical cities are similar to those noted above for 1950, for groups of cities of different sizes. As in that case, the apparent densities of the added areas are misleading; one knows from experience that the density on the periphery, which is the added area, is less than in the central part of the city. What is more probably true, as we have pointed out earlier, is that the real core of the larger cities decreased in density of residential population as their business districts expanded. But, between the real downtown business area and the city limits, population increased in density markedly each decade; and the added territory had a lower average density.

Bogue has attempted to measure changes in area of land used for urban purposes, by study of the data for standard metropolitan areas for 1929 to 1954.[30] He reasons that, instead of attempting to measure the impact of urban growth on agriculture by use of inadequate and somewhat questionable data on areas of land required or used for urban purposes, it might be possible to measure the changes in agriculture itself. He recognizes that about half of the SMA's were agricultural, but in our opinion he does not give sufficient recognition to the fact that the greater part of the remainder was also nonurban, and that therefore major shifts could occur in urban land use without necessarily affecting agricultural land use very much.[31]

Bogue finds that in 60 out of the 147 SMA's the area in agricultural use *increased* between 1930 and 1950, and that this increase more than offset the decrease in the agricultural area of the other 87 SMA's, so that for all of them there was actually an increase in area used for agriculture at the same time urban population was growing. He discusses three factors which "account for this apparent impossibility": land reclamation, transfer of grazing lands to private ownership, and allocation of land in farms to the county containing the farm or ranch headquarters. Each of these reasons is sensible, but Bogue omits what may be the most important factor of all—shifts in the use of some of the nonagricultural nonurban land. More than 40 per cent of the total land area of SMA's falls into this category.

However, based largely upon the relationships existing in those SMA's that lost farm area, Bogue calculates the area of farm land that

30 Bogue, *op. cit.*

31 Marion Clawson, Review (Donald J. Bogue, *Metropolitan Growth and the Conversion of Land to Nonagricultural Uses*), *Journal of Farm Economics*, Vol. 39, No. 1 (February 1957).

will be lost as urban population grows. He estimates that from 172 to 264 acres of land will be required for each 1,000 growth in population, or an average density of from 2,424 to 3,721 persons per square mile. These are average densities for the whole addition of population to the cities, not population density in the areas where additional land will be taken into the city, around the suburbs. That is, he recognizes that increases in population density will occur within cities, which will modify the density figures applicable only to the city fringe.

Since Bogue's work relates to the standard metropolitan areas, which include all the large cities and many but by no means all of the small ones, his results are not comparable either to the densities for the cities of 30,000 and over, shown in Table 13, nor to the density of all incorporated cities over 2,500 shown in Table 10. The density of all cities 25,000 and over in 1950 was 6,625 persons per square mile; the density for all incorporated cities was 4,360 persons per square mile. Bogue's estimates are thus considerably lower than actual population densities for average cities in 1950.

Measurement of the area withdrawn from other uses because of city growth would be worthwhile if acceptable definitions and procedures could be devised and appropriate data obtained. But the SMA is not a suitable unit of land use for the study of urban land use and urban expansion; too many factors other than expansion in urban population can affect land use within SMA's. The Bogue study may, in fact, have misled workers who have not fully understood its complexities.

A recent study of land use changes in Michigan during the 1930-55 period shows that urbanization has brought a decrease in agricultural activity in the first ring of townships around metropolitan centers, but an increase in agricultural activity in the second ring.[32]

Summary. This section about past urban use of land may be summarized as follows:

1. Information is seriously deficient. Prior to 1950, neither the area used for urban purposes nor the total area withdrawn by the city from other uses is known for all the cities and towns of the United States (and even for 1950 estimates are not entirely satisfactory).

2. There is some information in the census for the various years for

[32] Clarence W. Jensen, "The Effects of Urbanization on Agricultural Land Use in Lower Michigan," Doctoral dissertation, Michigan State University, 1958.

Table 16. Estimated area in cities 2,500 and over, by census periods, 1790-1950

Year	Total number	Total population (1,000)	Average population per city	Estimated average density[1]	Estimated total area[1] in square miles	Estimated total area[1] in acres (1,000)
1790	24	202	8,425	2,400	84	54
1800	33	322	9,760	2,575	125	80
1810	46	525	11,400	2,900	181	116
1820	61	693	11,360	2,880	241	154
1830	90	1,127	12,530	3,000	375	241
1840	131	1,845	14,080	3,150	586	375
1850	236	3,544	15,940	3,250	1,126	720
1860	392	6,217	15,840	3,320	1,873	1,200
1870	663	9,902	14,920	3,240	3,052	1,958
1880	939	14,130	15,030	3,250	4,340	2,785
1890	1,348	22,106	16,400	3,380	6,540	4,190
1900	1,737	30,160	17,350	3,480	8,670	5,545
1910	2,262	41,400	18,300	3,550	11,640	7,450
1920	2,722	54,158	19,900	3,640	14,880	9,535
1930	3,165	68,955	21,750	3,750	18,400	11,780
1940	3,464	74,424	21,480	3,720	20,000	12,800
1950—old	4,023	88,927	22,100	3,780	23,500	15,040
1950—new	4,741	96,468	20,320	3,680	26,200	16,750

[1] See text for method of estimation.

cities of over 25,000 or 30,000 population, showing area within city boundaries; this is assumed to be closer to "area withdrawn" than it is to "area used." Such data as there are show a closely similar relationship between city size and population density for all census periods from 1900 to 1950. Information on area is available for a group of larger cities.

3. If the 1950 relationship between city size, population density, and land area[33] is applied to data on numbers and total population of cities in earlier years, it is possible to estimate a total land area within cities for the entire national history (Table 16). On this basis, land within cities reached 1 million acres about 1860, 5 million acres

[33] See p. 87, n 22.

about 1900, 10 million acres shortly after 1920, and exceeds 15 million acres today. These figures more nearly conform to "area withdrawn from other uses" than to "area used for urban purposes," but they are not exactly either. They are entirely dependent upon the 1950 relationship between city size and density, a formula which applied fairly well to the larger cities for 1890 to 1950, but might not fit as suitably the smaller cities and longer time periods. If these data are representative, area withdrawn has not increased as fast as total urban population, because average density has risen as average city size has increased. This conclusion is contrary to much popular discussion of the subject, which assumes that population density in cities is decreasing, because suburbs are expanding rapidly and density there is generally lower than nearer the city center.[34]

FUTURE LAND USE BY CITIES

In a basically uncertain world, further increases in urban population in the United States seem about as certain as anything relating to the future. The total population of this country is growing rapidly, most of the net increase in population is within cities or urbanized areas, and more people in cities require more land to live on. These are the trends of the past and of the present; barring major catastrophe, they will continue. The real questions about future urban use of land are: How fast will our total population grow? What percentage of the total increase will be within cities? How much land will these additional city people need? How can we as a nation guide city growth, both to produce the best possible city and for the best over-all use of resources? These are primarily "how much" questions, not "whether" ones. The uncertainties as to the answers are in terms of specific figures, not as to general directions.

[34] No conflict is necessary between the easily observable fact that residential density is lower in the relatively new developments in suburbs and around the margins of cities than it is in older central city districts, and the analysis presented here which seems to indicate there has been comparatively little change in urban land density beyond that associated with increasing average size of city. Two different forces have been operating: on the one hand, some additional people have settled in the already densely settled parts of cities, thus adding appreciably to average urban density; on the other hand, as people have settled in the suburbs, greater urban densities have been attained there than would have been the case had they settled in the small towns and villages of the country.

Area of Land for Urban Expansion

When it comes to the land area needed for future urban growth, the ground is more uncertain. By any systematic definition of the terms, we know neither the present "use" of land for urban purposes, nor the present "withdrawal" of land by the city from other uses. The best available data are an intermixture of both ideas. We know even less about past trends in urban land use, by any definition.

Table 17. Estimates of future city and urbanized area, by regions, 1980 and 2000

Item	Unit	North-east	North Central	South	West	United States
1950[1]						
Total urban population	Mil.	31.4	28.5	23.0	13.6	96.5
Cities and urban places	No.	1,195	1,484	1,449	613	4,741
Average population per place	Persons	26,280	19,200	15,880	22,150	20,354
Estimated average density	Persons	4,100	3,660	3,370	3,840	3,710
Estimated total area	Mil. A.	4.9	5.0	4.4	2.3	16.6
1980[2]						
Total urban population	Mil.	46.9	49.0	55.8	34.1	185
Cities and urban places	No.	1,645	2,350	2,779	1,330	8,100
Average population per place	Persons	28,500	21,000	20,000	25,600	22,800
Estimated average density	Persons	4,200	3,760	3,690	4,040	3,900
Estimated total area	Mil. A.	7.1	8.3	9.7	5.4	30.3
2000[2]						
Total urban population	Mil.	61.9	64.0	75.8	54.1	255
Cities and urban places	No.	2,080	2,860	3,500	1,960	10,400
Average population per place	Persons	29,750	22,350	21,650	27,600	24,500
Estimated average density	Persons	4,260	3,840	3,790	4,120	3,980
Estimated total area	Mil. A.	9.3	10.7	12.8	8.4	41.0

[1] Using the new census definition of urban areas. Data on number and population of cities and urbanized areas in the United States from Appendix B, Table B-1; density and land area calculated from formula on relationship between size of city and density in 1950 ($Y = 3{,}295 \log X - 10{,}500$). Area is higher than in Table 10 because it includes an allowance for cities not reporting area and for other urbanized areas. It agrees with our conclusion on page 95.

[2] Population and number of cities and urbanized areas for the nation taken from projections in Appendix B, Table B-1; for regions, rough projections of past trends adjusted to yield national totals; density and area estimated from 1950 relationship between city size and density.

On the basis of the estimates presented in Table 17, the area within cities and other urbanized areas would rise from about 16½ million acres in 1950 to about 30 million acres in 1980 and to about 41 million acres in the year 2000. If the increased urban areas were to have a lower average density—say 2,500 persons per square mile, or the same average density that the typical city of 9,000 now has—about 39 million acres would be required by 1980 and about 57 million acres by 2000. Densities this low would seem to be improbable, but the figures may set an outside limit to the area in cities and urbanized areas.[35]

Each of the major geographic sections of the country would show an increase in urban population and in area within cities and urbanized areas, but the degree of the increase would differ. The South shows the greatest increase, over 8 million acres from 1950 to 2000, in absolute terms; the West shows the greatest percentage increase, nearly fourfold from 1950 to 2000. The Northeast and the North Central sections show roughly comparable increases, some 5 million acres each and a rough doubling in each, from 1950 to 2000. In general, the regional estimates are more speculative than the national ones.

Most of the increase in population and in urban land use will occur where there is now a town or city. But some new cities or urbanized areas will come into existence. The new major highways, especially at crossovers, provide excellent nodal sites. Factories, residential areas, shopping districts, and other urban land uses will often come into existence at such spots. The urban-like land use will be kept back a considerable distance from the road. New cities, in the legal sense, may be created; or the urban use of land may grow up under county or township government.

On the basis of present experience, it seems probable that at least one-third and perhaps more of the land thus within city boundaries and within urbanized areas will not be "used" for urban purposes—will be vacant and essentially unused. It is also assumed that all of the land within these boundaries will be withdrawn effectively from other uses, and that, in addition, other but unknown amounts outside of these boundaries will be withdrawn from other uses.

Part of the increased area within cities or other urbanized areas will come from farm land. It will be recalled that half of the SMA's

[35] For comparison, see somewhat larger estimates, on a different classification, by Raleigh Barlowe, "Our Future Needs for Nonfarm Lands," in U. S. Department of Agriculture, *Land, The Yearbook of Agriculture, 1958* (Washington: U. S. Government Printing Office, 1958), pp. 474-9.

are farm land. While probably few of the areas close to the cities within the SMA boundaries are of this character, yet, as the city spreads it will gradually withdraw land from farm use, even at distances well beyond the areas actually under conversion to urban use. Not all of the farm land will be in crops; some will be in pasture and some will be forested. Of the land converted to urban uses or withdrawn from other uses, that which was not farm land will in part have been used for grazing and in part be forested; in some cases it will have been idle or vacant. Much of the land under conversion at any time will have been idle in the immediately preceding years, because of the withdrawal effects of urban expansion. But some may have been idle over much longer periods. Unfortunately, the sparse and spotty information does not permit a reasonable estimate of how much land will come from each category.

Clearly, the extent and location of the urbanized areas of the United States in the year 2000 cannot be pinpointed. But, bearing in mind the uncertainties that hedge all long-term forecasts, it is possible to make some rough approximations. Figure 18 shows the major urbanized areas of the United States in 1950 as well as is possible on a small-scale map, together with our estimates of similar areas in 2000. In 1950 major urbanized areas (by our definition) were predominantly near the Atlantic Coast, around the southern edge of part of the Great Lakes, and in spots elsewhere especially on the Gulf and Pacific coasts. The general picture in 2000 is somewhat similar in its emphasis upon Atlantic, Gulf, and Pacific Coast and Great Lakes locations. A rash of areas conforming to our definition of major urbanized area now arises, especially across the South and to some extent in the Midwest. A large expansion in the middle Ohio Valley and neighboring areas is also apparent. There is an essentially continuous urbanized area from southern Maine to Richmond, Virginia—a "megalopolis," to use a word coming into use for such areas. Another such area begins at Rochester, New York, swings along Lake Erie to Detroit and beyond, across southern Michigan and northern Indiana to Chicago and Milwaukee, with extensions to Pittsburgh and the upper Ohio Valley cities. Smaller but relatively large congregations of cities are probable along the Texas Gulf Coast, in the southern California complex, and in the San Francisco Bay area reaching up to Sacramento and Stockton. Within these megalopolises there will be great variations in density of land use. The relative expansion of the cities and urbanized areas that are largely separate

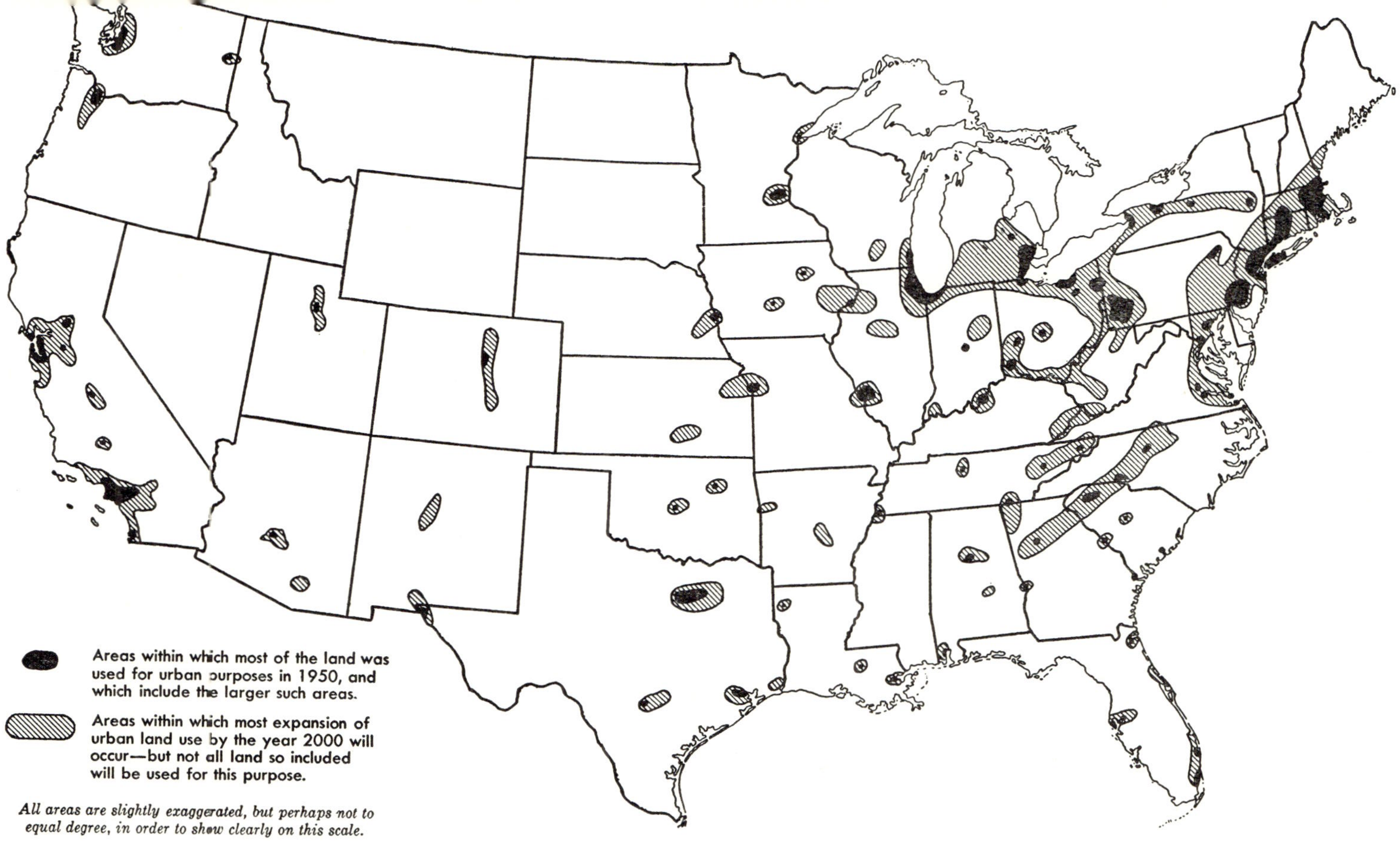

Figure 18. Major urban land use areas in 1950, and prospectively for 2000.

from one another will also be highly variable, and the chart can no more than suggest which ones are likely to cross the threshhold into major urbanization. Though urban growth will be a major land use adjustment in the next several decades, its impact will be different for the various major geographic regions and states.

The City of the Future

It would be interesting to speculate as to the form the city of the future may take, but to do so at length would lead us astray from the central purpose of this chapter. Here we are concerned with the shifts in use and competitions among various major land uses, and our purpose is served when land is assigned for urban use. Yet it must be recognized that the form of intra-urban land use may considerably affect the area of land needed for urban purposes.

An important factor affecting urban land use in the future may be the greatly increased per capita income. We estimated on page 15 that per capita income, in real terms, might about double the present figure by 2000. For an average family of four persons, this will mean an income (in 1955 prices) of roughly $15,000 annually. If past trends continue, a relatively greater proportion of the total population will have moderately large incomes, and fewer will have either very large or very small incomes, than at present. With incomes of this general level, people will be able to afford more for housing; and they are likely to want larger houses and more land.

It is wholly possible that considerably different forms of cities will arise in the future, possibly by the turn of the century. The three strongest centrifugal forces are increased speed and lowered cost of transportation and increased leisure time. In recent decades, there have been important changes in each, the full effects of which upon urban land use have not yet been felt. The private automobile is an indispensable factor in urban transportation today, but forty years ago its effect was minor. The transportation problem for large cities has not been solved, and no easy solution is in sight. Mass transport seems to offer the most logical answer, yet the trend increasingly has been toward greater use of the private automobile, with consequent increasing road congestion and parking problems in the larger cities.

If a satisfactory form of mass transportation could be devised and should prove economical, a new force toward spreading the city and reducing intensity of land use would come into being. But if mass transport is likely to prove economical, it will be only where traffic density is high, and hence where land use density also is high. If some form of personalized safe air transport should develop, this would be a major factor in urban dispersion. Further reductions in average work week, either in average hours per day, average days per week, or in longer paid vacations, seem highly probable. If the four-day, or even the three-day work week should become common—changes no more radical to consider today than the five-day work week of fifty years ago—the average worker might be willing to travel farther per work day than he does now. Either or both of these developments would tend to push people farther out from the city center, and result in a lower average density of urban land use and hence in a greater area of land for the same urban population.

The centrifugal forces operate not only for residential land use, but also for trade, business and industry. The suburban shopping center is a comparatively recent development, yet it is a major factor in retail trade. Shops and stores are built larger horizontally and lower vertically; and provision of parking space for patrons' automobiles requires extensive areas. The result is a much lower density of land use than in the old downtown shopping district. Some—but as yet not many—business firms have moved their offices from downtown to suburban or even open country sites; and far larger numbers of factories have been built there in recent years. The importance of efficient materials handling at lowest costs, and the much lower costs of horizontal than of vertical movement of heavy goods, has been at least one factor responsible. A shortage of adequate sites in the older city area, lower land costs in suburban or country locations, and more generous zoning provisions there, have been other factors. All of these developments might be accelerated in the future, with a consequent impact upon land use requirements.

While there are some centripetal forces operating to bring more people nearer city centers, their effects are relatively weak and are likely to remain so. If renewal of the decayed central portions of today's cities should proceed on a major scale and produce truly satisfactory living arrangements, a relatively larger proportion of total city population might live in areas comparatively close to city centers, at comparatively high land densities, and thus materially reduce the area of land required for urban purposes. However, the difficulties of urban renewal programs on an adequate scale are many, and it is difficult to estimate quantitatively the effect of such programs on total urban land use.

Some analysts think we may be building toward the multi-nucleated city, with many important business centers but without a single dominant center. The development of the suburban shopping centers, often by stores with their original store downtown, is sometimes cited as the beginning of this trend. Others have thought of cities built around airports, to reduce travel time from home to the airport—assuming that the dominant future form of travel will be by air. Other variants of these ideas contemplate great central parks in cities, partly to serve as recreation centers and partly to serve as pathways to the central airport for the incoming and outgoing planes. Many other ideas have been, or could be, advanced.

One common criticism to such advanced designs for cities is that their rebuilding costs too much. The fact is that every city is constantly being rebuilt; as homes, office buildings, and industrial plants become old or obsolescent, the need to build a new one arises. It may be rebuilt in the same site, or built in a new site. Over a period of 50 years or less, most of a present city will be rebuilt or replaced. By and large, the history of American cities has been that building in a new location has been much more common than rebuilding in the same location. Old sites have often proved too small or otherwise inadequate for the new uses; deterioration of the neighborhood has also been a factor leading to relocation elsewhere. Future city growth will surely lead to much migration of residence, factory, and business; it would be quite possible to plan the necessary rebuilding along new lines.

Yet there are serious economic and institutional obstacles to large-scale relocation of major city areas, especially of the central downtown area. Buildings become obsolescent one by one, not all at one time. The extensive and expensive service facilities—subways, other transportation lines, telephone and power lines, water and sewer pipes—available in an established area are a powerful argument for rebuilding there rather than in outlying districts where these facilities may have to be built. But the institutional obstacles to major city change are perhaps even greater. For the business district particularly, but to some extent for other types of major urban areas, location near service and competing businesses is important. Although a whole market might be better off in a new location, no single firm can afford to risk the loss of business that absence from the central market might entail.

Perhaps the greatest change in urban land use requirements could come from a relatively undramatic better utilization of the comparatively large areas, within or closely adjacent to cities, that are now idle. As we have noted, at least a third and probably half or more of the land the city withdraws from other use is not actually used by the city, but lies idle. Some is held speculatively, hoping for profitable rises in prices; other areas, speculatively held in the past, have accumulated such large charges and bad titles as to be worthless. Yet it would seem possible to reclaim these lands, usually by new institutions or procedures.[36] It would be physically possible to accommodate most of the increase in urban land use on land now withdrawn from other uses; but it will take much greater public responsibility and imagination than we have had in the past, if we are to do so.

Importance of Urban Planning

The foregoing analysis emphasizes, we believe, the great importance of urban planning, especially urban planning for the present suburbs and fringe areas. We have estimated that all the net population growth of the future will be in cities and urbanized areas. The welfare of the people located there will be affected vitally by the kinds of cities that develop over the next forty years. To a substantial extent, living conditions for them will have been determined by public and private urban land planners, who lay out the subdivisions, plot the streets and other improvements, and otherwise largely determine the character of the new urban area.

These additional urban areas will be the site of a major part of the additional investment in this country. By the year 2000 there will be over 150 million additional people in cities and urbanized areas, mostly in new areas rather than within the central part of present cities. Investment in homes, business districts, industrial plants, streets and other public facilities, and for other purposes connected with the land can hardly be as low as $10,000 per person. At this rate, a total investment in urban property of over $1,500 billion will be

[36] The article cited in n 16, p. 72, describes one set of procedures and cites court cases in which they have been upheld.

needed over the next forty years, or nearly $40 billion annually—less at first, more later. Even with the immensely productive American economy, these are large sums of capital. Comparatively small savings in the efficiency with which this capital is invested could produce very large gains—a 1 per cent total saving is still a large amount of money.

Planning for sound development of the new urban areas is both easier and more difficult than planning for good urban land use in developed areas. It is easier, because new capital will be needed, new buildings must be built, and other new improvements must be made in any event. If plans are available and accepted, the amount of capital and effort needed for good urban development may be no more and possibly less than for poor development, and the rate of obsolescence may be less. But it is difficult to make plans for a new urban area. Public bodies may be ill-equipped to do so; suburban developments today often precede establishment of cities as legal entities, and even where cities exist, they are often unable to provide sound planning for development. We have outlined earlier in this chapter why this is often true. Private real estate developers can hardly be expected to do more than provide sound planning, each within his own development. The sum of a number of individual real estate development plans, even if each is entirely sound, does not lead to a well-planned total urban area.

It should be stressed that this is not an issue of planning versus no planning, but rather one of good planning versus poor planning. Under any circumstances, the layout and character of a new urban area is planned—by the public bodies that lay out streets and highways, extend water and sewer lines, and provide or fail to provide public services such as schools, parks, and fire and police protection; by the private power and telephone companies; by the private real estate developers, who buy, subdivide, improve, and build upon land; by the merchants and other business people who come into a new area; and by others. The purchaser of a new home has little to say about these plans—he may reject some areas as unsuitable, or he may be unwilling to pay the asked-for prices for others, but he cannot directly make his desires felt in advance. A considerable degree of arbitrariness is necessary in the plans, public and private, and a large degree of risk, because it is impossible in advance to ascertain the desires of the potential customers.

Federal, state, and local governments possess two powerful tools to guide private use of urban land: the location of major arteries of transportation, and urban renewal. Highway, road, and street construction are some of our oldest recognized governmental functions. All too often, however, their construction is planned to serve the present needs along them. But when a new transportation route is opened it immediately creates a great new demand: more residences are built along it, or at one end, or nearby; and more factories and other businesses are attracted by it. The highway can be used as a major tool in guidance of land use. Urban renewal is a more recent governmental function, yet it has become firmly established. In the past, residential areas have gradually deteriorated, at least over large urban areas, and private capital has not found their revitalization profitable. By use of government legal powers, land is purchased in small tracts and combined into larger units for development. In the process, part of the costs are absorbed by government. The end result is often a higher density of land use, yet improved living conditions. Both transportation construction and urban renewal will continue on a large scale in the future, as far as we can see.

Some of the problems of planning wise urban development are also opportunities. For instance, one common complaint against private real estate development is that it "leapfrogs"—passes over some vacant tracts, to use other more distant ones, thus leaving areas of idle land. Street, road, power, water, sewer, and other services are more costly because of the scattered nature of the developed areas, under these conditions. But another common complaint is that private real estate development does not leave enough land for parks, schools, and other public uses. Why should not the leapfrogged areas be purchased for future park development? Instead of bemoaning leapfrogging, why not take advantage of it to buy the land as yet open? A far wiser course, it is true, would have been to plan the area in advance; bought the future park, school, and other public sites while the land was still at a price reflecting its undeveloped character; and then to have built roads and streets in locations to have influenced if not directed urban expansion along the lines planned. No small part of the difficulties of urban development in the past decade or more have arisen because the public agencies were not equipped and financed to keep abreast of private real estate development, much less to lead it.

In planning future urban development, perhaps more attention should be given to the small city and town. Most attention in city planning has been focused on the larger cities; it is true that most of the people are there. But, as we have shown, a much larger area of land is used by the small cities, and it is here that some of the greatest opportunities for saving land by good planning are to be found.

If the quality of city planning and development could be improved materially, substantial acreages of land could be saved for other uses. As a guess, we suggest that the area in the year 2000 might be only 35 million instead of 41 million acres. A saving of 6 million acres in a country of nearly 2,000 million acres may not seem large; but the saved acres would be located in highly important spots. The saving of land would be much less important, in monetary terms, than the further saving in transportation and public service costs it would make possible.

Water, Clean Air, and Air Space

Although our concern is with land and its use, air and water are closely associated with land; and cities are demanding in their requirements for both.

The city uses water. The amount of water that a city uses may be no greater than would be required to irrigate the same area of land in an arid climate, but in a humid climate the urban use of water is almost completely an addition to total water use. Cities use water to carry off their waste, human and industrial. As cities grow, they will use more water, probably more per capita. It will be necessary in most cases to reach farther and farther to get an adequate supply of water of acceptable quality, and this will mean higher water costs. An ample supply of water of acceptable quality is likely increasingly to become a major locational factor for industries which depend upon large water supplies. This, in turn, may require larger acreages of specialized municipal watersheds, or at least management of forest, grazing, and farm land with more regard to the effect upon urban water supply.

The city creates faster and more runoff, and thus adds to flood peaks and volumes. The modern urban area has a high percentage of impervious surface. Streets are likely to use a fourth of the total urban area; whether paved or gravel surfaced, they are almost impenetrable to water. House and garage tops and other areas are also hard surfaced. As much as 50 per cent or even more of the total area in densely settled parts of the city will be completely impervious to water. When it rains, runoff from these areas is virtually 100 per cent; total runoff from urban areas is much greater than from the same areas before urbanization. Storm sewers requiring large investments, or other means for carrying away large volumes of water very quickly, are necessary. Not only is total runoff increased, and hence the flood problem magnified, but infiltration into the ground-water basins is greatly reduced. In areas where pumping from ground water is a major source of water for urban needs—in the Los Angeles metropolitan complex, for instance—this may be serious.

Cities are vulnerable to floods. As cities expand and change, the hazard of floods and the damage caused by floods each may increase, although wise planning and sound development could reduce each. If we extend our indiscriminate use of flood plains as we have in the past, floods will certainly become a more serious problem. To a large extent "flood control" has so far proved illusory and certainly costly.

Clean air is a major attraction for living in any city. At the same time, the urban way of life pollutes the air above and around the city. Industry is likely to discharge smoke and other air pollutants; the burning of wastes of all forms does the same; and in some instances the greatest offender of all is the automobile. Gaseous wastes of various kinds are an unavoidable by-product of urban production methods and living patterns; the manner in which they are controlled or reduced in volume or rendered less objectionable may force major changes in the pattern of working and living. Nearly every large American city faces an air pollution problem, some to a greater degree than others; and none may be said to have fully solved it, though some have made more progress than others.

The modern city also needs air space for air travel. Wetmore has shown that the air space requirement for a major airfield may be 40 square miles, compared with a ground occupancy of 8 to 10 square miles.[37] The sound effects from jet plane operations may be large over an area as great as the air space, but not necessarily coincident with it in boundaries. In order to fully handle the incoming

[37] Louis B. Wetmore, "Future Demand for Urban Land," paper presented at Land Economics Institute, University of Illinois, June 25, 1958.

and departing aircraft, the space requirements may be several times as great as the approach zones themselves. The kinds of ground uses that are possible near airports are limited; and the effect of the airport on certain other kinds of use, such as residential, may be great.

Urban Growth and Agricultural Land Use

In recent years, many people with different professional and interest backgrounds have been disturbed at urban growth and its encroachment upon agricultural land. Most people have recognized that the total areas involved until now are small compared with the total area of agricultural land, and that no immediate shortage of agricultural land is in prospect. Some, however, have been disturbed about the immediate effects. Many have been much concerned about the long-run effects upon our ability to produce the crops we need, and most people have been concerned over the disturbance that urban expansion creates in the rural community. Some have written or spoken as though the city were a sinister thing, reaching out to strangle and engulf the rural area and its people.

In our view, much of this writing has been exaggerated. In the first place, most people are confused as to the nature of the force that is operative. It is not city growth, as such, that creates the problem, but population growth. Had the population growth of the last twenty or fifty years occurred without any major increase in city population—had it, in short, taken place in open country—the area required for site purposes would have been far larger. If it is use of land we are concerned about, the city is far more efficient than the open country: the larger the city, the less land is required for a given population increase. If we must have high rates of population increase and if our objective is to minimize the area of land needed for site purposes, then the city is the best place to put the extra people. This does *not* say that the cities that have arisen are the best that could be devised, or that the resulting pattern of land use within the cities has been perfect.

In the second place, most cries of alarm over urban expansion have been superficial in their level of economic analysis. The direct shift of land is from farm to urban use. But what are the indirect influences and results? American agriculture today is market-oriented and market-limited. The volume of agricultural output is definitely limited by available markets, and the quantity of resources used in agricultural production is also limited by market demand for agricultural products.[38] There is much land in the United States, physically capable of being farmed, that is not farmed.[39] In the southern Coastal Plain alone there are 36 million acres of good potential cropland.[40] As nonfarm populations grow, they constitute an increased demand for agricultural products, and this stimulates the use of land for agricultural purposes. It is true that, with present-day transportation methods and costs, the stimulus to land development is less local than it once would have been. Nevertheless, a considerable part of agriculture in the northeastern third of the nation has been directly affected by city demand within the region for fresh milk and other products produced locally. It should be recalled that Bogue found that the area in farms in the SMA's *increased* as urban population rose. He thought this was an "apparent impossibility," but we wish to suggest that this increase in farm area near cities was neither impossible nor unexpected. Rather, it was, at least in large part, a reaction to the stimulus of nearby increased demand for farm products.

In the third place, the discussion of the effects of urban expansion into rural areas ignores the technological, economic, and cultural complex of the present and recent past. As will be shown in Chapter IV, an agricultural revolution has taken place in the United States in the past forty years or so. If the total population of the United States today were 100 million, as it was at the time of the First World War, the area in crops would be *less* by at least 100 million acres. The growth of cities over the past generation has kept infinitely more land in agricultural production than it has taken out of it.

Urban expansion poses many serious land use and other problems; but solution is not facilitated by misunderstanding the nature of the problems and the facts that produce them.

38 For a stimulating discussion of this matter, see James Gillies and Frank Mittelbach, " 'Urban Pressures on California Land:' A Comment," *Land Economics,* Vol. 34, No. 1 (February 1958).

39 President's Materials Policy Commission (Paley Commission), *Resources for Freedom*, Vol. V (Washington: U. S. Government Printing Office, 1952), p. 64.

40 James R. Anderson, *Land Use and Development, Southeastern Coastal Plain,* U. S. Department of Agriculture, Agriculture Information Bulletin 154 (May 1956).

III

Land for recreation

In the United States today recreation is in many ways as important a factor as work, or as food, or clothing, or housing. In the future its importance will increase. By and large, we have met our most urgent basic needs for life. As Chapter IV tells, today we have nearly as much food per capita as we would use, no matter how high our incomes or how low the price of food. As many overfed as underfed people live in the United States today. There are limits to how much we can and will use of any specific goods; some of our rising level of living has taken the form of more leisure and more play. It will continue to do so. How we use our leisure time will affect the character of this country no less than how we work and what we produce.

"Play," "leisure," "recreation"—these are words with a large common element of meaning as well as important differences. Each involves a range of activities;[1] each also requires or implies the use of resources for fulfillment. This chapter is concerned with the latter. Further, it is limited to *outdoor* recreation, because it is this that

[1] The general sources found to be most helpful include the following: George D. Butler, *Introduction to Community Recreation* (New York: McGraw-Hill Book Co., 1949); George Barton Cutten, *The Threat of Leisure* (New Haven: Yale University Press, 1926); Foster Rhea Dulles, *America Learns to Play* (New York: D. Appleton-Century Co., 1940); Luther Halsey Gulick, *A Philosophy of Play* (New York: Charles Scribner's Sons, 1920); Harold D. Meyer and Charles K. Brightbill, *State Recreation: Organization and Administration* (New York: A. S. Barnes and Co., 1950); Harold D. Meyer and Charles K. Brightbill, *Com-*

makes a major demand upon land and water resources. As in other chapters, the concern here is with the land used primarily for the purpose under discussion. Much of the forest, farm, and grazing land of the nation has secondary, although sometimes quite large, recreational values. But these are disregarded; in this chapter we consider only those lands on which recreation is the sole or major land use.

Further, we emphasize the public recreation areas because, while a great deal of private land is used for recreation, little or no data exist as to its extent, location, and forms of use. There is a basic difficulty in that most privately owned land used for recreation has another primary use. Men may own forest tracts for recreational purposes, yet their land will ordinarily be classed as forest. The same is true of many farms. We focus on the public lands used solely or primarily for recreation. Some of these are called parks, but a wide variety of other names are used at times—recreation areas, forests, wildlife areas, waysides, and many others.

For this kind of outdoor recreation area, the decisions on land use are not made in the market place. Admission to most such areas is free, or at least not on a commercial basis. Calculations of costs, income, and profits do not provide the basic criteria for decisions on use of land for this major purpose. Rather, the decision to use or not to use land for outdoor recreation purposes is made by political processes; it is a decision for public, not private, action. In reaching it, calculations as to the value of recreation are seldom made, and they are not considered relevant, or at least not critical. The cost of establishing and maintaining the area is but one factor in the decision to use it for recreation. Often it is not easy to know accurately the value that people put on outdoor recreation, and their willingness to contribute to the public purse for it. Many factors enter into this type of public decision. Recreation, as a major use of land, differs in this respect from most or all the other major land uses.

Later in this chapter, we shall estimate the future demand for outdoor recreation and the area needed to satisfy it. But it should be

munity Recreation—A Guide to Its Organization and Administration (Boston: D. C. Heath and Co., 1948); Martin H. and Esther S. Neumeyer, *A Study of Leisure and Recreation in Their Sociological Aspects* (New York: A. S. Barnes and Co., 1949); Arthur Newton Pack, *The Challenge of Leisure* (New York: Macmillan Co., 1934); G. Ott Romney, *Off the Job Living—A Modern Concept of Recreation and Its Place in the Postwar World* (New York: A. S. Barnes and Co., 1945); Jesse Frederick Steiner, *Americans at Play—Recent Trends in Recreation and Leisure Time Activities* (New York: McGraw-Hill Book Co., 1933).

borne in mind that estimates for recreation are more than usually hazardous because they are necessarily and to a large degree forecasts of political action. Estimates of potential demand are possible, though with a rather wide margin of error; but how far the potential demand may be met is less clear.

RELATION OF RECREATION TO INCOME, EXPENDITURES, AND TECHNOLOGY[2]

One would expect certain rather clearly definable relationships between recreation and income, expenditures, leisure, and technology of various kinds. However, measurement of these relationships is hampered by the economic data which usually do not conform to the shape needed for this purpose. Recreation is a form of activity, or, more exactly, various forms of activity, undertaken for a specific purpose—that of enjoyment. The usual classifications of economic activity, especially income and expenditure data, are in terms of types of goods—food, clothing, shelter, etc.—or in terms of activities or forms of production. But recreation as a human activity cuts across all these. What is spent for sport fishing tackle is clearly an expenditure for a commodity usable only for recreation. But a large part of the expenditure for private transportation, especially for automobiles, is for recreation. Some forms of housing, such as mountain cabins or seashore cottages, are primarily for recreation. A considerable amount of clothing is bought expressly for recreation. Even food may cost more than it otherwise would, when it is bought for recreation use.

For all of these reasons it is difficult to define recreation satisfactorily as far as expenditures are concerned. A narrow definition could be adopted which would include only those goods and services clearly and solely usable for recreation—sports equipment, admissions to sports events and places of entertainment, etc. But this definition would exclude other expenditures made mainly to provide recreation for the spender and his family. The motivation and the relation be-

[2] Sources for this section include Frederic A. Dewhurst and others, *America's Needs and Resources—A New Survey* (New York: Twentieth Century Fund, 1955); and the editors of *Fortune, The Changing American Market* (Garden City, New York: Hanover House, 1955).

tween income and expenditure are almost surely different for housing in the usual sense and housing as part of recreation activities, for instance. However, an attempt to broaden the definition to include a much wider range of activities encounters all sorts of difficulties, not the least of which is confusion with other and well-accepted economic categories. Neither is it clear where one would draw the line between expenditures for recreation and those for other purposes: What about expenditures on alcoholic beverages, for instance?

There are other and special difficulties as far as outdoor recreation is concerned. Activities of this kind are often engaged in because they are relatively cheap to enjoy. A day in the park may involve very little cash cost to a family other than what would have been incurred at home. The family may take its own low-cost lunch and may drive the family car at small gasoline expenditure; the result may be a happy outing for a low cash outlay. The indirect costs to the family, as for car depreciation, may be much larger; and the whole affair is possible only because some unit of government has provided the area involved, often at considerable expense. Even for longer trips and more ambitious outdoor activities the direct cash costs may be fairly low. In part this is because the full costs do not show up as cash costs at the time, in part because some of the costs are borne by government and others, and in part because the participant himself provides the many services required.

As a result of all this, if the expenditures for outdoor recreation, especially the direct ones, are compared with the expenditures for other types of recreation or for other types of activities, then a false idea may easily result. Outdoor recreation may appear in a less important role than perhaps it really occupies.

The differences in definition of recreation are revealed by three estimates, each of which uses the same basic data (Table 18). The United States Department of Commerce collects and analyzes data basic to national income and national expenditure estimates. It calculated that in 1952 total expenditures for recreation were $11.4 billion. Dewhurst and associates used the same data; except that they excluded 58 per cent of the expenditure for reading, because they considered this as private education, they used the same definition. However, they used earlier estimates published by Commerce, whereas those in Table 18 attributed to Commerce are later revisions; the differences in the two series are thus revisions, and give some idea of

Table 18. Personal consumption expenditures for recreation according to different definitions

(million dollars)

Item of expenditure	Expenditure (in current dollars) as estimated by: Dewhurst[1] (1952)		Department of Commerce[2] (1952)		*Fortune* magazine[3] (1953)
Sports equipment:		2,153		2,279	3,330
nondurable toys and sports supplies	1,284		1,164		
wheel goods, durable toys, etc.	790		1,115		
boats and pleasure craft	79				
Radio, television, and musical instruments:		2,800		2,576	2,710
purchases	2,324		2,100		
repairs	476		476		
Participant recreation:		867		839	
parimutuel net receipts and coin machines	419		[4]327		
billiards and bowling	129				2,200
golf	169				
other	150				
Spectator amusements:		1,876		1,691	
motion picture theaters	1,134		1,284		
legitimate theater, miscellaneous	474		185		
spectator sports	268		222		
Organizations and clubs		556		511	
Reading:		888		2,120	
books and maps			504		3,280
magazines, newspapers, sheet music			1,616		
Flowers and plants		836		641	
Other		513		717	
Dining out					1,030
Alcoholic beverages					8,860
Vacations, weekends and foreign travel					9,190
Total		10,489		11,374	30,600

[1] Dewhurst and Associates, *America's Needs and Resources—a New Survey* (New York: Twentieth Century Fund, 1955).

[2] *Survey of Current Business,* U. S. Department of Commerce, July 1956.

[3] Editors of *Fortune, The Changing American Market* (Garden City, N. Y.: Hanover House, 1955).

[4] Parimutuel only.

the statistical stability of data on this subject. The editors of *Fortune,* on the other hand, have a total nearly three times as large as the others, because they include alcoholic beverages and vacation costs, items the others exclude. Some other differences also are evident, even though they, too, use the basic data of Commerce. But the variations in estimate stem from differences in definition as to what constitutes recreation as well as from differences in the basic data themselves.

Although the general trend of expenditures under each set of estimates is similar, there are some variations (Appendix D, Table D-8). The Dewhurst estimates are in terms of 1950 dollars as well as of current dollars. On this basis, total expenditures for recreation (as Dewhurst uses the term) rose more than three times from 1909 to 1952. A doubling occurred from 1909 to 1929, after which expenditures fell off during the severe depression years to almost the level of 1909. They rose again during the 1930's and the war, but the rate of most rapid increase in recent years has been since the war.

On logical grounds, as pointed out above, one would expect per capita expenditures for recreation to rise as average income of the population or the individual rises. There is some evidence that this is true, other evidence that it may not be. According to Dewhurst, the percentage of consumer expenditures made for recreation purposes increased irregularly over the 1910-52 period (Figure 19). In the earlier period, about 3 per cent of all consumer expenditures was for recreation, as these authors define the term—rather narrowly, as we have seen. This percentage rose rather rapidly to a peak of nearly 5 per cent in 1930, declined during the depression and war years to about 4 per cent, and since the war has climbed, though irregularly, to over 5 per cent.

The editors of *Fortune,* with their broader concept of recreation, use two definitions, one to include alcoholic beverages, TV, radio, and dining out; the other to exclude them. But even their narrower definition includes vacations, week ends and foreign travel, reading, gardening, etc. As a result the *Fortune* definition shows expenditures for recreation of the order of 6 to 8 per cent on the narrower basis, and of 11 to 13 per cent on the broader basis, in each case measured against total consumer outlays. The *Fortune* comparisons begin in 1929. In percentage terms they show some modest declines during the depression, a recovery in the late 1930's, a sharp decline during the war, and a later recovery; but for the whole period covered by

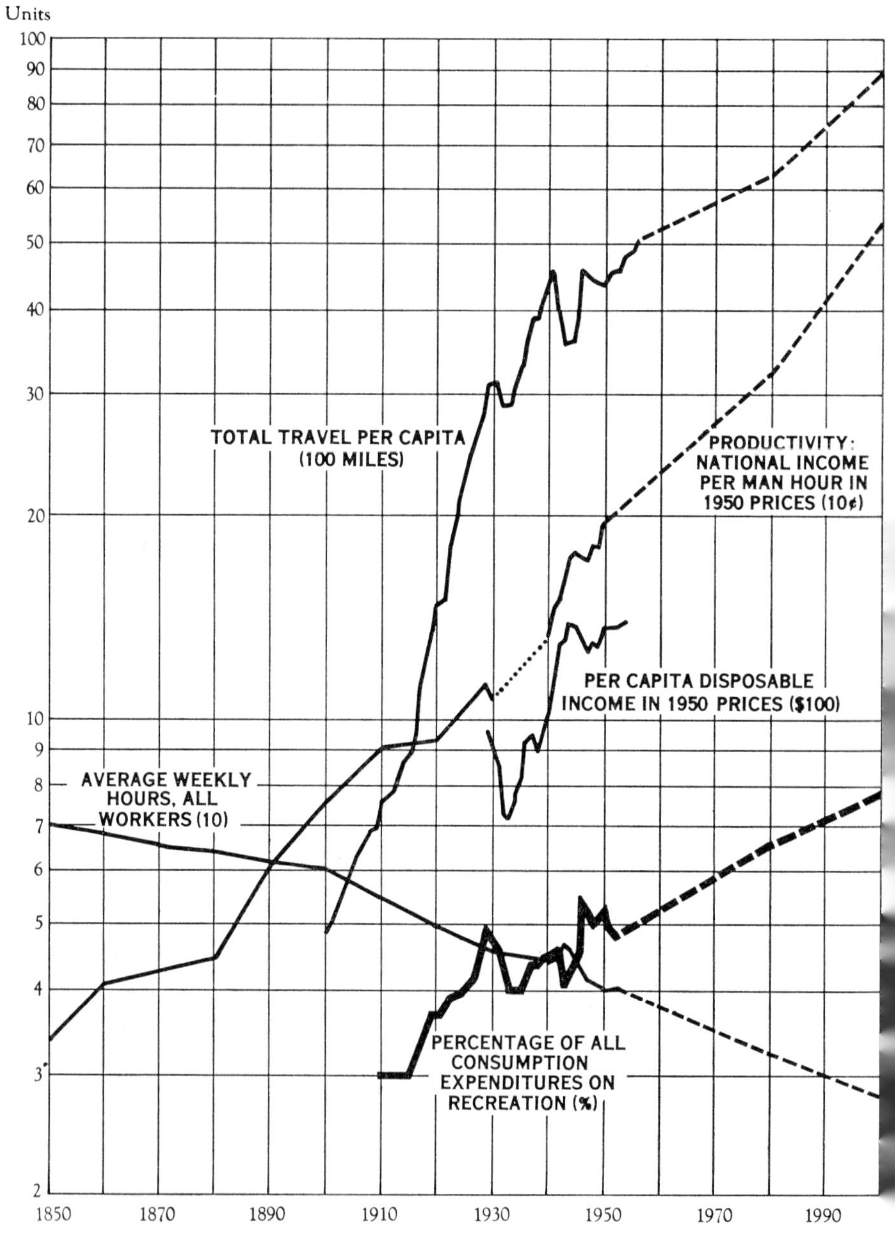

Figure 19. Income, travel, and leisure factors affecting the use of outdoor recreation facilities, 1850–2000.

these data, there is no clear trend in either direction; actual dollars spent reflect changes in income.

Fortune also shows the relationship between the income of the family and the expenditure for recreation in 1953, using both definitions of recreation. The amount spent for recreation is comparatively high, percentagewise, when family income is low; drops considerably (roughly 25 per cent) as family income rises to the $3,000-$4,000 range; and then rises slowly and steadily to new high levels as family income reaches $7,500 or $10,000. This relationship is due to the fact that many of the low-income family units are made up of retired elderly people, not dependent entirely upon current income for expenditure, and that other family units in this income range are of young people who obtain parental assistance to maintain a level of expenditures above their incomes. Among those family units that are limited in their expenditures to their income, there is almost certainly a trend toward greater expenditures for recreation as income rises. It seems reasonable to expect that as *average* income for the whole population rises, per capita expenditures on recreation may also rise, even proportionately to income.

But should this not happen, the total sum of money spent for recreation would nevertheless increase greatly as *total* population and *total* income rose. The *Fortune* analysis emphasizes the lags in expenditure patterns, especially for recreation. According to this thesis, expenditures for recreation in any period tend to reflect the income-expenditure relationship of the previous few years. This is believed to be true when incomes decline, as in a depression, or when they rise. At the least, there is good reason to think that expenditure for recreation will not be a lower proportion of total income in the future than in the past; it may well be higher. It seems likely that outdoor recreation will share in the increased total expenditures, possibly at a more than proportionate rate. But there is little direct evidence on this latter point.

Real incomes per capita have been rising in the United States for many decades, more or less regularly however one measures them.[3] This is evident in Figure 19, which shows per capita disposable income (measured in constant prices) rising considerably, but irregularly, for the period 1929-53, and shows also productivity per man-hour rising over a much longer period and at more or less the same rate. If the evident trend in the latter is extended to the year 2000, a

[3] Reference is made to the discussion of this point on pp. 14–16.

productivity of roughly $5.50 per hour (in 1950 prices) is obtained, compared with about $2.00 in 1952 and a forecast by Dewhurst of $2.40 in 1960. These trends may not continue, but they have persisted for at least a hundred years; no major error may be introduced if they are extended fifty years into the future. This would give a slightly higher rate of increase than was assumed earlier (page 15).

The amount of leisure enjoyed by the average person of working age has also risen greatly over recent decades. But no direct measures of leisure are available; they must be estimated from the time spent in work. While part of the decrease in the latter has been absorbed by increased travel to and from work, a good share of the reduced work week has produced more leisure. Dewhurst shows the hours worked per week by the average worker to have declined from seventy in 1850 to forty in 1950. His data include an allowance for overtime, for time lost through shutdowns of various kinds, and for vacations; to a considerable extent one balances the other, so that the average work week is only modestly shorter than the typical work week. His data include agricultural workers, notorious for their long hours of work, as well as urban workers. The decline has not been steady over the entire period, but breaks into rather definite periods as follows:

	Decline in average work week			
	from	to	absolute	per cent
1850-1900 (50 years)	70	60	10 hours	14
1900-1920 (20 years)	60	50	10 hours	16
1920-1950 (30 years)	50	40	10 hours	20

In considering the future, it makes considerable difference whether the past trends are extended on a logarithmic or an arithmetic basis. Using the former, a twenty-eight-hour average work week by 2000 is probable; this might involve a work schedule of roughly five days of six hours each, or a four-day seven-hour work schedule, or more days and more hours but with a longer vacation schedule. It is not improbable that some of the increased leisure will come about through all three forms of work-time reductions; but the proportions may vary in ways not now clear. If one takes the arithmetical approach, it can be pointed out that the average work week was shortened ten hours in twenty years between 1900 and 1920, and ten hours in thirty years between 1920 and 1950; if it were to be shortened another ten hours, even in another thirty years, this would mean a thirty-hour average work week by 1980, declining almost to a twenty-hour work week by the year 2000. It would take a rash prophet to assert that such changes will not occur, yet on the whole it seems more reasonable and more conservative to base estimates of the future on logarithmic relationships.

Fortune estimates average working hours in a somewhat different way but with similar results. The total weeks of paid vacations have risen from 17.5 million weeks in 1929 to 70 million weeks in 1956; considering the number of employed workers in different years, this averages .37 week per worker in 1929 and slightly more than one week per worker in 1956. Many workers, especially the casual and those in certain activities such as agriculture, typically get no paid vacation; others receive more than the average. The increase in paid vacations was modest until after the war, but it has doubled in the past ten years. Further major increases appear highly probable.

Another big factor in the growth of recreation over the past decades is the increase of travel in the United States. The data are not perfect, but it is clear that average travel per capita has vastly increased, especially between the years 1900 and 1930 when the automobile was becoming the common method of personal transportation. The smaller but still significant increase since 1930 is partly obscured by the depression and the war, each of which limited travel, although in different ways (Appendix D, Table D-9). With the past as guide, one can foresee further increases in travel per capita, even with the same means of travel that we know today. The super-highway, especially as a system covering the whole country, will encourage more people to drive their cars greater distances. And tomorrow's travellers may very well take to the air for their personal transportation in far greater number than they do today. Although new and radically different transportation technologies are not predictable, neither should they be excluded from consideration of future expanding travel.

Some rough approximations of the total effect on expenditures for recreation of more people, higher real incomes per capita, more

Table 19. Factors related to outdoor recreation in the future

Item	Unit	1956 Amount	1956 Index	1980 Amount	1980 Index	2000 Amount	2000 Index
Total population	Million	170	100	240	141	310	182
Personal income:							
per capita	1955 $	[1]1,630	100	2,525	155	3,660	224
total	Bil. $	[1]269	100	605	225	1,135	422
Expenditures on all recreation:[2]							
per cent of income	%	5.2	100	6.6	127	7.8	150
total	Bil. $	14.0	100	39.9	285	88.5	632
Time:							
average work week	Hours	40	100	32	80	28	70
discretionary leisure per week[3]	Hours	30	100	38	127	45	150
paid vacation: per worker[4] .	Weeks	1.0	100	2.5	250	4.0	400
total	Million Weeks	70	100	240	343	496	709
Travel:							
per capita, total	Miles	5,000	100	7,000	140	9,000	180
per capita, for recreation[5] . .	Miles	2,000	100	3,500	175	5,000	250
total for recreation	Billion Miles	340	100	840	247	1,550	456
Recreation visits:[6]							
user-oriented:							
total[7]	Million	1,000	100	2,000	200	3,750	375
per capita	Visits	5.8	100	8.3	143	12.1	209
intermediate:							
total	Million	312	100	1,200	384	5,000	1,604
per capita	Visits	1.8	100	5.0	278	16.1	894
resource-based:							
total	Million	116	100	750	647	5,000	4,310
per capita	Visits	.7	100	3.1	443	16.1	2,300

[1] 1955.

[2] Using Dewhurst definition of recreation.

[3] Assuming 8 hours for sleep and 6 hours for eating, miscellaneous personal activities, and travel to and from work in 1956. Similar assumptions in 1980 and 2000, but some reduction in travel time as work days per week are reduced.

[4] Dividing total paid vacations by total labor force.

[5] In 1957 approximately 40% of all trips were for vacation and pleasure (other than visiting relatives). *Travel Survey, 1957,* U. S. Bureau of the Census, September 1958.

[6] To public areas only.

[7] Incomplete estimate all years, chiefly an index of actual use.

leisure, and more travel per capita, and the effect upon outdoor recreation are found in Table 19. It should be emphasized that projections of this kind are rough in the extreme, and at best will yield only general ideas of comparative magnitudes. All too little is known of past relationships between recreation and other activities. For the future, even if accurate historical information were in hand, there still would be large uncertainties. Shifts in recreational activity over recent decades have reflected changes in the social structure and ways of life in the United States. Some of these changes may have been unique to the period in which they occurred; nothing comparable may occur in the future; or it may be argued that even greater changes will come about. In any case, mere extension of past trends is no guarantee of accuracy, and yet there is little else to use even for a rough approximation.

The estimates of total population and of income per capita in Table 19 are the same as those developed on page 15. The average work week, average miles traveled per capita, and percentage of expenditures on recreation are rough extrapolations of past trends. They lead in total to an expenditure on recreation by private individuals of about $40 billion in 1980 and about $88 billion in 2000, using in each case the Dewhurst definition of recreation; if either of the *Fortune* definitions is used, the resulting figure is much higher. Although one cannot be sure that the expenditure for recreation in 1980 and 2000 will be as estimated, there is fair assurance that the total expenditure will be greatly in excess of what it was in 1950. The effect of the increases in the various related factors seems to be roughly *multiplicative*—population is doubled by 2000, per capita incomes are more than doubled, expenditure for recreation is a 50 per cent larger part of total income. When all these increases are multiplied together, they result in a very large increase. How much of this total will be spent for outdoor recreation is unknown—as is the proportion so spent in 1956—yet it is fairly certain that outdoor recreation will share at least proportionately in this increase in total private expenditures. If this is the case, expenditures on outdoor recreation in 1980 may be in the rough order of three times those of 1956, and by the year 2000 they may reach six times the expenditures for 1956. Even these amounts would exclude outlays for transportation, food, clothing, shelter, and some other items basic to the recreation. In addition, substantial sums of public funds will be spent for outdoor recreation.

A BROAD CLASSIFICATION OF OUTDOOR RECREATION

There are so many kinds of outdoor recreational activities, so many types of areas upon which they are carried on, so many different times of use, and so many organizations providing recreation, that the whole situation becomes confusing unless there is some basis for classifying the areas. This has been attempted in Table 20.

One major type of recreation area is user oriented. The essential

Table 20. General classification of outdoor recreational uses and resources

Item	Type of recreation area		
	User oriented	Resource based	Intermediate
1. General location	Close to users; on whatever resources are available	Where outstanding resources can be found; may be distant from most users	Must not be too remote from users; on best resources available within distance limitation
2. Major types of activity	Games, such as golf and tennis; swimming; picnicking; walks and horse riding; zoos, etc.; playing by children	Major sightseeing; scientific and historical interest; hiking and mountain climbing; camping, fishing and hunting	Camping, picnicking, hiking, swimming, hunting, fishing
3. When major use occurs	After hours (school or work)	Vacations	Day outings and week ends
4. Typical sizes of areas	One to a hundred, or at most to a few hundred acres	Usually some thousands of acres, perhaps many thousands	A hundred to several thousand acres
5. Common types of agency responsibility	City, county, or other local government; private	National parks and national forests primarily; state parks in some cases; private, especially for seashore and major lakes	State parks; private

characteristic of these areas is that they be located close to their users; the physical characteristics of the resources are secondary to the location factor, although good physical characteristics are important here as well as elsewhere.[4] At the extreme in this general class are the tot-lots, provided usually in densely settled areas where backyards for play are not available. There are also children's playgrounds, playing fields for games of all kinds, swimming pools, picnic grounds, local parks, and the like, as well as zoos, museums, and other similar improvements. Unless areas of this description are readily accessible to the people most interested, they are to all intents and purposes nonexistent. They are used primarily after school or work hours. The areas themselves may be small, often of only a few acres; and they must be provided primarily by local government, and to some extent by private business and voluntary private associations.

At the other extreme are the resource-based recreation areas whose essential characteristic is their superior natural features. It is not always easy to measure the quality of a recreational area, for quality must be judged by subjective standards. Yet few people would deny that most of our national parks, many of our national forests, some state parks, and many of the better lakes and ocean beaches have recreational qualities superior to those of the average wooded area even if the latter has small lakes within it. Experience has shown that if a recreational area has truly superior natural qualities, people will be attracted to it, assuming that reasonable methods of transportation exist. Even such remote and unusual areas as Mount McKinley National Park and Katmai National Monument, in Alaska, attract a few people in spite of the difficulties of time, travel, and cost. The resource-based recreation area is used primarily by people on vacation. The activities enjoyed there are quite different from the activities of a more organized kind, enjoyed in the user-oriented area. For the most part, the resource-based recreational area is included in a federal landholding such as a national park or national forest, or is in private ownership, such as an ocean beach or a major lake.

Between these two extremes is a type of recreation area which, for lack of a better name, we call, "intermediate." The intermediate area must be within reasonable distance—perhaps one to two hours' travel time—of most of its users, and it is intermediate in that its physical

[4] Many of these areas fall within cities; hence their area is likely to be included in data on urban land use. Further consideration of this matter is found in Chapter VIII.

characteristics are important but not all-dominant. Subject to these influences, however, it may be located in one of perhaps many possible locations. Once located, it is as permanent in location as either of the other types. Typically, a recreation area of this kind is a state park, possibly a large municipal park, located outside of but not too distant from a major city; it is used primarily for all-day and weekend recreation; ideally contains attractive forests and waters, and is used for somewhat more extensive recreational undertakings than are the city parks and playgrounds.

It should be obvious that these are not completely watertight categories. Although the resource-based areas are classified as the vacation spots, for example, some people may spend their vacations in a local city park or in the state park a few miles away. Similarly, a few people may be in a position to use a national park or forest for a workaday evening picnic. But these are only variations from the general pattern.

One would expect more substitutability among the types of recreation within each of these major categories, than between the types in one category and those in another. The time of use and location factors alone would operate in this direction. A youth may decide to play tennis because the swimming pool is not open, or an adult to play golf because his softball league is not yet organized; but neither can very well visit Glacier National Park in the same available time period. Likewise, a family planning a vacation may be undecided between the mountains and the seashore, and between locations at each; but most families will not be satisfied with an evening picnic in the local park each night, if they have a paid vacation and can afford to go somewhere.

The substitution between these various kinds of outdoor recreation and various other types of recreation, including commercial amusements, also differs. Movies, television, and reading will readily substitute for user-oriented outdoor recreation; foreign or extensive domestic travel to urban locations will more readily substitute for resource-based outdoor recreation. The degree of substitutability between certain types of outdoor recreation and certain types of non-outdoor recreation may well be higher than substitutability within the outdoor recreation groups.

The relation of the distribution of leisure time to the kind of recreation in demand should also be obvious. If shortened average work weeks take the form of fewer hours per day, then the user-oriented recreation areas will feel the major brunt of more leisure; if the number of days worked per week are shortened to four—or conceivably some day to three—then it is the intermediate type of recreation areas that will feel the major increased demand. On the other hand, if longer vacations should absorb most of our increased leisure, the demand for resource-based recreation will rise the most rapidly. It seems probable that greater leisure will take each of these forms, and hence that all three types of recreation areas will feel the impact; but the proportions may still vary considerably.

Each of these areas could be classified further, according to intensity of use. This may be measured on the basis of total annual or seasonal use, or of maximum numbers of persons during any day, or even hour, of use. The two measures would not necessarily give the same results, and each would be important, but for different purposes. The classification might be based upon an entire administrative area, such as a national forest or a state park, or upon a use classification of areas within such larger administrative units. Typically, a recreation area of almost any kind is used heavily in some parts, and lightly in others. This is true even of a playground: its blacktop may be well trampled, but not usually its shrubberies and borders. As for national parks, it has been estimated that nearly all their use occurs on 5 per cent of their area.[5]

It is possible to classify the use intensity of recreation areas roughly as follows:

1. Very heavy, with perhaps 1,500 or more total uses or visits per acre per year. A large share of the municipal parks will fall into this use classification, in whole or in part, and perhaps some highly popular areas within state or national parks.

2. Heavy, with typically 50 to 100 uses or visits per acre per year. Camps and picnic areas, popular lakes and stream edges, and some other types of areas would fall into this use class.

3. Moderate, with typically one to three uses or visits per acre per year. Many hunting and fishing areas, hiking and riding areas, and others would fall into this class.

4. Light, with typically one-tenth use or visit or less per acre per year. This class would include the "back country" of national and state parks and forests; sometimes even the more remote corners of large municipal parks would qualify.

[5] Marion Clawson and Burnell Held, *The Federal Lands, Their Use and Management* (Baltimore: Johns Hopkins Press, 1957).

HISTORICAL DEVELOPMENT OF OUTDOOR RECREATION

In considering the development of outdoor recreation in this country from the earliest settlement to the present, it is essential to distinguish between informal and the more or less formal types of recreation. The kinds of informal outdoor recreation that people have engaged in for centuries past can be readily undertaken in almost any open area of suitable terrain. In contrast, a gradual, but not uninterrupted, development of specialized recreation areas, mostly publicly owned and improved, has brought with it a wide variety of activities, many of which, like the conventional sports, involve quite large groups and are highly formalized. Paralleling this development has been the gradual shrinkage, and in some localities the virtual disappearance, of open and unused areas suitable for more casual activities, and the rise of commercial amusements of all kinds.

Dulles has expressed very well the changes in the place of recreation in American life:

> ". . . Two important factors, I think, stand out from the record [of the past three centuries]. The first is the continuing influence of an inherent puritanism, both arising from and enforcing a dogma of work born of economic circumstance, which may be traced from the seventeenth century to the twentieth. Until recent times it has frowned severely upon what the early settlers called any 'misspense of time.' If today this attitude has somewhat changed, the American tradition still insists that amusements should at least make some pretense of serving socially useful ends. The businessman plays golf to keep fit for business; the woman's club emphasizes its educational program; and reformers would have all popular entertainment directed toward the establishment of higher cultural standards.
>
> "The second factor is the paramount influence on recreation of the gradual transformation of our economy from the simplicity of the agricultural era to the complexity of the machine age. No field of human activity has been more deeply affected by this change and the concomitant growth of cities. The machine has greatly increased the leisure of the laboring masses, and it has at the same time made life less leisurely. The traditional patterns of everyday

living have been completely altered with an ever-growing need for play that can effectively compensate for the intensity under which we must work. If many of the forms of recreation that have evolved under these circumstances appear far from ideal, the question is nevertheless posed as to what the urban masses, granted the conditions of modern life, would be doing if they did not have their commercial amusements and spectator sports."[6]

The readily observable changes discussed later in this chapter—acreage of parks, volume of user participation, and the like—are merely the tangible expression of changes in popular attitudes toward recreation stemming from the more basic changes that have taken place in our personal values and ways of living.

The Period before 1875

In the earliest colonial days, recreation and play of all kinds were severely suppressed. Idleness was a social and moral sin. Given the precarious nature of life in the early colonies, this viewpoint was natural and in the interest of group survival. Attitudes of this kind are commonly ascribed to the Puritans; but in the earliest days they existed also for the middle and southern colonies and punishments for their violation were as severe in Virginia as in Massachusetts. In the South, however, the social climate changed in a comparatively few decades. By and large, by the end of the seventeenth century recreation in almost any form was an accepted part of southern living. And in New England, while social pressures against such activities continued, a majority of the people had loosened their ties with the church, and were either quietly disregarding or openly flouting Puritan attitudes and laws. Nevertheless, while Puritanism failed to suppress either the instinct to play or its release in convenient forms, it left its mark in terms of basic social attitudes.

In the first half of the eighteenth century, the colonies enjoyed a wide variety of diversions and sports. There was hunting and fishing for sport, as well as for food; there were occasional fairs, corn-huskings, barn-raisings, and other gatherings to which people from miles around would come for play and emotional release from loneliness;

[6] Dulles, *op. cit.*, pp. viii-ix.

organized sport, including horse racing and cock fighting, was common, especially in the South; dancing, shooting matches, and many other diversions took place. Drinking had always been a release from work and monotony, and the tavern had been a part even of the New England society from the earliest days. These recreational activities were indulged in by the well-to-do and the rich to a far greater degree than by the poor, of course, but even the latter had their means of relaxation to the extent that their money and their leisure permitted. In the South, especially in Virginia, there grew up in this period perhaps the most outstanding leisured class this country has known—a landed aristocracy which took its fun without apology or shame, and produced some outstanding political leaders.

The American Revolution brought a sharp change. Both government and public opinion forced a sharp curtailment of recreation and amusement. In part this was dictated by the stringencies of the times. But in large part it also grew out of the class nature of the struggle for independence: a revolt of the masses against the privileged classes. In the years that followed, the severe strictures and restrictions of the churches against amusements of various kinds were reimposed to a large degree. On numerous occasions through the first half of the nineteenth century, religious and educational leaders denounced the theater, card playing, profanation of the Sabbath, and other activities which once had been equally denounced and then relaxed. The saloon and grog shop often became the source of release for those to whom other enjoyments were denied. To a considerable extent the saloon was the poor man's club. Play and godlessness were again equated, at least to a large degree. This was the period of the frontier, with its infrequent but lusty and unrestrained pleasures. This was also the beginning of the rise of commercial amusements in the towns and cities.

There were few parks as we know them today in these earlier times. The typical New England village had its town square or common, originally used for pasturing cattle and providing an open space for walking, sports, and shows. Philadelphia had open squares or parks from the beginning, and in 1828 developed a park of 24 acres surrounding the new water works. In 1853 New York began to purchase land for Central Park. By the time of the Civil War, a few stirrings could be heard for formal outdoor recreation areas. But while some cities were beginning to feel the pinch of crowding, there was still plenty of open space in most of the towns or nearby.

The Last Quarter of the Nineteenth Century

For recreation, the last portion of the nineteenth century was a period of beginnings. A number of activities, relatively unimportant at the time, in later years came to have immense significance.

One of the chief actions just prior to this period in 1872, was the establishment of Yellowstone National Park.[7] Sequoia, General Grant (now a detached section of Kings Canyon) and Yosemite National Parks were established in 1890 and Mount Rainier National Park was established in 1899. The Forest Reserves, now national forests, began in 1891, and by the end of this period extended nearly to 40 million acres. This same period also saw the beginning of state-owned forests and parks on a continuous basis. These reservations of public land for recreation were largely a matter of individual instances; the concept of systems of national parks, national forests, and state parks was not yet generally accepted. Most of the areas established were used only lightly; they were too remote, and travel to them was too expensive, difficult, and time-consuming.

During this period, also, organized sport spurted forward. Horse racing, boating and sailing, cock fighting, foot racing, and a few others had long been popular. Baseball had been played from comparatively early times, but in 1845 the first formal rules were adopted for it. During the Civil War the game was played by the armies on both sides, and afterward gained rapidly in popularity. Croquet became popular for both women and men. Roller skating had begun about 1863, and by about 1876 bicycling began to provide the personal mobility which the automobile was later greatly to extend. College sports began on a large scale during this period, with matches between schools as well as play within them. Swimming became popular with women as well as men. Sports usually began as an activity of the upper classes, but soon spread to lower income groups as well.

This was a period when the summer vacation became more common. Hitherto travel and vacations had been confined largely to the

[7] Yellowstone is usually considered the first national park, and indeed it was, as a national park. However, the great mineral springs of Hot Springs, Arkansas, had been reserved from disposition to private ownership as early as 1832, and later, in 1921, this area was made a national park. A considerable part of Yosemite Valley was transferred to the state of California in 1864, for a state park. Later, in 1890, this area was transferred back to federal ownership and became a national park.

wealthy; even before the Civil War, canals and the new railroads had provided a means of travel for pleasure. Vacations were especially noteworthy in the northeastern third of the nation, at mountain and seashore spots. But during this period, travel for pleasure extended to the "wild West," and the published accounts of the few who had made this exciting journey aroused wide interest.

There was some beginning of city parks during these years. Of 103 cities with 100,000 or more people in 1950, 66 could lay claim to one or more parks in 1880.[8] These parks were not improved and managed in the way we believe city parks should be today, and they often constituted little more than unimproved open spaces. Nevertheless, their reservation at this early date is significant.

The Period from 1900 to 1932

Great social and economic changes took place in the years between 1900 to 1932—changes that were to have far-reaching effects upon outdoor recreation. Of perhaps greatest importance was the tremendous increase in travel, from about 500 to over 3,000 miles per capita; 91 per cent of the increase is attributable to the automobile. By the end of this period, many more automobiles and all-weather roads throughout much of the country had made the typical individual infinitely more mobile. His range of outdoor recreational opportunities had been enormously increased; the more distant areas were no longer the almost exclusive preserve of the wealthy. During this same period per capita income rose sharply, working hours were reduced considerably, and paid vacations made some progress. However, the driving force was still personal advancement, and the typical successful businessman rarely converted much of his success into increased leisure and outdoor recreation.

During this period seventeen national parks were added.[9] Under

[8] The statistics on recreation used in this chapter are drawn mainly from data compiled by Resources for the Future, in Marion Clawson, *Statistics on Outdoor Recreation* (Washington: Resources for the Future, Inc., 1958). There were 106 cities with 100,000 or more people in 1950, but information about their parks was available for only 103 of them.

[9] Crater Lake, Platt, Wind Cave, Mesa Verde, Glacier, Rocky Mountain, Hawaii, Lassen Volcanic, Mount McKinley, Acadia, Grand Canyon, Zion, Hot Springs, Bryce Canyon, Grand Teton, Carlsbad Caverns, and Great Smoky Mountains.

authority of the Antiquities Act of 1906, the President created several national monuments. The National Park Service was established in 1916. The idea of a national park system was firmly established by the end of this period, although that system had not yet acquired its present dimensions.

The national forests experienced a major expansion during this period, to about 160 million acres. Perhaps more important, their management was placed on a sounder and more intensive base. Recreation came to be regarded as an important use of the national forests.

Some progress in the provision of state parks was made during these years, although the physical evidence was modest. In 1928 only twenty-six states reported parks and over 80 per cent of the acreage was in New York. Only nine states had provided annual appropriations in any considerable amount for state park and related work.[10] Nevertheless, the idea and the ideal of state parks had become generally accepted.

There was a material expansion in the number and acreage of city parks during this period. At its end all but one of the cities of 100,000 people and over in 1950 supported one or more parks; the number and acreage of their parks had each roughly quadrupled. Comparable data for smaller cities are unavailable, but it seems probable that, for these, park numbers and acreages increased at a relatively faster rate. The idea of park provision and management as a proper municipal activity was firmly established during this period.

The period saw an enormous growth in both participant and spectator sports. Some of these, such as golf, baseball, and tennis were predominately users of land areas. Others often being indoors activities required less land. The larger participation of women in sports of many kinds contributed to the demand for recreational space.

Attitudes toward recreation underwent certain changes in these years. During the First World War the need of recreation to build morale and the need of physical training to develop muscular strength and stamina were impressed upon several million men. In the light of more recent experience, the methods used were ill designed, but they did represent a step forward. The Puritan heritage still cast its shadow over recreation; most recreation was defended in terms of its value for building sound bodies, or for keeping young people out of mischief. But whatever their motives, more and more people began

[10] U. S. Department of the Interior, National Park Service, *A Study of the Park and Recreation Problem of the United States,* 1942, p. 93.

actively to take part in sports, to take trips for recreation purposes, and to have planned vacations.

Depression and New Deal

The Great Depression of the 1930's cut a wide swath into the amounts people had to spend on recreation. But for those who could afford any recreation at all, the outdoors could and did often provide the cheapest source. There was a profound upheaval in popular thinking during these years—a concern as to the purpose of life, and a questioning of material goals. How leisure should be used occupied many people's thoughts. Because productive employment was hard to obtain, it was believed that means would have to be found to keep people satisfied with inexpensive leisure-time activities. Concern for the aged increased with the realization that their relative numbers would increase, and that such people need satisfying activities.

At the beginning of this period public expenditures for recreation declined greatly. States, cities, and other units of government, faced with falling revenues and sometimes with mounting costs, sought to economize wherever they could; and as a result park and recreation expenditures were often sharply reduced. The public programs launched with the coming of the New Deal, however, were of unprecedented variety and magnitude. Of most direct concern to outdoor recreation were the Civilian Conservation Corps, the Works Progress Administration, and the Public Works Administration, each of which provided labor for resource development in a volume previously unknown; the submarginal land purchase program, which led to the acquisition of many millions of acres of land, much of which was used for forest, park, or wildlife purposes; and planning activities of various kinds, often financed by the federal government.

The national park system expanded in these years. Shenandoah, Mammoth Cave, Olympic, Isle Royale, and Kings Canyon were added to the national parks. Management of various military parks, battlefield sites, and similar areas was transferred from the War Department to the National Park Service. The system itself was rounded out in several important ways and became more firmly entrenched in the acceptance of the American public.

The national forests also changed somewhat in this period. Although little land of national forest caliber could be added from the public domain, several million acres were purchased from private owners. A great many improvements were made in the forests, either directly or indirectly for the benefit of recreation. Roads and trails were built, opening up new areas for ready use; campgrounds and other facilities were constructed; and in general the abundant labor supply was used productively.

Some national wildlife refuges had been created in the early years of the century and later, but their number and area was small. Now, however, land purchase funds were used to make major additions to the wildlife refuges, and other funds were available for their improvement for wildlife purposes.

During this period, too, the number and area of federally built reservoirs increased greatly. The Tennessee Valley Authority, the Bureau of Reclamation, and the Corps of Engineers each built big-scale dams with large bodies of water behind them. Recreational use of the water areas began and quickly mounted to relatively large figures. Increasingly, recreation was accepted as a recognized use in the planning and management of such projects.

It was in this period that the state parks really became established on approximately their present scale. Shortly after 1930 almost all the states had some form of park administration.[11] The acreage of state parks, excluding the two big New York parks, increased fourfold in twelve years. The federal government gave subsidies for park planning by state agencies. This encouraged as well as aided the states in developing both immediate and long-term programs, and in taking advantage of the various public programs available. Another major factor encouraging the states to expand their park areas was the availability of essentially free labor, through the various federal programs mentioned above. Land was relatively cheap in these depression years, and with much labor available at very low cost, the states reaped the benefit. The facilities within state parks were increased, and the quality of management improved.

The development of municipal parks was more modest. For one thing, the cities carried a disproportionately heavy financial burden because of unemployment and other demands arising out of the depression. During the early years, when the depression was the most severe, even park administration suffered greatly, and staffs were reduced in many cases. Nevertheless, some increases in numbers and area of municipal parks later occurred, and also some increase in facilities and improvements.

[11] *Ibid.*

War and Postwar

For the third successive major period economic and cultural changes had a major impact upon outdoor recreation. During the war gasoline rationing and other restrictions on travel barred the use of the national parks and national forests to many people. Hours of work increased and total employment reached new high levels. Incomes increased greatly, and were more evenly spread through the whole population than during the depression. There were fewer goods on which to spend the larger income. One result was a greatly increased expenditure on commercial amusement. In the postwar years employment has remained high, incomes and total population each have continued to rise, and the amount of paid leisure has risen considerably.

In this period comparatively few changes have taken place in the areas of the national park system, the national forests, and the wildlife refuges. During and after the war appropriations for these areas, especially for improvements, were at lower levels of purchasing power than before the war. In addition, the various special sources of labor, such as the CCC, were no longer available. The greatly increased use of these areas has been made on facilities constructed before the war; severe crowding and overuse has resulted.

During these years, the number and area of federally built reservoirs have continued to increase, and much greater recreational use has been made of most of them.

Some limited expansion of the area of state parks has taken place during these years, but at a declining rate. Much of the increase in the earlier years had been due to transfers of recreation demonstration areas from federal to state ownership. Since the prewar period the area added annually to state parks has generally declined, while the cost per acre of the land purchased has risen sharply—much more so than the price of farm land in general.

Although there has been some expansion in municipal park systems since the war, this has definitely been less than the rate of population growth within cities. If only the central cities are considered, the disparity is not great. It is in the suburbs where the rate of population growth has been most rapid, and where expansion of park acreage has lagged most seriously.

The war and postwar years have seen a continued and accelerated rise in both participant and spectator sports, and in commercial recreation generally. With more people, higher incomes per person, and more leisure, it could hardly be otherwise.

USE OF LAND FOR OUTDOOR RECREATION TODAY

It would be useful if we could analyze recreational uses of the land today and in the recent past according to the scheme of classification presented in Table 20. However, such data are lacking or are not presented in a way that makes this possible, and some approximations are necessary. Because data are not available as to the recreational use of privately owned land, our consideration must be limited to the use of the publicly owned areas. Even here, the data are assembled by broad types of areas, primarily according to type of administering agency rather than kind of use.

A general view of outdoor recreation may be a helpful introduction to a more detailed discussion of the various kinds of areas used for recreation (Table 21). Municipal parks are taken as typical of consumer-oriented outdoor recreation areas. The best examples of the resource-based recreation areas in public ownership are the national park system, the national forests, and the federal wildlife refuges. The basis of establishment has been their natural and historical resources; they are not so located as to be available to many people except by rather considerable travel.

Intermediate areas, in public ownership or regulated by government include state parks, TVA reservoirs, Corps of Engineers reservoirs, and hunting and fishing generally. A few units of each are so located as to be comparable to consumer-oriented areas; perhaps more units are comparable with resource-based areas. Certainly, some state parks closely approximate national parks and national forests. Some units, especially the reservoir areas, are located because of particular resources available; but, it will be recalled, the definition of "intermediate areas" assumes that the best areas within certain time distances will be chosen. In general, it is believed these units or these activities more nearly belong here than elsewhere, and are more nearly representative of this type of area than any other broad group for which we have data.

Table 21. Area, overnight capacity, and attendance by major types of outdoor recreation areas

Major type of outdoor recreation	Kind of areas representative of each type	Present area (mil. acres): Total	Present area (mil. acres): Primarily for recreation	Overnight capacity (1,000 persons)	Attendance: 1955 or 1956 (mil.)[1]	Attendance: Average annual percentage increase in post-war years
1. User-oriented	Municipal parks	0.7	0.7	?	1,000 plus	4
2. Resource-based	National park system	24.4	24.4	77	55	8
	National forests	188.1	14.0	400	53	10
	Wildlife refuges	17.2	?	?	8	12
3. Intermediate	State parks	5.1	5.1	195	201	10
	TVA reservoirs[2]	0.98	0.2	12	40	15
	Corps of Engineers reservoirs[2]	4.7	3.3	36	71	28
	Hunting[3]	[4]	[4]	[4]	[5]14.5	2.7
	Fishing[3]	[4]	[4]	[4]	[5]18.7	4.5

[1] Most recent years of record: data are in terms of visits, except for wildlife refuges and TVA reservoirs, which are in terms of visitor days.

[2] The chief attractions of these areas are their water resources; however, their use seems to be primarily of the single-day type typical of intermediate areas.

[3] Although some hunting and fishing is carried out on relatively distant areas, most of it is on a one-day basis, and is on areas conveniently located.

[4] Not applicable.

[5] Licenses; number of times hunted or fished may have been much greater.

SOURCE: Marion Clawson, *Statistics on Outdoor Recreation* (Washington: Resources for the Future, Inc., 1958).

The municipal parks are the smallest in total area, the largest in total use (assuming that "visit" is a reasonable measure of use),[12] and hence have much the heaviest use per unit of area. The data for these areas are particularly incomplete, but it appears that their use is growing at a rate slower than that for most other kinds of areas. The number of uses per acre for this whole system is probably in excess of 1,500, or, in the terms of our earlier classification, "very heavy"; this undoubtedly includes many areas with over 5,000 uses per acre annually, as well as some areas of lighter use—perhaps with 50 to 100 or fewer annual visits per acre ("heavy," in terms of our earlier classification).

At the opposite end of the scale are the resource-based areas: largest in total area, smallest in total use, and hence lowest in intensity of use. They seem to be experiencing a medium growth rate, compared with either the user-oriented or the intermediate types of areas. For this group of areas, average use is about half a visit per acre per year; for the national park system as a whole, it is only slightly more than two visits per acre annually. This we have considered "moderate" in our classification of use intensity. However, these averages undoubtedly conceal great variation in intensity of use. The most popular spots within some national parks and national monu-

[12] The usual units to measure recreation use are (1) the visit, or the number of individuals who enter the specific recreation area, each entrance by each person being counted once; (2) visitor-days, similar to the foregoing, but with the added dimension of the length of time the users stay; and (3) visitors, or the number of different persons who enter each area. For a more detailed discussion of these measures, see Clawson and Held, *op. cit.,* pp. 69-71. The number of visits is the most easily determined figure; it is roughly comparable to the number of admissions to movies, athletic events, and the like. The total recreation load upon an area is perhaps better measured by number of visitor-days, especially when there are considerable differences in the length of average stay, as between one area and another. It is usually more difficult to estimate this figure. The number of different individuals who use an area, or a system of areas such as the national forests, or all outdoor recreation areas within some geographical radius (as within a state), is also an important figure, particularly when considering social policy for the provision of recreation areas and the sharing of their costs. Each measure has its usefulness; the difficulty comes when they are confused. Many popular releases and statements use data on visits but refer to them as visitors; to the extent this connotes that an equal number of different individuals was involved, it is definitely wrong and misleading. Unfortunately, there seems no ready way to convert the different kinds and lengths of visits into common units.

ments probably have annual visits running up to a few thousand per acre, while millions of acres of remote country are visited by only a very few in total.

The intermediate areas, as we have termed them, are indeed intermediate—in total area, in total use, and in intensity of use per unit of area. However, the rate of growth of use of such areas in recent years has been the highest of all. For all these areas, use averaged about 36 visits per acre annually, varying from about 21 on the Corps reservoirs to about 67 on the TVA reservoirs, with the state parks at about 39. This is midway between moderate and heavy, according to our scale. Undoubtedly there is also considerable variation within these averages; the most heavily used areas probably approximate use in the most heavily used areas of the national park system and the average of municipal parks. However, the places where use is light are probably not as extensive as in the more remote parts of the national parks, national monuments, and national forests.

Because of their popularity, we have included hunting and fishing in the "intermediate" category, although there are many drawbacks to considering them along with area-based data. These activities occur on land of both public and private ownerships, and some part of them are included in the visits to certain types of publicly owned recreation areas. The fact that the data regarding them are in terms of licenses, rather than of "visits" is also a disadvantage, since the typical license holder goes several times during the year.[13] In spite of these deficiencies, the data on these activities have interest and value. While hunting and fishing have gained in popularity, the increase in numbers of licenses issued is at a far lower rate than the increase in use of recreation areas. However, we know very little about the increases in number of times that license holders actually went hunting or fishing.

If we divide the total number of visits to the different areas by the total number of persons in the United States, we get a use of public recreation areas by the mythical "average" person as follows:

[13] U. S. Department of the Interior, Fish and Wildlife Service, *National Survey of Fishing and Hunting, 1955,* Circular 44, Washington, 1956. According to this survey, the typical (median) fisherman fished on 9½ days during the year and 32 per cent of those fishing went on a fishing trip lasting more than one day; the typical (median) hunter hunted on 8½ days, and 21 per cent of those hunting went on a hunting trip lasting more than one day.

	Per 100 persons
Visits to municipal parks	600 plus
Visits to the national park system	32
national forests	31
wildlife refuges	5
Visits to state parks	118
TVA reservoirs	24
Corps of Engineers reservoirs	42

In addition, there would be hunting and fishing under license, on all types of ownership as follows:

	Per 100 persons
Fishing licenses	11
(at the typical rate of fishing, this would be equivalent to 104 days of fishing)	
Hunting licenses	8½
(at the typical rate of hunting, this would be 72 days of hunting)	

In practice, of course, this average recreation pattern would apply to relatively few people. Much of the use of municipal parks is by children and young people; most of the hunting is by adult males; and various other age, sex, income, location, and taste factors operate to make the recreation pattern of different individuals vary greatly. To the extent that some persons participated little or not at all in these recreational opportunities, others enjoyed them the more. When people of all ages, from babes to octogenarians, and of all locations, from backwoods to the heart of the largest cities, are included, it seems that an average per capita use of more than six times in municipal parks, nearly one day in resource-oriented areas, two days in intermediate areas, and two days fishing and hunting is really a rather large amount of outdoor recreation.

User-Oriented Outdoor Recreation

The available information on user-oriented outdoor recreation is largely limited to municipal and county parks. This is unfortunate, for in user-oriented recreation, perhaps more than any other type of outdoor recreation, private recreation activities are important. It is a matter of common knowledge that business concerns have provided outdoor recreation activities of various kinds when the investment has seemed desirable. Commercial amusements probably compete more with user-oriented than with other kinds of outdoor recreation, primarily because of the time requirement for enjoyment of each.

The data available on municipal and county recreation are in themselves severely limited. Most of the data have been assembled by the National Recreation Association by means of questionnaires addressed to the cities and counties. Reply was voluntary and many cities and counties made no response, among them, doubtless, some possessing parks. An unknown, but possibly not serious, degree of under-reporting thus existed, and we are not sure that it was constant from period to period.

The reported acreage of all municipal parks in 1925 was about 249,000, and of county parks, about 67,000–316,000 in total. This was about one-half acre per 100 of the 1925 urban population. Each type of acreage increased in the next five years, so that by 1930 urban parks amounted to about 309,000 acres and county parks to 108,000 acres, or 417,000 acres in all. The rate of increase of county parks was greater than that for city parks, a relationship that has persisted to the present. In the five depression years after 1930, when most forms of municipal activity were shrinking because of restricted revenues, the acreage of parks increased again—to about 381,000 acres for city parks, and 159,000 acres for county parks, or to a total of 541,000 acres. A further increase took place in the five years ending in 1940, when the city parks had about 444,000 acres and the county parks about 197,000 acres, or 641,000 in total. In the ten years between 1940 and 1950, the acreage of city parks apparently shrank slightly, to about 431,000 acres. The extent to which this was an actual shrinkage, due to loss of park acreage for other purposes, or a paper shrinkage, due to differences in reporting, is not clear. The area of county parks increased slightly, to about 213,000 acres in this same period, slightly more than offsetting the loss in city park acreage. Data for 1955 are not available for each separately, but the total acreage was up to about 749,000 acres. A large percentage of the county parks is in counties near cities of some size; if the total city and county park

---

Figure 20, opposite. Number and acreage of parks compared with population for all cities of 100,000 population or more in 1950, 1880–1950.

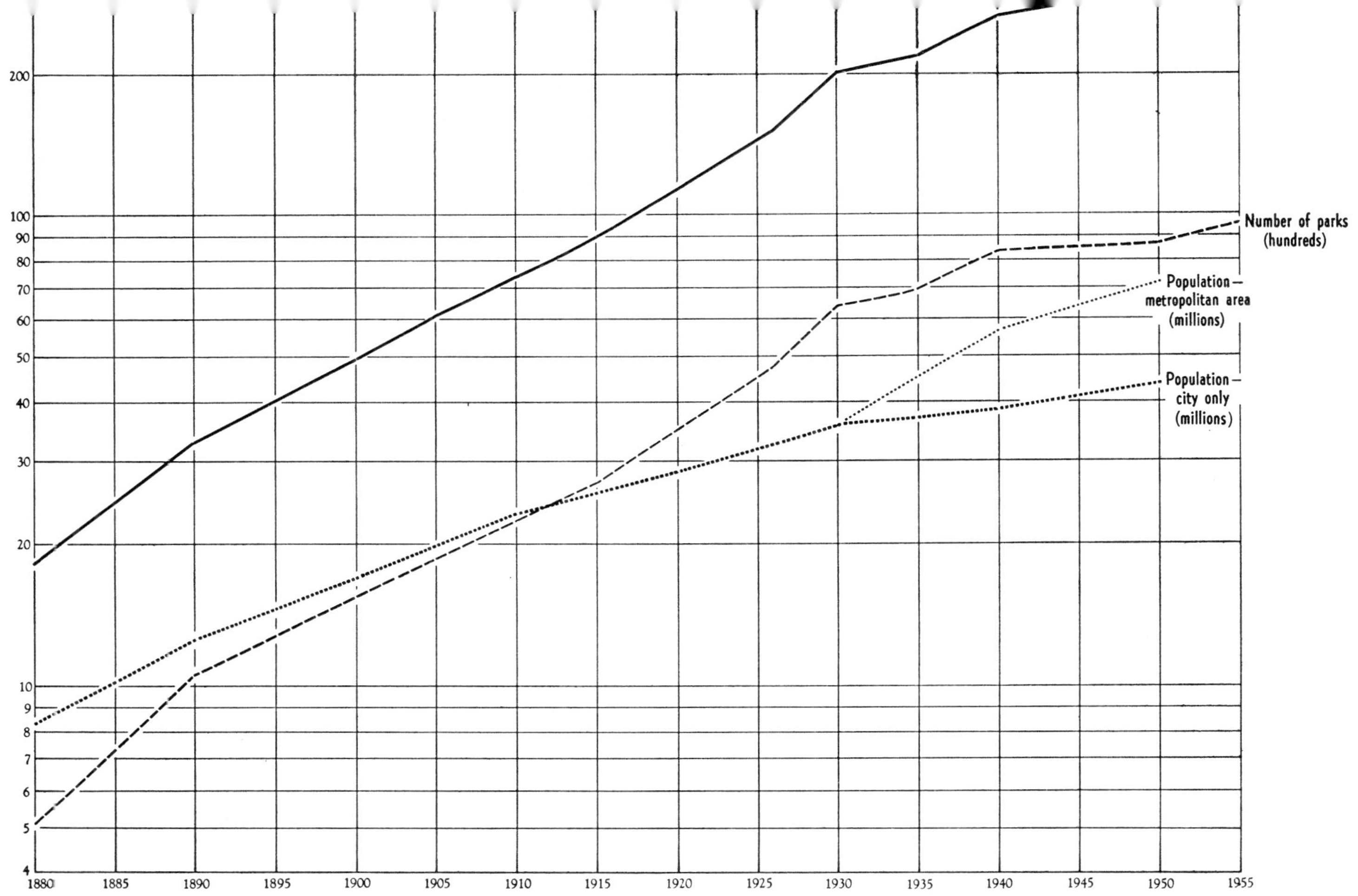
Number of parks (hundreds)
Population—metropolitan area (millions)
Population—city only (millions)
200
100
90
80
70
60
50
40
30
20
10
9
8
7
6
5
4
1880
1885
1890
1895
1900
1905
1910
1915
1920
1925
1930
1935
1940
1945
1950
1955

acreage is related to the total urban population, an average of about 0.7 acres per 100 urban population is evident. This represents an increase in area per capita of roughly 40 per cent compared with the situation thirty years earlier. However, this comparison for the entire thirty-year period is somewhat misleading. Municipal and county park acreage per capita of urban population rose fairly rapidly and regularly until 1940, when it was about .85 acre per hundred population, or nearly 70 per cent higher than in 1925; but since 1930 it has declined about 17 per cent.

A somewhat similar, but not identical, relationship exists for all cities with 100,000 or more population in 1950 (Figure 20). From 1880, the first year for which estimates are available, until perhaps 1930, the number and acreage of parks increased somewhat faster than did the population of these cities; in the decade of the 1930's the rates were about the same; and since 1940 the expansion in parks and acreage has definitely lagged behind that of population growth. During the later years, these data, which relate to cities as governmental units, do not fully represent the situation for the entire metropolitan area of which each of the cities was the core. To assemble accurate data on the recreation area available for metropolitan areas is a task of no mean size; from general observation, however, it is clear that expansion of park areas in suburbs has been much less rapid than in central cities; the total situation, therefore, has worsened more than the data indicate.

The amount of municipal and county park acreage in relation to urban population differs from state to state (Figure 21), but the differences conceal great variations among cities within some of the states. Moreover, acreages are sometimes misleading; the very high acreages for Arizona and Colorado are due to city and county ownership of desert and mountain parks of types not possessed by most cities elsewhere. Nevertheless, some interesting inferences can be drawn from the data. The Mountain and northern Plains states show up especially high; it may well be that much of this area consists of relatively unimproved acreage, in some instances outside of city limits; nevertheless, land has been set aside and is available for later development in these states. The North Central states nearly all show up well, as do scattered states elsewhere—New Hampshire, Florida, and Texas, for example. The Pacific Coast states also seem to be doing relatively well. The poorest showing is in the South. There is some tendency for states with large metropolitan centers to lag be-

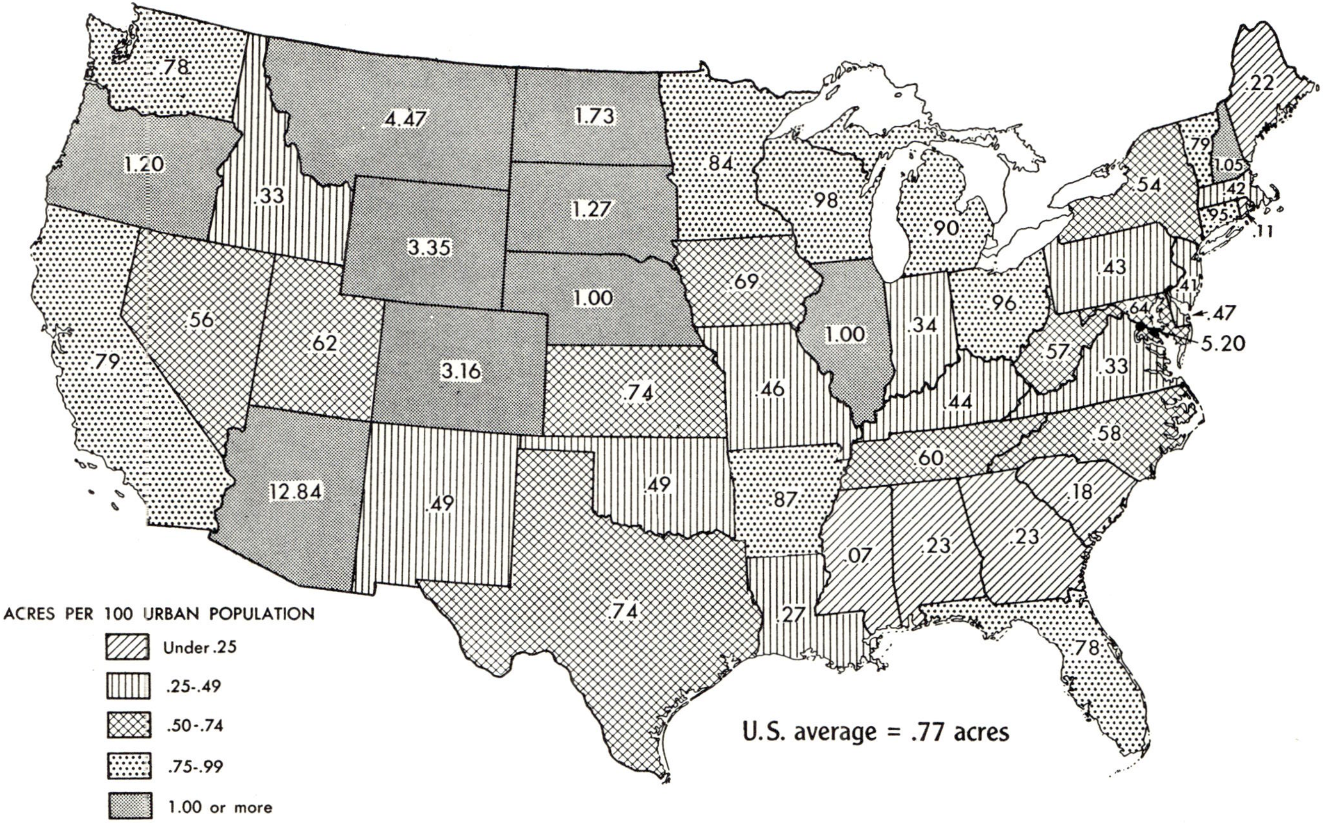

Figure 21. Area of municipal and county parks in 1955 in relation to urban population in 1950.

hind—this may be the explanation for the relatively small acreage in Massachusetts, New York, New Jersey, and Pennsylvania. But, on the other hand, Illinois, Wisconsin, Ohio, and Michigan all show up well above average. Cities in the older states along the Atlantic seaboard may have more difficulty obtaining suitable park acreage than cities in the newer states of the West.

These data on total park acreages and on acreage per hundred urban population are not very meaningful as absolute numbers; but they would be more significant if they could be related to some standard of "adequacy." It is not easy to define what is an adequate number and acreage of local parks; no completely objective and verifiable standards have been established, and possibly none could be. The estimates made by park administrators and others long familiar with recreation work are, after all, subjective in that they are based on their own experience. Moreover, statistics on total numbers of parks are perhaps no more important than is information on the distribution of parks in relation to the potential user public; and statistics on total acreage may be no more important than is information on the physical character of the areas, their layout and development, and the skill with which recreation is managed on them.

Butler, reflecting the experience of the National Recreation Association, suggests the general standard of at least one acre of publicly owned park and open space within cities for each hundred of population, and an additional area in parkways, large scenic and forest parks, and similar larger areas within and adjacent to the city.[14] Smaller cities will, he believes, need much more land in relation to population. He stresses that the balance among areas, as to kinds, relative size of each, and location with respect to population, is at least as important as total area. He describes the following special types of areas:

PLAY LOTS—primarily for housing projects and underprivileged areas, to serve as a substitute for the backyards of more prosperous areas —from 2,000 to 5,000 square feet per unit.

NEIGHBORHOOD PLAYGROUNDS—primarily for children of 6 to 14 years of age; the areas must be within one-fourth to one-half mile to be of service to these children; such areas should usually be from 4 to 7 acres in extent, and can often be located at or near schools; 1 acre per 800 of the total population is the standard suggested.

[14] Butler, *op. cit.*

LARGE PARKS—should be of not less than 100 acres in size, often cannot be over 300 acres for lack of available tracts; every city of any size needs at least one; the larger cities should have one for each 40,000 population.

NATURAL RESERVATIONS—comparatively large areas, usually of 1,000 acres or more, kept largely in a natural state, perhaps outside of the city proper; need for these depends greatly on availability of state and federal park areas.

SPECIAL RECREATION AREAS—such as golf courses, camping areas, bathing beaches, swimming pools, athletic fields, stadiums, etc.; some of these may be included in the foregoing types of areas.

NEIGHBORHOOD PARKS—not for active recreation but for quiet informal enjoyments; should be one in each square mile of city, in the 2- to 50-acre size range.

These areas and acreages are in addition to reasonable playgrounds in connection with schools, and on the assumption of some play areas in private yards (except in the case of play lots).

In a careful study which considered varying climatic conditions, and hence varying recreational patterns, in different parts of California, and which provided for many kinds of activities, estimates were made of the acreages required, as follows:[15]

NEIGHBORHOOD RECREATION CENTERS—when adjoining elementary schools, roughly 0.3 acre per 100 of total population; when a separate center, roughly 0.6 acre.

COMMUNITY RECREATION PARKS—when adjoining junior or senior high schools, roughly 0.1 acre per 100 total population; when a separate park, nearly double this acreage.

CITY-WIDE RECREATION FACILITIES—roughly 1 acre per 100 total population.

These estimates are based upon an efficiency in area layout and use that would not often be attained in practice; they also assume an ideal distribution of the different types of areas in relation to population. They may thus be considered as minimum adequate areas to provide the various types of recreation services.

The reasonableness of these estimates of recreation experts may be judged by the experience of those cities that have made the most

[15] California Committee on Planning for Recreation, Park Areas and Facilities, *Guide for Planning Recreation Parks in California* (Sacramento: Documents Section, Printing Division, 1956).

generous provision of parks and recreation areas. Of the 1,465 cities reporting parks in 1940, 370 or 25 per cent reported one acre or more of park for each hundred of population.[16] On the other hand, 339 cities, nearly half of them above 5,000 population, in the same year reported that they had no park acreage; this was in addition to the nearly 50 per cent of all cities over 2,500 population making no report. (Most of these were comparatively small and probably had no parks.) The citizens of some cities have been willing to provide park acreage well in excess of the common one acre per hundred population standard that is so often used. This is not necessarily proof that such cities have "adequate" park systems; the earlier comments about distribution of parks and use of acreage within parks should be borne in mind. Moreover, some of the larger city parks fall in the categories which Butler has suggested require an additional acre per hundred population—the larger scenic and forest parks, and parkways. Attainment of an acre per hundred population is, in itself, not always enough.

It is difficult to estimate how far municipal and county park acreage falls short, in toto for the entire country, of any standard of adequacy. If we accept the standard of one acre of local park and playground per hundred population, plus another acre per hundred population of scenic and forest park and parkways (thus assuming that the problem of proper location and spacing of parks and of proper use of land within parks is one of efficiency in administration rather than of land use *per se*), the urban population of the country in 1955 would have required about 2.10 million acres of parks. Some of this need probably was met by relatively nearby state parks, even national forests and other federal land areas in some cases, and by parkways. The area reported in municipal and county parks was 0.75 million acres; to this would have to be added some acreage, perhaps small, for parks in the unreporting cities and counties. On the other hand, some of the county parks reported may have served primarily rural rather than urban people.

Thus, municipal and urban parks in 1955 were probably less than half as large as was then needed, and certainly still smaller in comparison with future needs. In this connection, the lag in park expansion in relation to growth of urban population since about 1940, and the scarcity of parks in the growing new suburbs, assumes seri-

[16] George D. Butler, *Municipal and County Parks in the United States, 1940* (New York: National Recreation Association, 1942).

ous proportions. While large relative to the area now in parks, the total deficiency in urban park acreage is small in an absolute sense. Something on the rough order of a million acres in a country of nearly 2,000 million acres, is comparatively small. Even as compared with the estimated 17 million acres now in cities and urbanized land uses, the urban park deficiency is not large. However, the deficiency is usually critical as to location; to be most useful, any added park and recreation acreage should be close to people—where little idle land now exists or none that is not very expensive.

More striking than the relationships between urban park acreage and population are the changes that have taken place in intensity of use of park acreage (Figure 22). These must be measured indirectly, but the evidence seems clear. The number of paid recreation leaders has increased much more rapidly than has either the area of parks or the numbers of urban people. Numbers of recreation buildings and indoor centers have also increased rapidly—at a somewhat more rapid rate in fact, than have numbers of playgrounds (including among the latter only those under definite leadership). Numbers of selected facilities also have risen; this understates the increase in facilities, because it is based upon the same kinds throughout, when in fact a constantly increasing variety of facilities has been provided.

All of these factors clearly point to increased use per unit of area. The available data on recreational use of municipal parks show a relatively slowly rising trend—something in the general order of 4 per cent increase annually in the years since the war. This rate of increase was more rapid than the increase in area in these years, but not quite as rapid as the rate for area increases before the war. However, it is difficult to get accurate attendance data for urban parks, considering their ready access to users and the lack, in most areas, of any formal system for recording use. The reported uses may, therefore, be much too low, not only as to level but also as to rate of increase. The considerably larger number of playgrounds, buildings, and specific facilities would certainly lead to greater use; and the additional playground and other paid leadership was surely not provided except as greater use made the expenditures necessary.

Summary of problems. The major present problems in user-oriented recreation areas are their small size and poor location. Their total area is probably half or less of what would be adequate for today's needs. Some cities, and some parts of many cities, are tolerably well

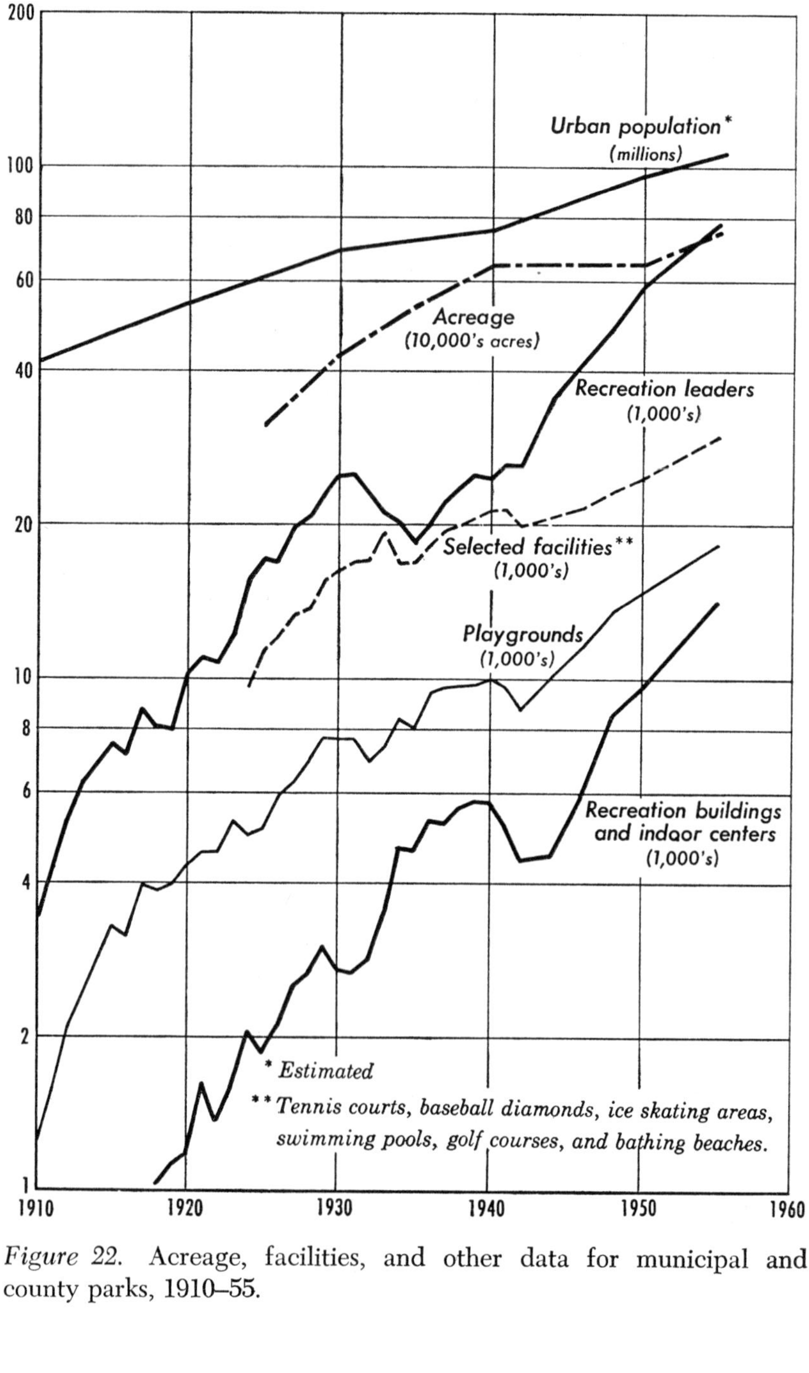

Figure 22. Acreage, facilities, and other data for municipal and county parks, 1910–55.

supplied, while other cities and other parts of most cities are very short. The oldest parts of many large cities, which tend to be slums or at least seriously decadent, generally lack adequate parks—sometimes they lack parks entirely. At the other extreme, many of the fastest growing suburban areas have thus far made extremely inadequate provision for parks—and with the available land taken up in large-scale developments, there are neither vacant lots nor vacant sites upon which parks may be constructed later.

Resource-Based Outdoor Recreation

By definition, the resource-based recreation areas include those areas and resources of such unusual quality that people are willing to travel considerable distances, if necessary, to visit them. Largely for historical reasons, a great deal of the best resource-based recreational opportunity is in public, usually federal, ownership. There are, however, some excellent resource-based recreational opportunities in private ownership, especially seashore and lake areas, and to a lesser extent mountain areas. In Table 21, we used the national park system, the national forests, and the federal wildlife refuges to illustrate resource-based recreational opportunities. These three types of federal land areas provide a substantial part of this kind of recreation and, in addition, they offer a considerable body of information for study and comparison.

The national park system includes the national parks, national monuments, historic sites and battlefields, and other areas with unusual recreational value. Each unit of the system has been placed in its category because, defining recreation in the broadest possible terms, it represented a major recreation resource worthy of preservation for public enjoyment. Each unit is of national, rather than regional or local, significance. Superlative scenery, outstanding scientific value, national historical importance, and similar qualities characterize the system. With few exceptions, its areas are not available for commercial uses such as grazing, timber harvest, mineral development, and the like.

The national forests were established to preserve valuable forested, grazing, and watershed areas, and are administered under a program of multiple-use management. Timber harvest, grazing, mineral development, and other commercial uses are permitted and encouraged, but under such conditions that the productivity of the resource is preserved. Recreation is a recognized use, and one for which some areas have been set aside as the principal or sole use. There are specialized campground and recreation sites which, although comparatively limited in area, derive a large part of their attractiveness from the large surrounding areas of national forest which serve as buffer zones to keep out disturbing uses and to lend general attractiveness. There are in addition comparatively large areas of wilderness, primitive, roadless, or other areas in which commercial use of resources is prohibited or kept to a minimum and which have exceptional value for a special type of recreation.

Table 22. Area of federally owned resource-based recreation areas

Item	Unit	North[1]	West[2]	South[3]	United States[4]
National forest area (1957)					
total	Mil. A.	11.4	136.6	12.3	160.3
reserved for recreation	Mil. A.	—	—	—	14.0
National park system area (1955)	Mil. A.	0.8	13.7	2.8	17.3
Federal wildlife refuge area (1955)	Mil. A.	1.1	6.8	1.7	9.6
Total population (1950)	Million	83.9	19.6	47.2	150.7

[1] New England, Middle Atlantic, East North Central, and West North Central regions, as defined by U. S. Bureau of the Census.

[2] Mountain and Pacific regions.

[3] South Atlantic, East South Central, and West South Central regions.

[4] In addition, other areas are found in Alaska, Hawaii, Puerto Rico, and the Virgin Islands.

The main purpose of the federal wildlife refuges is to provide a suitable habitat for certain species of wildlife, especially for scarce species or those threatened with extinction, and especially at certain seasons for migratory wildfowl. The presence of the wildlife enhances the areas for recreational use; the animals and birds may be watched, photographed and, in some areas, shot at certain seasons; and fishing is often permitted. There are in addition certain general-interest recreational opportunities, such as camping and picnicking.

These federally owned resource-based recreational areas are predominantly in the West (Table 22): the West, in fact claims 85 per cent of the national forests, 79 per cent of the national park system, and 71 per cent of the wildlife refuges. A much larger proportion of the historic sites and other relatively small areas of high value are in the East and South, so that the gross acreage figures somewhat distort the regional distribution as far as importance of areas is concerned. Nevertheless, resource-based recreation areas are heavily concentrated in the least populous regions of the nation. The location factor in such areas is secondary; yet is it unfortunate for many that the regional distribution is so extreme. For a very large proportion of the total population, long travel and associated high costs are unavoidable if the areas are to be visited.

For the country as a whole, federally owned resource-based recreation areas are available to the extent of 124 acres per hundred population, if the entire national forest area is included. If only the area reserved for recreation in national forests is included, the average is 27 acres. In either case there is obviously a vastly greater area of resource-based than of user-oriented recreation. If only the population resident in the region is considered, the total federally owned resource-based areas range from 16 acres in the North (including New England and the Middle Atlantic States) to as high as 803 acres in the West, for every hundred people.

The national forest system began in 1891; by 1905, when the Forest Reserves, as they had been called up to that time, were placed under the management of the Forest Service, the area was up to 75 million acres (Figure 23). In the next few years acreage increased very rapidly, as public domain was withdrawn for this purpose, to a total in excess of 160 million acres in 1909. The area declined somewhat in the following decade, and since has been followed by a slow rise in area to a present total slightly in excess of 181 million acres (including Alaska, and territories). The changes since 1909 have been such that the whole period can be characterized as one basically of stability in area.

In contrast, the national park system has grown more slowly. The first national park was Yellowstone, established in 1872. Other parks were added slowly, so that by 1900 the total area was slightly more than 3 million acres. National monuments were provided in the Antiquities Act, passed in 1906, and other units of the system have been added in later years. Additions tend to be in fairly large blocks, as new parks or monuments are added. However, a generally steady upward trend, at an average rate of nearly 5 per cent annually, was evident from 1900 to about 1942. At that time the total area reached

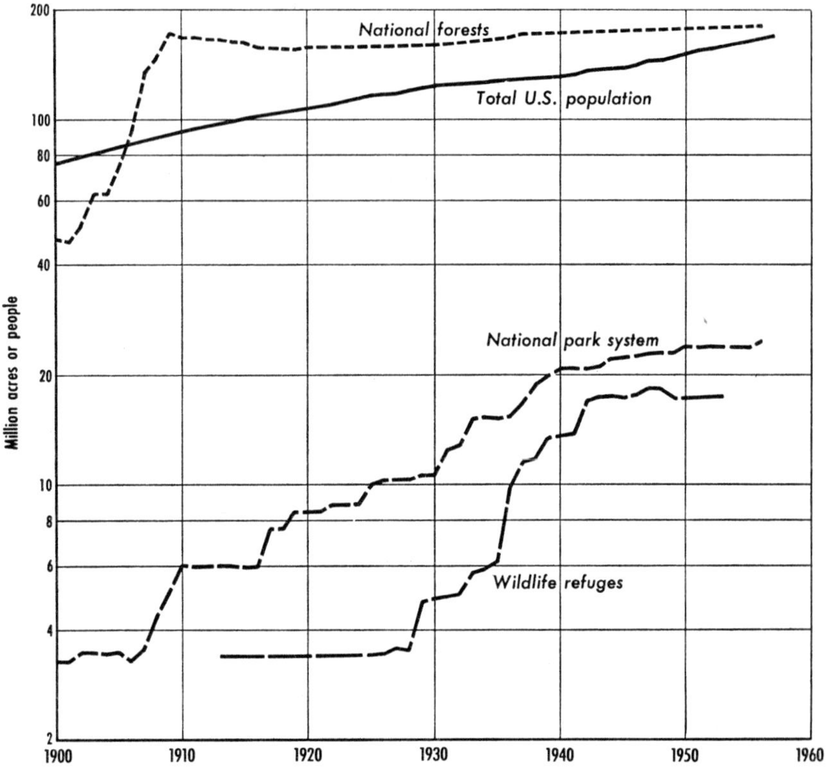

Figure 23. Area of national forests, national park system, and wildlife refuges in the United States and possessions, in relation to U. S. population, 1900–1956.

about 21 million acres. Some expansion has occurred since then, but at a markedly slower rate: had the 1900-1942 rate continued, total area today would have been in the neighborhood of 40 million acres, rather than the 24 million acres actually found. Much of the expansion of the national park system has come about by transferring federally owned land from public domain or national forest status to national park or other status. Also several significant donations were made by states and individuals.

Although wildlife refuges were first established as early as 1903, their area until 1913, was too small to show on the scale of Figure 23; at that time the addition of a very large acreage in Alaska brought the total area up to more than 3 million acres. There was almost no further increase in their area until 1929; from then until 1942 total area rose very rapidly to about 17 million acres. Since 1942 the expansion has been small.

The rate of increase in national forest acreage from 1890 to 1910 was very much greater than the increase in population; since 1910 it has been less. Up until about 1942 the rate of increase in acreage of the national park system was greater than the increase in total population; since 1942 it has been roughly equal. The rate of increase in acreage of wildlife refuges has been very irregular.

The basis for establishing resource-based recreation systems and for selecting individual areas to fit into them, in general has been related to physical characteristics, not to demand for their use. Idealistic foresters led the country into the creation of a system of permanently reserved federal areas to ensure a future timber supply, before popular demand had required such a step.[17] Of land in federal ownership, that with the best forests, or most valuable for watershed purposes, or of critical importance for grazing, to the extent of the available knowledge, was withdrawn. The first units of the national park system were chosen because of their outstanding quality and a conviction that they should remain in public ownership rather than be submitted to commercial development. Later units have been selected on the same basis; only incidentally has demand for the areas' use entered into discussions as to the wisdom of their establishment. Similarly, the units of the wildlife refuge system have been selected because of the need for protecting certain key species that otherwise were faced with elimination or decimation, not because of their popularity as recreation areas.

As far as we can ascertain, no attempt has ever been made to establish standards of "adequacy" for a resource-based recreation system. There is nothing here comparable to the rough standards of adequacy for user-based recreation. The limiting factor has not been demand or need, but their supply. As the areas have become better known, interest in their preservation and use under the protection of the national park system has increased.

For this reason, it is impossible to say whether the present areas are, or are not, adequate. Other areas in the United States might

[17] Marion Clawson, *Uncle Sam's Acres* (New York: Dodd, Mead and Co., 1951), especially pp. 104-11. See also, John Ise, *History of the United States Forest Policy* (New Haven: Yale University Press, 1920).

merit inclusion in the national park system; the wildlife refuges should be extended, if they are to provide adequate protection to wildlife; and some areas might be added to national forests. At the same time, there is considerable opposition to a material expansion of federal land ownership.

The relationship of the national parks as such, of them and the national monuments, and of the whole national park system is clearly seen in Figure 24. Visits to each have increased partly because of the addition of new units to each part of the system. Typically, visits to a new park grow very fast in the first years after establishment, and later grow at a slower but often steadier rate. In 1956 about 20 million visits were made to the national parks, nearly 9 million to the national monuments, and about 26 million to other units in the system—55 million visits in all. For the entire period of record the annual rate of growth has been close to 10 per cent; for the postwar years, about 8 per cent. There is no convincing evidence of a major decline in rate of growth. A constant rate on a ratio or semi-logarithmic basis means that the actual numbers of increase have been greater in recent years, and also that greater and greater absolute numbers of increase may be expected in the future if past growth rates continue.

Increase in recreational use of the national forests has been similar. Before the war use of these areas grew less rapidly than for the national park system; since the war the reverse seems to be true. But the differences are not great, and may be due, in part, to differences in methods of collecting the statistics. Since 1936, total recreation visits to the national forests have been nearly the same as total visits to the whole national park system, but from two to three times the visits to the national parks alone. Except during the recovery from war restrictions, the rate of growth of recreational use of the national forests has been faster in the 1952-56 period than in any past period of equal length.

Data on recreational use of the wildlife refuges are available only since 1951. In these few years, this use has grown more rapidly—on the order of 12 per cent annually—than that of either the national

---

Figure 24, opposite. Trends in visits to different parts of the national park system 1910–56, to national forests 1924–56, and to wildlife refuges 1951–56.

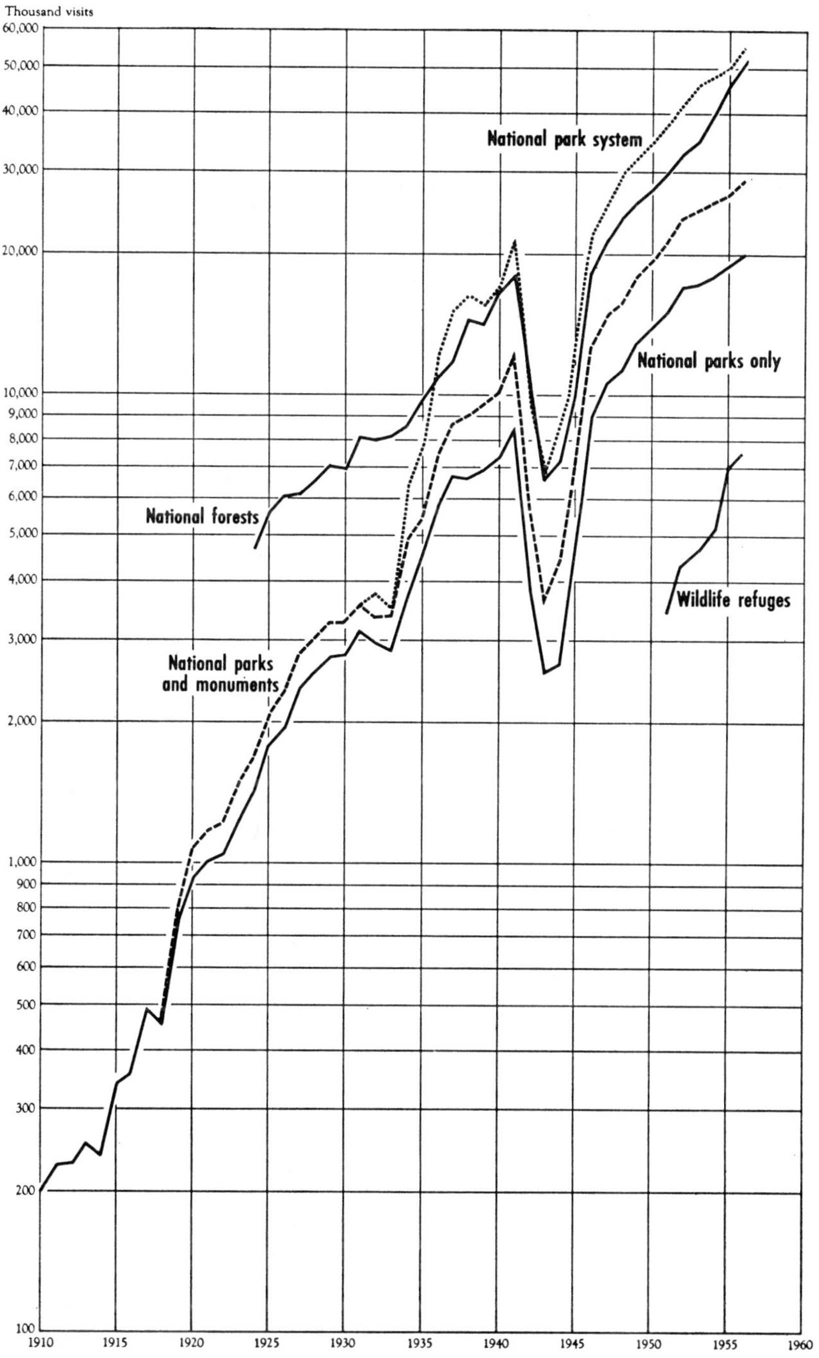

Thousand visits
60,000
50,000
40,000
30,000
20,000
10,000
9,000
8,000
7,000
6,000
5,000
4,000
3,000
2,000
1,000
900
800
700
600
500
400
300
200
100
1910
1915
1920
1925
1930
1935
1940
1945
1950
1955
1960
National park system
National parks only
National forests
Wildlife refuges
National parks and monuments

forests or the national park system; but the actual amount of recreational use is still fairly low.

It has been pointed out in connection with Table 21 that recreational use of the resource-based areas is at the lowest average intensity of any type of areas, and also at the lowest level of total use. However, some parts of the areas are heavily used. The most popular spots within the national parks, such as the areas around Old Faithful geyser or Yellowstone Falls in Yellowstone National Park, or at Canyon Rim in Grand Canyon National Park, or in the floor of Yosemite Valley in Yosemite National Park, are probably used with an intensity approaching that of the most heavily used municipal parks—several thousand visits per acre annually. Likewise, some of the most popular picnic and campground areas within national forests have extremely heavy use, probably of the same general magnitude.

Of those who visited national forests for recreation in 1956, 27 per cent went for general enjoyment of the forest environment, 24 per cent primarily for picnicking, 18 per cent primarily for fishing, 8 per cent primarily for hunting, 7 per cent for camping, 5 per cent for winter sports, 3 per cent for swimming, 3 per cent for riding and hiking, and the remaining 5 per cent for miscellaneous purposes. Of those who visited wildlife refuges for recreation in 1956, 6 per cent went primarily for hunting, 37 per cent went primarily for fishing, and the remaining 57 per cent went for other unspecified purposes. Presumably many of the latter were interested primarily in wildlife observation.

The average length of recreational visit to national forests in 1956 was 1.33 days; there had been a rather regular decline in average length of visit since the war, amounting to 28 per cent in total. For purposes of measurement, any recreational visit of ¼ hour to 3 hours in length was considered as one-fourth day in length; visits of 3 to 5 hours, as one-half day; visits of 5 to 7 hours of three-fourths day; and visits 7 hours or more as a full day. The longest visits were to summer homes on national forests—6.50 days; privately owned organization camps had an average of 5.28 days; government owned organization camps, 3.66 days; and wilderness areas, 3.04 days. In contrast, most other types of uses had visits only slightly exceeding one day in average length.

One would expect visits to national parks to extend over longer periods than these, but in fact the average visit to some of the major national parks was not much longer. Of those staying in Yosemite

National Park overnight, the average length of stay was 3.1 days; the average length of time spent on the whole trip, of which the visit to Yosemite was a part, was 10.6 days. The average number of days spent in Grand Canyon, Glacier, and Crater Lake national parks ranged between 1 and 2 days, as part of trips ranging from 14 to 22 days in total length.[18] It seems clear that the majority of the visitors limited their contact with the park to what could be seen from the highway or at a few well-patronized special attractions, and that the typical visitor saw little of the qualities and character of the area—the special qualities which had brought about its establishment as a national park and which, in fact, had prompted him to visit the area.

Information as to the residence location of visitors to four of the national parks is shown in Table 23.

Table 23. Per cent of visitors to four national parks, by distance

Airline distance from home	Grand Canyon[1]	Glacier[2]	Yosemite[3]	Shenandoah[4]
	(Per cent)	*(Per cent)*	*(Per cent)*	*(Per cent)*
up to 100 miles	1	3	18	25
100-300 miles	3	20	28	38
300-500 miles	23	20	29	20
500-1,000 miles	19	19	5	13
1,000-2,000 miles	40	31	10	3
over 2,000 miles	15	6	10	1

[1] Data for 1954.
[2] Data for 1951.
[3] Data for 1953.
[4] Data for 1952.

SOURCE: These data come originally from studies of the National Park Service and co-operating agencies; the analysis by distance zones is made and presented in Marion Clawson and Burnell Held, *The Federal Lands* (Baltimore: The Johns Hopkins Press, 1957), pp. 119-20.

Yosemite draws 74 per cent of its visitors from California, where the large urban centers in the San Francisco Bay region and in the south form a large potential user group. Many of them flock to Yosemite over week ends. To this extent, Yosemite resembles an inter-

[18] For this purpose, every entry into a park was considered as one day's use, every overnight stay as two days' use, and so on.

mediate recreation area for Californians. However, 65 per cent of all visitors come during the summer, the vacation season, although the spectacular waterfalls of Yosemite are at their best in the spring. Shenandoah also draws a major part of its visitors from a comparatively short radius, chiefly from the large cities of the Middle Atlantic seaboard. Nearly half of all visitors to Shenandoah come in the summer, and more than one-fourth in the fall, when the area is famous for its autumn foliage.

In contrast, Glacier and Grand Canyon, two parks of national and even international fame, are located in regions of relatively small resident population. As a result, well over half of their visitors travel more than 500 miles to reach the parks, and a large proportion travel over 1,000 miles—clear evidence, if evidence is needed, of the drawing power of outstanding resource-based recreational areas. The situation as to these two parks is typical of many others in the system.

The comparatively short time the average visitor spends in the national parks is the more remarkable when one considers the length and cost of the trip necessary for him to visit the park. The average visitor to these relatively distant parks spends a good deal of money in order to reach them—$508 (in 1954) for the average party visiting Grand Canyon, $373 (in 1951) for the average party visiting Glacier, and $243 (in 1953) for the average party visiting Yosemite. It is true that for many the visit to the park may have been only one reason for the trip. For example, only 20 per cent of the visitors to Grand Canyon, 31 per cent of those to Glacier, and 56 per cent of those to Yosemite, made the visit to the park their principal goal. However, it would seem that, having spent the time, energy, and money necessary for the journey, a larger proportion of the visitors would have stayed for a longer while to become better acquainted with the parks they were seeing so casually.

Comparable information is not available for national forests. Some of their visitors are identical to those who visit national parks; they camp in or otherwise use the forests on their way to and from the parks. Many picnickers in the national forests are local residents who use the forest as a user-oriented or intermediate recreation area. And many people fish and hunt there. It seems probable that a higher percentage of forest than of park visitors come from the 100- and 300-mile radiuses, and far fewer from the over 500-mile distance.

As far as we can learn, no effort has been made to measure the present or potential total recreation capacity of the federally owned resource-based recreation areas. Many factors would be involved: willingness of people to use what are now the more distant parts of the areas versus their insistence upon congregating at the spots of greatest attraction; distribution of visitors through time, as to days of the week and seasons of the year; physical facilities for accommodation; the desire of people to look and quickly go on versus their desire to stay and appreciate the areas more fully. It seems probable that the potential capacity is infinitely beyond present use.

However, the capacity of specialized recreation areas within the larger areas has been estimated. The Forest Service, for instance, has calculated that the capacity of its 4,742 camp and picnic areas in 1955 was 41,400 family units or 277,200 persons, for which 64,770 acres had been set aside. This is an average of about 1½ acres per camp or picnic unit; it is doubtful if most other agencies have allowed as much room. In the layout of any intensified or developed area within the national forests or the national park system, some estimate of the capacity of the area is essential. So long as unimproved sites exist, the need for more capacity can usually be met rather easily with further expenditures for roads and facilities.

Summary of problems. The major present problems in the use of the resource-based recreation areas arise out of a flood of visitors, inadequate facilities to care for them, and inadequate funds for administering the areas. The great increase in visits in the postwar period has put an enormous strain upon facilities constructed in the postwar years and designed to serve half as many people. Serious overcrowding of the more popular spots within forests and parks has been evident. It was in recognition of this that the National Park Service developed its Mission 66 and the Forest Service its Operation Outdoors.[19] In each case, a program for several years was formulated and funds were sought to implement it. If carried out, these programs will meet present needs and demands for a few years, but still greater needs and demands will occur in the longer-term future.

Another major set of problems arises out of popular use of resource-based recreation areas in ways that do not fulfill their potentials. The short-stay character of the visits has been noted; far too many visits

[19] U. S. Department of the Interior, National Park Service, *Mission 66 for the National Park System* (processed), Washington, January 1956; U. S. Department of Agriculture, Forest Service, *Operation Outdoors, Part I, National Forest Recreation,* Washington, 1957.

to the major national parks resemble a Sunday afternoon visit to the zoo—quick entertainment, not appreciative contemplation and study. If the national parks, in particular, are to fulfill the objectives of their original sponsors, the nature of public use must undergo rather basic changes.

Intermediate-Type Outdoor Recreation

Intermediate-type outdoor recreation looks both to user-based recreation, which it must resemble in providing a degree of convenience to users, and to resource-based recreation, in that the best sites within the allowable radius are chosen. There is more leeway both as to location and as to character of resources than is possible with either of the other kinds; yet, in the nature of things, intermediate-type recreation involves something of a compromise. In some cases, highly convenient location and superb natural resources are found on a single site, but, more commonly, some degree of compromise of one characteristic or the other, or both, is unavoidable.

There is a wide range of area within the state parks and other units we have chosen to represent this broad group. Some state parks are little different from city parks—located a little farther from the center of the city, requiring a little longer travel, but with similar characteristics and uses. On the other hand, other state parks are rather similar to national parks in character. Of the groups for which data are available, however, the state parks best illustrate the intermediate type of recreation.

Another common type of intermediate recreation area is the reservoir recreation area—TVA, Corps of Engineers, and Bureau of Reclamation reservoirs. It is true that these areas take advantage of particular resources; yet they are man-made rather than primarily natural resources—a characteristic that applies to some of the state parks also. Park specialists and enthusiasts will not, as a rule, regard their qualities as equal to those of the outstanding natural areas set aside as national parks. The common use of state parks and reservoir areas is for single-day outings—a characteristic of the intermediate type of recreation.

We have included fishing and hunting among intermediate recreation activities, even though, properly, these activities are not area based. We have reasoned that while some hunting and fishing is of a vacation or extended type, the greatest part of it lasts only for a day, and is likely to take place on the intermediate areas. There is some overlap of the data for fishing and hunting with the data for recreation on certain types of areas, but the availability of the data plus the popularity of these activities make it worthwhile to consider them explicitly.

Slightly over 5 million acres were reported in state parks in 1955. However, this acreage, in total and in relation to population, was most unevenly distributed around the United States. Over 2½ million acres were in New York; of these, Adirondack Park comprised 2.2 million acres, and Catskill Park 0.23 million acres. Thus, over half of the total was in a single state. Sixty-four per cent of the total area of all state parks was in eight large parks of more than 50,000 acres each in five states; but 35 per cent of all areas had less than 50 acres each. Three-fourths of the total was east of the Mississippi and north of the Ohio rivers. The acreage was small in the West, where the comparative abundance of national parks and monuments and national forests may make state parks somewhat less necessary. The outstanding exception to this generalization is California, which has an important state park system with, next to New York, the largest acreage of any state. Oregon and Washington also have good state parks. The total area of state parks was also low in the South where, unfortunately for the people of that region, federal areas cannot supply the deficiency.

The acreage of state park land per capita varies greatly among the states (Figure 25). For the nation as a whole, in 1955, there were 3.1 acres of state parks per hundred of population. But this is a prime example of how misleading "averages" can be; only seven out of the forty-eight states had average acreage or above, but some of these parks were so extensive that they lifted the nation's average considerably. Three of the seven states were in the northeast and three in the Far West; the only state in the center of the country which exceeded the average was South Dakota.

The data on state park acreage are somewhat impaired because in most instances, they exclude state forests which, in some states, have major recreational values. As with other types of recreational areas, acreage data alone fail to provide a measure of the quality of the land for recreational purposes.

Apparently, no measure of adequacy for state park acreage has ever been devised. We suggest that a satisfactory standard would

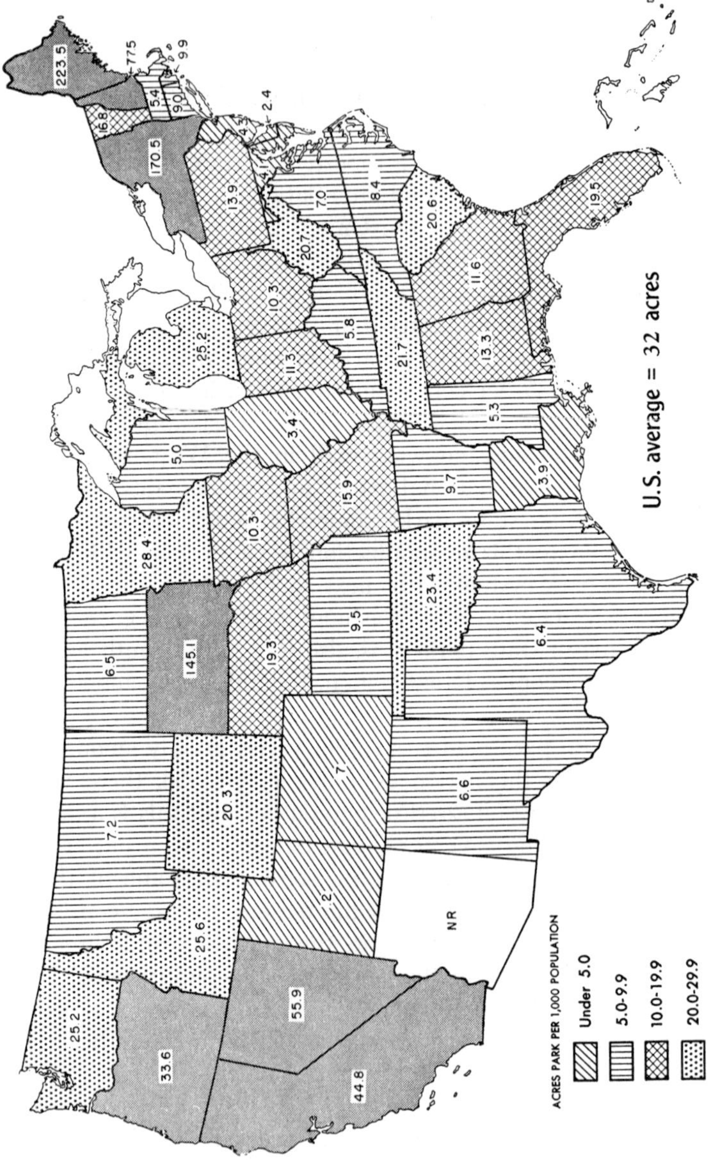

Figure 25. Acreage of state parks per thousand of state population, 1954.

be the provision of one or more state parks within two hours' driving time for 90 per cent of the total population.[20] If state parks are to serve primarily as a place for day outings, they cannot be much beyond two hours' distance. If possible, some should be much closer. The actual miles of distance would depend on the character of the roads, particularly on the provision for through-ways where traffic would be at a minimum.

The acreage required would depend on how intensively the areas were developed, how much buffer area was desirable around the more intensively used zones, and what proportion of the population wished to use state parks. Considering some of the states, such as Oregon, Michigan, and Oklahoma, where use per capita is comparatively high but acreage is relatively highly developed, it would appear that something in the neighborhood of 3 acres per hundred of population might be the minimum for adequacy. Where parks are of a relatively large and relatively undeveloped character, as in New York and South Dakota, a much larger acreage would be needed. On this basis, something like 10 million acres of state parks, compared with the present 5 million acres, would be the minimum present acreage to meet even these modest standards of adequacy in the states that now have less area, and to retain the larger areas in the states now possessing them. This acreage would suffice only if the additional area were carefully located and fully developed.

The gross area of the TVA and Corps of Engineers reservoirs is of the same general magnitude as that of state parks, namely, slightly more than 5 million acres. The TVA reservoirs are located along the Tennessee River and its tributaries, and thus are confined to one major region of the country. Although some Corps reservoirs are to be found in all parts of the country, a majority are on the Mississippi, Ohio, and Missouri river systems, and through the central and southern parts of the country generally. Although a few of the smaller dams of the TVA system were closed before 1933, this system really began in the 1930's and the greatest area of reservoirs is in those closed during the early 1940's. More than half of the Corps' total reservoir acreage is in reservoirs completed since the war. Most of the

[20] In Massachusetts most people are within a half-hour to an hour driving time of major state parks and beaches. *Report of an Inventory and Plan for Development of the Natural Resources of Massachusetts, Commonwealth of Massachusetts.*

reservoirs of the TVA system have been built, but, for the Corps, a great many reservoirs have been proposed or authorized so that further increases in the area of Corps reservoirs seem probable.

The numbers, locations, and size of the reservoirs were not determined in order to provide some desired level of recreation. Therefore, it is difficult to apply any standard of recreation "adequacy" to them; more properly they are evaluated in terms of their adequacy for flood protection, navigation improvement, or power generation. Nevertheless, recreation is often an important additional use.

The Bureau of Reclamation has built several major and many smaller reservoirs that have significant recreation values. These include Lake Mead behind Hoover Dam, Franklin D. Roosevelt Lake behind Grand Coulee Dam, Shasta Lake behind Shasta Dam, Millerton Lake behind Friant Dam, and others. The administration of these areas for recreation, in general, has been turned over to other agencies—to the National Park Service in the case of three dams, to the Forest Service in the case of most reservoirs lying within national forests, and sometimes to state and local agencies. In such cases, the data on use of these areas have been included in the statistics and discussion relating to the other kinds of areas.

Although complete data are lacking, the trend toward greater use of state parks was evident prior to World War II. Visits may have reached 45 million in 1930, 61 million by 1933, and 75 million by 1938. The earliest reasonably complete data are for 1942, when total visits were 70 million. Use in that year was probably below the prewar peak and, with travel restrictions, it later fell off sharply. Since the war, use has been rising rapidly. The annual rate of increase in recent years has apparently been about 10 per cent, or the same rate as the increase in recreational use of the national forests. The total number of visits to the state parks was over 200 million in 1956, or nearly twice as great as visits to the national park system and the national forests combined. Since their area is much less, it is clear that state parks have a higher intensity of use. In 1956 about 39 visits per acre was averaged; this is midway between what we have designated as "moderate" and "heavy." It seems clear that, especially in their most popular parts, the use of the most heavily visited parks must run close to that of many well-used municipal parks, while use of the more remote parts of the larger state parks, which have extensive buffer areas, is probably not much different from that of the more remote parts of national parks and national forests.

Attendance at state parks in relation to total population differs greatly among the various states (Figure 26). In eight states there were more than two days of use per state resident, but in Oregon, South Dakota, and New Hampshire, this heavy use may have been due in part to a comparatively large number of tourists. Tourists are probably important in Michigan also. But there were also states, such as Rhode Island, Iowa, and Oklahoma, where use per capita was well above average and where tourist use was not likely to be a major factor. Some states of heavy use per capita are not states of high acreage per capita. Use of state parks was especially low in the Mountain states where, it will be recalled, state park acreage was also low. Similarly, state parks in most of the southern states were patronized less than average.

Although some state parks have camping, lodge, and other facilities for overnight stay, the typical state park has no such facilities. Roughly 6 per cent of total visits to state parks in recent years have been overnight; the rest have been single-day or part-day visits. Overnight use has increased from about twenty days annually per unit of overnight capacity in 1942 to nearly sixty days in 1956. This would seem to indicate that the overnight facilities are becoming better known and more fully patronized. However, the heavy predominance of single-day use is in keeping with our characterization of these areas as areas of intermediate outdoor recreation.

Although no recent study has been made of the characteristics of state park use, an excellent study was made just before the war.[21] At that time, about 40 per cent of the total attendance was on Sunday, with no marked preference for other days of the week. In nearly all the parks studied, 40 per cent or more of the visitors arrived in the afternoon; in a comparatively few parks, more visitors came in the morning or in the evening than during the afternoon. Curiously enough, there was almost no correlation between the number of people in what was estimated on a distance basis to be the potential service area of a park, and the actual attendance. This indicates either that many people were unaware of the parks and their attractions, or that the logical service areas were for some reason not the actual ones. The median distance travelled for a day's outing varied from 25 miles for the lowest income groups to 42 miles for the highest income groups; possibly this reflected differences in

[21] U. S. Department of the Interior, National Park Service, *Park Use Studies and Demonstrations,* Washington, 1941.

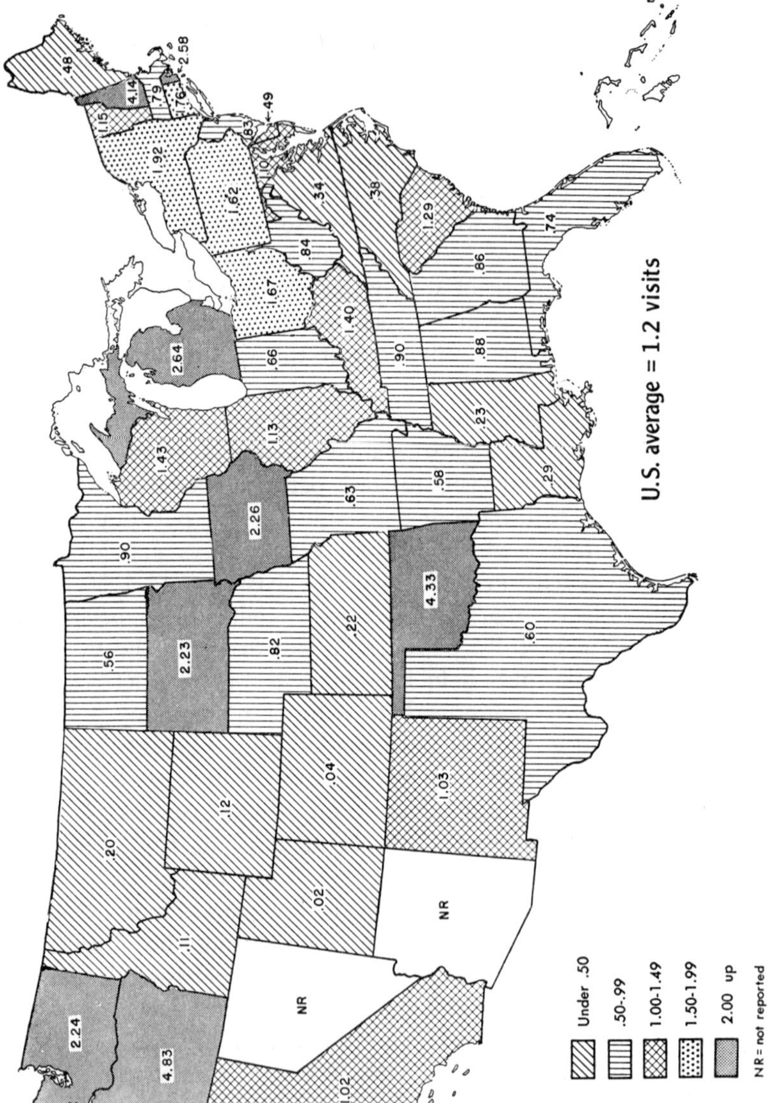

Figure 26. Attendance at state parks (guest days) in 1956 per capita of state resident population in 1954.

quality of automobile as well as of ability to undertake greater travel expense.

Some additional information on annual vacations was obtained from some of the same people: about half had two weeks and less than 20 per cent a longer vacation. Interestingly enough, the distance travelled for vacation bore almost no relationship to income, the median running close to 260 miles for each income class. The several thousand families interviewed were apparently spending, on the average, between 5 and 6 per cent of their annual income for outings. At this time, the overwhelming proportion of the park visitors were adults. The common size of party was four—a comfortable load for the average automobile—and nearly half of these parties were without children. Less common but still numerous were parties of two persons. Nearly all the users of state parks were city people, with far less use by rural people than their numbers would suggest. A disproportionately large number of users came from the white-collar class. Swimming, picnicking, fishing, group camping, and playground activities were the activities most commonly enjoyed.

Since this study included 135 parks in twenty-five states, it presumably reflects reasonably well the typical use of state parks in the prewar period—the data cover 1938. At that time state parks were not reaching all groups of the population that might have been expected to benefit from them. Many of the parks were comparatively new, and even on week ends they were incompletely used.

From the immediate prewar situation until 1956 the acreage of state parks has increased about 20 per cent, total personnel engaged in their administration has increased fourfold, and attendance has trebled. Of the funds spent for capital improvements, about two-thirds have gone into improvements and only one-third into land acquisition. The combined result of these trends is a greatly intensified management and use of the land area of these parks, not an expansion of the system to meet increased present demands or to provide capacity for meeting still larger future demands.

The annual rate of increased use on TVA reservoirs in recent years has been about 15 per cent and on Corps reservoirs, about 28 per cent. Part of the latter has been due to large-scale additions to the Corps system of reservoirs. Total amount of use on these two groups of reservoirs is roughly the same as total use of the national park system and the national forests, and somewhat more than half the total reported use of state parks. Since their area is roughly the same as that of the state parks, average use per acre is less. However, use of key areas around the shores of the reservoirs, for camping, picnicking, and boat launching, is heavy, while use of some parts of the water area is probably much lighter. Roughly one-fourth of the visits to the TVA areas are to the dams, and the rest to the reservoirs.

Hunting and fishing, increasingly popular as outdoor sports, are among the activities enjoyed on several of the federal and state areas already discussed, as well as on private lands and waters. The available data relate to fishing and hunting under license only; in some states young people, especially those under 12 years of age, and older ones, sometimes over 60 or 65 years of age, are not required to have licenses; in some states landowners and their tenants are not required to have licenses when fishing or hunting on their land; and in some states men in the armed services and some classes of veterans are not required to have licenses. Despite these loopholes, license data appear to provide a fairly accurate index to total hunting and fishing activity. Roughly 11 or 12 per cent of the adult population hold resident hunting or fishing licenses. The number of license holders has grown remarkably in recent years, especially in the South and to a lesser extent in the West; in the North the growth has been slower. In 1950 more than one-fourth of the population 15 years of age and over had fishing licenses in four states, and almost half in one state. More than one-fourth of the same age class had hunting licenses in six states, three of which—Montana, Wyoming, and Idaho—also had an equally high percentage of persons with fishing licenses. When there are reasonably good hunting and fishing opportunities within reasonable distances, it appears that at least one-fourth of the population 15 years of age and older will buy licenses. Considering that, even in the states of highest percentages, some people live far from localities of good fishing and hunting, the proportion may even be as high as 40 to 50 per cent.

Summary of problems. The major present problems of the intermediate type of outdoor recreation can be briefly listed. For the state parks, inadequate total area and insufficient numbers in many needed locations rank high on the list of problems. For the reservoirs, location according to hydrological considerations rather than recreation is a limitation. Recreationists perhaps should be thankful that the dams engineers have provided for flood control and other uses also have a high degree of recreation usefulness; but the recreation need has not been met in many parts of the country. Users apparently have not yet fully realized the potentialities of state parks, any more than they have those of the national park system and the national forests. This, which was evident before the war is still apparently true, especially as to quality or satisfactions. The funds spent on the parks in many states have been insufficient funds for best current operation. And, as a system, there has been inadequate provision for the further demands that the future will certainly bring.

Summary of Outdoor Recreation Today

The user-oriented, intermediate, and resource-based recreation areas all form parts of a total recreation system, the different uses of which are closely interrelated. The user-oriented areas tend to be used predominantly after school and after work, the intermediate areas for all-day outings, and the resource-based areas for vacations; yet there are many exceptions. More importantly, use of one kind of area at one time is both a substitute for other areas, and a stimulant to their greater use. Recreation must be viewed in relation to the total culture and the whole lives of people in a particular culture and economy; and outdoor recreation must be studied in relation to other ways of enjoying leisure and having fun.

All the forms of outdoor recreation are used heavily today—more heavily than at any time in history—and the trend of use for most of them is steeply upward. Even if use of the municipal parks climbs at the rate of only about 4 per cent annually, as the admittedly inaccurate figures seem to indicate, this is still twice as rapid a rate of increase as the rate for total population. For other types of recreation, the rate is faster—typically of the magnitude of 10 per cent annually, and for the reservoirs built by the Corps of Engineers a whopping 28 per cent increase annually. There is no clear sign of a diminution in the rate of growth; and growth at a constant percentage rate over a period of years quickly mounts to very large figures.

The outdoor recreation system, in toto, is inadequate to meet the demands of today. There is insufficient acreage, or the units are not properly located, or they are inadequately improved, or there is a combination of all these deficiencies. The situation varies from city to city, state to state, and region to region. In many instances, funds are not sufficient for properly administering the present park areas.

Even though complete data are unavailable, one cannot escape the conclusion that the full recreation potential of even the present areas is not being realized. Too many people make only superficial or fleeting use of them. Better recreation and educational programs within the different types of parks might well lead to different types of use, which would better distinguish outdoor recreation areas from commercial amusements.

Lastly, the inadequate provision being made for the future is disquieting. For the past fifteen years the acquisition of park acreage at all levels has been at a rate definitely slower than the rate of growth of population, to say nothing of the rate of growth in demand for recreation. While some federal and state organizations are making and trying to implement "long-range" plans, most of these deal with periods of less than ten years; and most outdoor recreation agencies have no plans, apparently, for even such short periods.

FUTURE USE OF LAND FOR OUTDOOR RECREATION

What will be the demand for land for outdoor recreation in the year 2000? At this stage of knowledge and experience, any figure is little more than a guess, but the general direction seems clear, and possibly the magnitude can be indicated.

In this book we estimate that by the year 2000 there will be 310 million people in this country, compared with 170 million in 1956; that per capita income will be $3,660 (in 1955 dollars), compared with $1,630 in 1955; that the average work week will be 28 hours, compared with forty in 1956; and that travel per capita will be 9,000 miles, compared with about 5,000 miles in 1956. Thus, population, per capita income, and travel will each roughly double its respective figure, and leisure will increase. What will these massive changes in the causal factors of recreation mean, in terms of recreation demand? As we have pointed out earlier in this chapter, their effect seems to be multiplicative—twice as many people times twice as much income per person times twice as much travel times considerably more leisure.

These forces will affect the demand for all kinds of recreation, indoor and commercial as well as outdoor. Moreover, they will affect the demand for outdoor recreation on all types of land, private as well as public, and on land where recreation will be but one use as well as land where it will be the dominant or sole use. The estimates of recreation demand and acreage required given in this chapter apply to the latter type of land—that which is used solely or dominantly for recreation. For areas of multiple use management, such as national forests, this means that the areas set aside in one way or another for recreation must be increased; this will have repercussions on the larger surrounding areas that serve as buffer zones for recreation and also produce wood and other commercial products.

When it comes to estimating the future demand for outdoor recreation and the areas that will be required to meet that demand, it is well to face frankly the difficulties and the slender analytical base upon which any projections must rest. First of all, we really know very little about past changes in demand for outdoor recreation. Some data are available on total visits or admissions to different types of areas, but even this is not complete. We do not know how many different individuals the total admissions represent; and as admissions have increased, we do not know to what extent this means more people and to what extent it means more visits by the same people. Presumably it means some of each, but in what proportions? We know little or nothing about the age, family composition, occupation, income, and other characteristics of users of different types of outdoor recreation, and of how these may have changed over the years. We know very little about the distances people now travel, or are willing to travel, for certain types of outdoor recreation.

As a result, we lack firm coefficients or factors for relating future demand for outdoor recreation to future changes in these factors. Even if it is assumed that our projections of future population, income per capita, leisure, and travel are each highly accurate, we still do not know how to convert this information into valid estimates of demand for outdoor recreation. And, obviously, there is a margin of error on projections of these basic parameters, which may in itself be large.

In any case, projections of demand for outdoor recreation involve long-term extrapolations into the future from a relatively short basis of past experience. The record of use of the national park system is the longest of any statistical series on this matter; it extends from about 1910 to the present, or less than fifty years. In this case extrapolation to 2000 involves a time period almost as long as the historical record. For the federal wildlife refuges the extrapolation must be six or seven times the record of the past. It can well be argued that economic and social conditions have changed so much that only the postwar experience is relevant; if so, then extrapolation for all areas must be for three to four times as far into the future as the past usable record. At best, long extrapolation from a short historical base is dangerous; many unforeseeable factors may intervene to change the relationships.

Moreover, in basing estimates of future demand for outdoor recreation on past trends in use, it must be realized that the past trends were possible only because the supply of outdoor recreational areas expanded so greatly. Had the number of areas, their acreage, and their access facilties remained unchanged, use of the areas could surely not have increased as it did. Just as surely, we cannot expect future use to increase at the same rate as past use unless more areas, larger acreages, and better roads are provided. Even without development, use will increase on present areas, but a ceiling may be reached quite soon in many instances.

There is no clear evidence of a slackening in the past rates of growth in use of the major types of outdoor recreation areas. Yet a mechanical continuation of past trends leads in a few decades to total recreation figures which are improbable, even absurd. A slowing down in growth rate at some stage seems likely; after all, the capacity to enjoy outdoor recreation, even of the most outstanding kinds, is limited. It would be enormously helpful if there were solid evidence of a slowing down or satiety of demand on the part of any substantial group of the population. This would provide us with some clue as to the future behavior of the entire population with regard to recreation. As it is, we can only speculate on when, at what level, and to what degree a slowing down will occur in the rate of growth of demand for outdoor recreation.

The estimates of future demand for outdoor recreation, which follow, are rough at best. One must assume that a wide margin of error attaches to them. Where it is estimated that total demand for all outdoor recreation may rise in the year 2000 to ten times its level in 1956, this should be interpreted as something between five and fifteen times. With the information available at the time this study was written, no more precise estimate can be made. No definite formula can be applied to the problem. The estimates have been arrived at from trend extension modified with a liberal dosage of "judgment."

Table 24. Recreation use and area, 1956 and 2000

Use and area	Type of recreation: User-oriented[1]	Intermediate[2]	Resource-based[3]
1956			
Million visits	1,000 plus	312	116
Actual area—million acres	0.7	[4]9	[5]45
Adequate area—million acres[6]	2.0	[7]15	[8]
2000			
Million visits	3,750 plus	5,000	5,000
Adequate area—million acres[6]	5.0	[9]70	[10]60

[1] Using city and county parks as an index of this type.

[2] Using state parks and federal reservoirs as a measure of this type.

[3] Using the national park system, national forests, and federal wildlife refuges as a measure of this type.

[4] State parks, 5.1 million acres; remainder, federal reservoirs.

[5] Includes area of national park system, federal wildlife refuges and national forests used primarily for recreation; additional areas are available for recreation and add value to specialized recreation areas.

[6] See text for discussion.

[7] Assuming area of state parks doubled and of federal reservoirs unchanged.

[8] No estimate made.

[9] Assuming reservoir areas of 20 million acres and state parks of 50 million acres.

[10] Assuming some increase in federal areas used primarily or solely for recreation.

A comparison of the actual situation in 1956, the estimated adequate area in that year, and the projections for 2000 are given in Table 24. The projections measure potential more than actual anticipated figures. While they involve a substantial reduction from a mere trend extension, and thus are perhaps lower than a potential defined in this way, the use estimates assume that adequate areas will be available to satisfy a demand of this magnitude. If actual areas fall short, some curtailment of use or some greater intensity than that assumed will be necessary. In Chapter VIII we return to this matter; there our estimates of probable recreation area are contrasted with the area desirable to meet the potential demand.

The base for these estimates requires explanation. The 1956 data are those previously given in Table 21, summarized now by type of recreation. The user-oriented potential recreation demand in the year 2000 is based upon the assumption that 250 out of the 310 million people will live in cities; that the full two acres per hundred urban population, which is desirable, will actually be available; and that the average urban person will use a municipal park fifteen times or more during the year. The latter is admittedly a crude estimate, but the acreage figure does not depend upon it. In the case of user-oriented recreation areas, in addition to the difficulties of projecting so far into the future we have very poor knowledge as to their present use.

The estimate of use of resource-based areas is based upon extension of present trends, but reduced (arbitrarily) to about half of what a mere trend extension would produce. While the trends seem firmly established and very steady over a long period of years, undoubtedly there is some ceiling to the use of these areas; an annual growth rate of 10 per cent cannot continue forever. It has been assumed that the publicly owned areas of this type would not be much larger than at present; not only is there fairly strong sentiment against further major extensions of federal ownership of land, but, more basically, there is little land physically suitable for some of these categories that is not already federally owned. Much of the seashore and lake-shore land, desirable for this category, is in private ownership and its acquisition for public use is improbable. Some additions are likely to national parks, but some of these are likely to come out of national forests. Extension of the wildlife refuge system to areas that specialists consider necessary would require only a comparatively few million acres. A larger part of the total area of federal lands might well be used primarily for recreation in the future.

The potential recreation demand in intermediate-type areas is built upon the assumption of fifteen or more visits per capita annually. In a few states it is now about four to state parks alone, to which must be added some more to reservoir areas. An average of fifteen, under the income, travel, and leisure conditions we have assumed, does not seem unreasonable as a potential demand. It is far less than an extension of past trends would indicate. As will be shown in Chapter VII, the area of federal reservoirs will expand considerably, and primarily for purposes other than recreation. To meet the increased use, state parks should expand tenfold, from 5 million to 50 million acres. It has been further assumed that the use per acre for all areas of this type would be somewhat heavier than now—over seventy visits per acre annually, compared with the current thirty-nine in state parks and thirty-two on federal reservoirs. For comparison, the present use of state park systems is 195 days per acre annually in Connecticut, 144 in Oregon, 105 in Michigan, but only 23 in California and 11 in New York.

Some further light on the reasonableness of these projections can be gained by reference back to Table 19, where the changes in all the factors are in relative terms, with 1956 as a base. The number of visits per capita in the year 2000, compared with 1956, is about twice as high for user-oriented areas, about nine times as high for intermediate areas, and about twenty-three times as high for resource-based areas. The latter is a much greater increase than that in travel, leisure, or income per capita, and hence might seem too high. On the other hand, as these personal factors become more favorable, a disproportionate percentage of the increase will be used for this type of recreation. If our conclusion is correct that the effect of income, travel, leisure, and related factors are multiplicative, then the amount of recreation per person will rise greatly—perhaps even more than we have estimated. Relatively long trips—long in miles and in time—to visit several relatively distant areas may in another generation become as familiar as today's picnic outing.

A few comparisons between types of areas can be drawn. User-oriented areas will experience the smallest relative increase in use, and in fact may be outnumbered by use in each of the other two types of areas; but the area of user-oriented recreation must be increased relatively a great deal to meet the projected needs, partly because it is deficient now and partly because virtually all the growth of population will be in the cities. The area of resource-based recreation cannot increase much, but use will grow many fold. Use of intermediate areas will reach the same total, but now starts from a higher point; and the acreage of intermediate-type areas will increase most of all, absolutely and relatively. These are the potentials; the reality may be different.

If these estimated areas of land and future demand actually develop, the intensity of use will, in the year 2000, differ considerably from what it was in 1956. For the consumer-oriented areas, intensity of use will actually decline by half; the increase in area, to meet the standards set up, is greater than the projected increase in use. For the intermediate areas as a whole, intensity of use will about double; for most areas, the increase will be far greater than this. Little of the increase in use will be in the large type of area now found in the largest state parks, such as are in New York, where on much of the area use is comparatively light. More of the increase in use will be in moderate and small-size areas, where use will average heavier than now. The greatest increase in intensity of recreational use will take place upon the resource-based areas—more than thirty times the present use, we estimate. A large part these areas are now used very lightly for recreation; but increases of this character will bring about much greater use of the more popular areas and a large expansion of the areas of relatively heavy use.

If these potential future demands for outdoor recreation are to be met, the distribution of recreation areas will be somewhat different from what they are today. The user-oriented areas must be near the users; and the location of these users, in a general way, has been suggested in Chapter II. The increase in area of intermediate areas must be mostly within the same states and general areas, in order to lie within the assumed travel distances from most users. This means a great increase in the area of such parks in the more populous states. On the other hand, the increase in resource-based areas will be relatively much smaller and will probably not differ too greatly from their present distribution. A large increase in seashore and lakeshore areas is greatly to be hoped for, but unfortunately is not very probable. The greatest increases of all are needed in the South, where areas are now so deficient.

The physical character of the additional acreage of each type of recreation area will differ also. The resource-based areas may include some mountain areas, but most of these that are suitable are in federal ownership now, or are unattainable; some areas may be shifted from national forest to national park status; some seashore and lakeshore areas may be added; further swamp areas may be added, partly for wildlife as well as for recreation; similarly, one or more large areas of prairie or grassland might be established. This about exhausts the possibilities of additions of resource-based types; and each area would have to meet stiff standards if it were to be selected.

The user-oriented areas will have only a narrow range of selection as far as location is concerned. Within that range, many physical types of land will be adequate. Playground and other areas of heavy use should have well-drained soils but be moderately sloping or level, to be of greatest usefulness. But flood plain areas that are subject to occasional flooding, within or adjacent to cities, might be more suitably used as parks or playgrounds than developed for residences or industry. Sometimes areas too steep for economical residential or industrial development will do very well for city parks, particularly the larger ones. If adequate urban recreational areas are to be acquired, then advantage must be taken of the by-passed or "leapfrogged" areas ignored, for one reason or another, by private developers.

The greatest possibilities of choice exist for intermediate areas. Relatively unproductive farm land and forest land may be entirely suitable for intermediate recreation areas. Dams can be constructed and ponds or lakes developed; trees can be planted, and in time forests grown—not the most magnificent forests in the world, perhaps, but adequate for the purpose. In short, intermediate recreation areas can be "made" as, in fact, some state parks and many recreation demonstration areas of the past have been made. It is true that naturally developed sites may be more attractive, especially for the first decade or two; but it also is true that such sites may be expensive.

If the foregoing represents the potential demand for outdoor recreation in the year 2000, what are the major problems involved in its attainment?

First, this potential will cost a great deal of money, which may not be forthcoming. Our rough estimate is that state parks will require something in the range of $30-$50 *billion* over the next forty years or so; city parks, $22–$30 *billion,* and federal investment possibly $5 *billion.* Someone will immediately say that total investments for recreation of $60–$80 billion, even over a forty-year period, are totally unrealistic. But let us look at it in the following way: This sum is one-third to one-half of the national net income in 1950, but it would be roughly only 10 per cent of it in the year 2000. If most of the expenditure were concentrated in the next twenty years, the annual outlay would be only $3–$4 billion—something on the order of 1–2 per cent of our total national income. Twice as many people with more than twice as much income per person yield an enormously greater national income, which can make possible many things that are today considered improbable.

A second and related problem in attainment of the land needed to meet the potential demand is the matter of timing. The need for expansion of recreation acreage is now, or soon. For cities, ideally the park acreages should be added well ahead of urban expansion, while land is still available and relatively cheap. At the least, expansion of park acreage should move concurrently with the spread of cities. In urban planning provision should be made for recreation areas, and lands should be reserved for this purpose. If acquisition of land lags, in many cases it will be impossible to get the needed land in blocks of adequate size and in the desired locations; at the best, the cost will be higher. Yet, as we have seen, it is in its period of expansion that the new city or suburb is least able, financially and politically, to acquire land for future need. Possibly some means could be devised whereby interested private parties would acquire unimproved land for later resale to public agencies. For state parks, the problem of land acquisition is perhaps easier, because locations can be more flexible and, in many instances, it is possible to make an attractive park out of land relatively unproductive for agriculture or forestry. Early action is needed to acquire the comparatively small acreages of resource-based recreation areas that are desirable, before they are diverted to other uses and become unavailable.

Third, meeting the recreation potential demand will require action at all levels of government. We have assumed that private recreation facilities also will expand. To some degree, they are competitive with public recreation facilities, but to a greater extent they are complementary: when state parks are built, for instance, they create more new opportunities for private recreation facilities than they displace.

The consumer-oriented facilities would be provided chiefly by the cities, the intermediate type chiefly by the states, and the resource-based areas chiefly by the federal government. The comparative burden of meeting the potential demand is perhaps not proportionate to the ability to bear it; the federal government would be required to make the least contribution, and the states the most; but few students of fiscal matters would rate their revenue-raising abilities in this order. It may well be necessary for the federal government to make grants-in-aid for recreation to the states, and possibly to the cities as well.

The fourth and perhaps most serious problem is the quality dimension of outdoor recreation—both the ability of the areas to supply the quality of recreation demanded, and the desires of people for a quality of experience markedly less than the areas are potentially capable of supplying. Accurate data on this point are admittedly lacking, but there are strong indications that a substantial proportion of the users of various types of recreational areas experience only the minimum of the recreational opportunities offered. The brief and superficial character of much of the use of the various areas has been noted. Perhaps more emphasis should be placed upon the quality of individual experience rather than upon quantity in terms of more and more visitors who by their very numbers destroy the experience for each. More educational and leadership services for the areas would at least make possible a less superficial type of use. The basis of the publicity or appeal to potential visitors might be shifted to emphasize the longer stay and the closer look. Perhaps materially higher entrance fees for some types of areas would discourage the casual visitor, without frightening away those with serious intent.

The actual development of land for recreation in the future is difficult to predict; it will almost surely be less than the potential demand would indicate, perhaps far less. But to the extent it is smaller, this nation will be poorer in opportunities for the good life. The degree to which the potential demand for recreation is met will depend largely upon how quickly the people of the nation wake up to the potential needs of the future, and how quickly they begin to prepare for them.

IV

Agricultural land use

If ours is an urban age, if an era of greater leisure lies ahead, it must be remembered that humanity has been a long time in coming this far. Nor can we forget that while values shift and objectives change, man, wherever he lives or whatever he does, must still make provision for those needs that are most basic to his existence. But the American today is not often satisfied to merely meet these basic requirements. Although he does not want to devote too large a part of his income to food and clothing, he wants more than a cheap source of calories and fiber. The refinements, the services, the concessions to variety and taste, and the luxury to which he has become accustomed as part of his consumption of food, shelter, and clothing can now be easily satisfied from the land and other resources devoted to agricultural production.

Projections of real per capita income for the next twenty-five and fifty years make it reasonable to assume that Americans will be able to *afford* the same relatively expensive diet they now enjoy; moreover, with no major increase in the relative price of food, the average family may well be in a position to demand a diet that only the higher income groups now enjoy. This, plus the projected increase in population and the increased competition from other uses for the agricultural land available raises the question of whether the agriculture of the United States will then be able to contribute to the world market, and even whether it will be able to meet the increased domestic demands.

The average American is apt to take our present agricultural riches for granted, failing to appreciate fully the amounts of land and hours of human effort required until comparatively recent times to provide even the basic needs. In many parts of the world today most people are still tied to the soil, committed to relatively unrewarding toil, because their agricultural productivity is so low. It is one of the paradoxes of our times that those economies in which a high proportion of the labor force is committed to agriculture are likely to be those with the lowest per capita consumption of agricultural products and all other goods and services. Such a commitment of human effort is not characteristic of the most productive agricultures of the world.

The United States today is primarily an urban nation, yet agriculture is still of major importance. The urban and agricultural sectors of the economy are complementary, each dependent upon the other and each contributing to the other. Agriculture feeds the city dwellers and supplies industry with raw materials. Its great increases in productivity have freed a major part of our total labor force for work in the cities. As a result, the nation's position is certainly stronger and it is able to provide the American people a higher and more varied level of living.

As a user of resources, agriculture is now of relatively less importance than in 1910, when 42 per cent of the nation's privately owned tangible assets were used for agricultural production. In 1955 agricultural production required only 22 per cent of these assets, but the dollar value of these assets per farm worker in 1952 was six times greater than in 1910 while the value of assets per worker in nonfarm occupations was slightly more than three times greater.[1]

Without question, the cities have stimulated agriculture, primarily through the provision of a market for farm products. From urban industries have come consumer goods for the farm family as well as the new tools of farm production, the application of which, together with the contributions from the emerging agricultural sciences, have made possible the tremendous showing which one man on the land can now make.

Even with land and the labor force declining, the state of agricultural technology is such that production capacity still exceeds market

[1] John D. Black, "Agriculture in the Nation's Economy," *American Economic Review*, Vol. 46, No. 1 (March 1956), p. 4.

requirements by a small though important margin. Because of the still unrealized potential of existing technology as well as the promise of even further gains, agricultural economists today are in general agreement that, for at least the next twenty years and with the population growth now anticipated, surpluses of food and fiber are a more real continuing problem than shortages. American agriculture could even provide food and fiber for a much larger population, but the price might be the sacrifice of some of the developed tastes and amenities related to food and dress.

The possibilities of revolutionary developments in the processing of raw agricultural products, in the art of preparing food for the table, and the development of synthetic foods by the chemical industry similar to its development of synthetic fibers must be seriously considered. But short of such developments in the next twenty-five years, to what extent can American agriculture meet the probable food and fiber requirements of the period of years between 1980 and 2000?

Agricultural pursuits differ greatly in their use of land. While an extremely extensive use of land, such as grazing, is no less an aspect of agriculture than a use as intensive as the production of mushrooms, we shall include with cropland only improved and rotation pasture together with farmstead and other miscellaneous land. We exclude at this point land which may lie within farm or ranch boundaries but which is devoted principally to the production of timber or to native grasses which are grazed.

It is not possible to adhere to these definitions strictly. The unit of management in the production of crops is the farm, and the data used are primarily oriented in terms of farms. It is accordingly impossible to consider the use of land for crops aside from a discussion of farms as such.

Cropland, improved pasture, and such miscellaneous farm land as roads and farmsteads accounted for 27 per cent of the total land area of the nation in 1954. This was exceeded only by the land used for permanent pasture and grazing, some 37 per cent of the total area. Far more than half of the total land area is devoted to these two uses. Taking in addition the 25 per cent devoted to commercial forestry, some of which is also on farms, we thus account for 89 per cent of the total land area. Of these three, far more people and capital are employed with each acre used for cropland than either of the others.

Agricultural use of land shows the impact of four major forces long in operation:

1. Appropriation of virgin land to private ownership and use.
2. Development of improved and lower cost transport, bringing markets closer to the farm and various agricultural areas into competition with one another.
3. Growth of cities and of the urban market, to which changes in agriculture have also made significant contributions.
4. Changing technology in agriculture and the whole economy which has affected the type of resources required in agricultural production as well as the quantity and quality of goods and services required and the goods and services available.

Our concern is to trace in some detail the effect of these broad phenomena upon the use of land for agricultural purposes. Farm people have also been affected and although we do not explore those effects, they may be inferred from the changes in land use. Some people have moved from farm land, either to other farm land, or to the cities. This process helped to build the cities and to form the new nation in its pattern.

The form of much of our data permits "still" pictures at intervals in what is essentially a continuous process; we catch particular postures and positions, which were not the same the year before or the year after. We wish to emphasize neither the details at any date, nor the detailed changes between specified dates, but rather the broad process and sweep of change.

The past two to three hundred years have seen a vast sorting of land take place on a trial and error basis. Much land has been taken into agriculture, either for a subsistence type of agriculture or under mistaken notions as to its productivity, which is entirely unsuitable for this use in the light of today's needs, technologies, and opportunities. Successive generations of farmers have largely determined the areas of land in the United States best suited to agricultural use. This is not meant to imply that no further changes are possible or desirable. Some land now in crops may well be shifted out of them, either because of demands for other uses or because it is not suited to continued crop use; and some land not now in crops could be used for this purpose on a continuous basis but its preparation for crops is not now profitable. Experience has taught us a great deal about the suitability of land for agriculture, but land use is dynamic, not fixed.

HISTORICAL DEVELOPMENT OF AMERICAN AGRICULTURE

Even with the heavy dependence on it, land cannot be considered apart from men, and the machines and other items they use with it in crop production. A number of possibilities present themselves to the man who wants to expand his production. He could do so by increasing one, two, or each of the three broad categories of production factors: land, labor and capital goods. His decision will depend upon a number of things, but of major importance will be a comparison of his costs to produce an additional unit of output under each of the alternatives he can consider.

Although the other factors may substitute for land, they are still less than perfect substitutes. Land cannot be eliminated from crop production and it is strictly limited in total area. While at present land is not a sharply limiting factor, circumstances might arise such that land would restrict the expansion of crop production, particularly after the intensive margins of production have been exploited at a given level of technological progress.

The intensity of the use of land for agricultural purposes varies with time and place depending upon the nature and combination of its physical, economic, and institutional environments. The conditions affecting plant life and growth—temperature and sunshine, rainfall and evaporation, topography, and soils—vary from site to site. Any of these may be excessive at some sites, deficient at others, but optimum at still other sites.

With effort and ingenuity and the necessary materials, equipment, and tools, a particular site can sometimes be made more favorable to plant growth. Because such changes can work both ways, the manipulations, deliberate or incidental to other ends, may lead to adverse results. Fortunately, because drastic changes in the physical characteristics of the site may be physically, economically, or institutionally impossible, the requirements of plants of economic value vary and may also be altered genetically. The cost of selecting crops that will tolerate given environmental conditions is usually considerably less than that of changing the environment to suit the crop. Thus, genetic variations make possible five major classes of wheat: some that are more resistant to cold weather than others, some for which a humid climate is most favorable and some that thrive on a minimum of moisture.

Man's ever-increasing understanding of the world has enabled him to push back the physical frontiers of agricultural production, perhaps only ever so slightly, when it has become economic to do so. Within narrow limits he has learned to control temperatures and moisture, but the era of weather control still lies ahead. He has developed the machines and means that enable him to "move mountains." He knows enough about soils and fertility to increase the productive capacity of soils. The real limits on the supply of land are economic and institutional in nature. Even where physical conditions are such that the cultivation of crops would be feasible, the cost of producing and marketing the crop may prevent its production. Land tenure arrangements, tax laws, foreign relations, and trade arrangements, acreage allotments, customary farming practices—these are only examples of other restraints on the use of land.

Biological factors impose an upper limit on the intensity of land use for agricultural purposes, lower than other uses, and usually lower for crop or primary production than for such secondary production uses as poultry and livestock feeding enterprises which can compete more effectively with nonagricultural uses for the required space.

Resources Devoted to Agriculture

The total area of land in farms grew rapidly from 1850 to 1900 as farm numbers increased, but has increased more slowly since then (Figure 27). The growth in total area within farms was both a change in land ownership and control and a change in land use. Land passed from public to private ownership or from large private owners, such as railroads, to small farmers. During much of this time, total area within farms in the Northeast declined slightly.

Data on changes in farm or crop area by states or regions mask important localized changes within the different states. After all, land is peculiarly a locational factor, and *where* the change in land use occurs may be as important as the extent of the change in use. The shifts in location of improved farm land acreage or in acreage of harvested crops over the decades graphically trace the westward movement of agriculture within the United States.

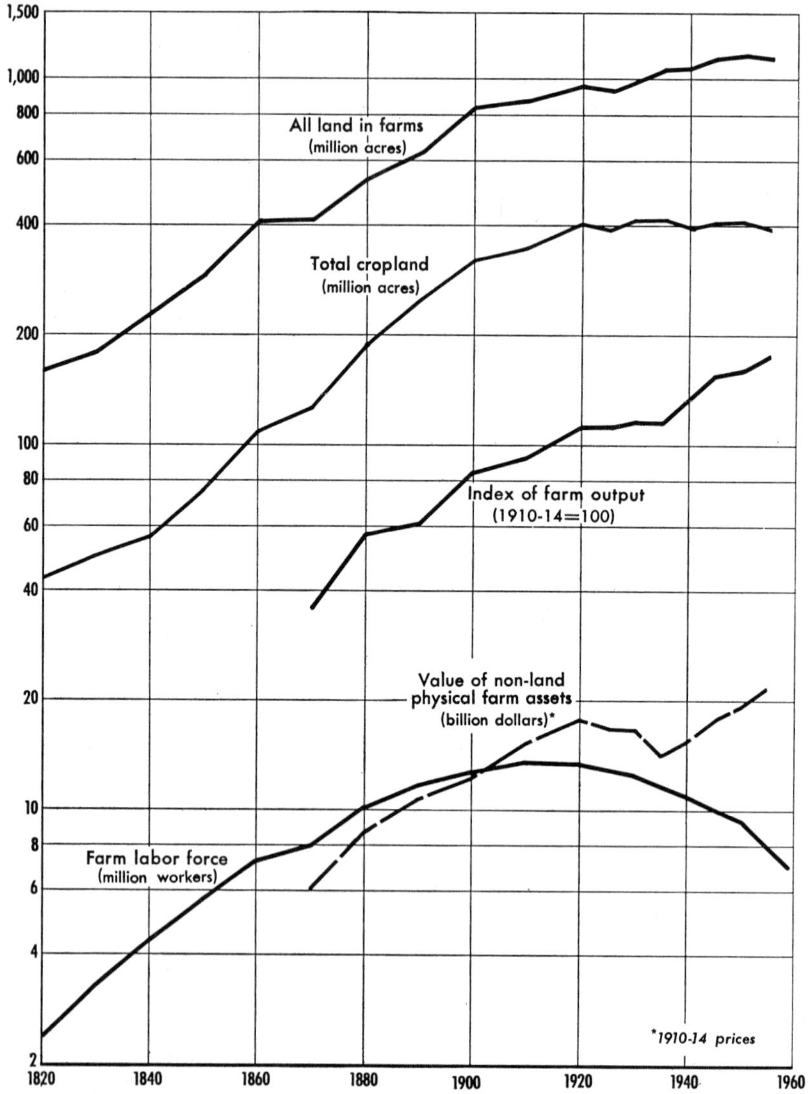

Figure 27. Land, other physical assets, labor supply, and agricultural output, 1820–1954.

In the decade before the Civil War, the great central Corn Belt area showed the greatest increase. Expansion in the decade after the war was greatest on the western edge of the Corn Belt. In the 1870's the expansion was still largely concentrated in these regions with some in Texas and in California. The 1880's saw the first major move onto the Great Plains as such.

Changes in total cropland acreage since 1910 have been relatively minor. Significant increases occurred up to 1920, but since then there has been no consistent trend, either upward or downward. An upward trend in cropland area in both the Mountain states and the northern plains is still continuing. A similar trend in the southern plains reached its peak in 1935. Small but long continued declines in cropland area have occurred all along the Atlantic Coast and in the Corn Belt since 1920. The over-all picture of cropland use is thus one of small changes in total area within and between regions. This contrasts sharply with other and greater changes occurring in agriculture. The greatest concentration of cropland, however, remains in the Corn Belt (Figure 28), while the farm land of the Western United States is largely pasture and range and is dealt with in Chapter VI.

The continuing increases in the land reported in farms by the agricultural censuses from 1930 through 1950, small but in sharp contrast to the relatively static position of cropland and its ultimate decrease, is explainable largely by changes in tenure rather than by changes in use. The percentage increase in farm land made by the Mountain and Pacific regions between 1930 and 1954, 66 per cent and 26 per cent respectively, was unsurpassed by any other region. However, except for the cropland added by reclamation projects, most of these increases represent no actual change in use.[2] The land, usually open range, was already being grazed and because of changes in census definition as well as changes in the tenure arrangements for its use, came to be included as farm land.[3]

Abandonment of land has accompanied increases in improved or harvested cropland. Some abandonment had occurred on the tobacco

[2] Outside of these regions, Florida, South Dakota and Louisiana showed total farm land gains of an order of 261 per cent, 23 per cent, and 22 per cent respectively. Cropland increased by only 16 per cent in the Mountain states, 15 per cent in the Pacific states, and in the three individual states above only Florida, which gained 29 per cent more cropland, added to the area under cultivation. South Dakota and Louisiana actually lost cropland.

[3] U. S. Bureau of the Census, *U. S. Census of Agriculture; 1954, Vol. II, General Report,* p. 17.

lands along the Tidewater areas even before our nation was founded, and later, but as early as 1830, in New England. These areas were comparatively small. The rate of abandonment rose faster later. While the areas abandoned have not been large in total from a national viewpoint, they have been major in some localities.

Most of the past decreases of improved or harvested crop acreage have come about, not because the land was in such demand for other uses that agriculture could not compete against them, but because its productivity for agriculture was so low that satisfactory incomes could not be earned on it. This type of farm abandonment generally meant that alternative employment opportunities existed elsewhere to absorb the people who left, although some drought-driven exoduses from the Plains were forced by the very harshness of conditions there.

The chief competitor to some agricultural areas in the past has been other agricultural areas, not other land uses. In many instances the land abandoned by agriculture did not immediately go into other land use in the sense of a consciously planned and directed use. It grew up to trees or grass, depending upon its location, and in time came to have some use, perhaps a very extensive one.

In terms of value, land is now and long has been the largest single investment item in American agriculture. The land itself, exclusive of buildings, constituted as much as 66 per cent of all physical assets used for production in 1910, and was as low as 50 per cent in 1950. A more typical level for the last eighty-five years would be about 58 per cent.[4]

Land values have increased over the years as investments have been made to clear, drain, or irrigate land, thus directly affecting its productivity. They have also capitalized the gains anticipated or realized from the economic development and growth of the community. Landowners have captured some of the gains in the productivity of other input factors and the resulting economies of the new technologies. Similarly, they have, to an extent, been the beneficiaries of price

[4] Alvin S. Tostlebe, *Capital in Agriculture: Its Formation and Financing since 1870* (Princeton: Princeton University Press, 1957), p. 66, Table 9; *The Balance Sheet of Agriculture, 1957,* Agriculture Information Bulletin 177, U. S. Department of Agriculture, 1957, p. 11.

Figure 28, opposite. Principal cropland areas of the United States. (Map from *Agriculture Handbook No. 153,* U. S. Department of Agriculture, 1959.)

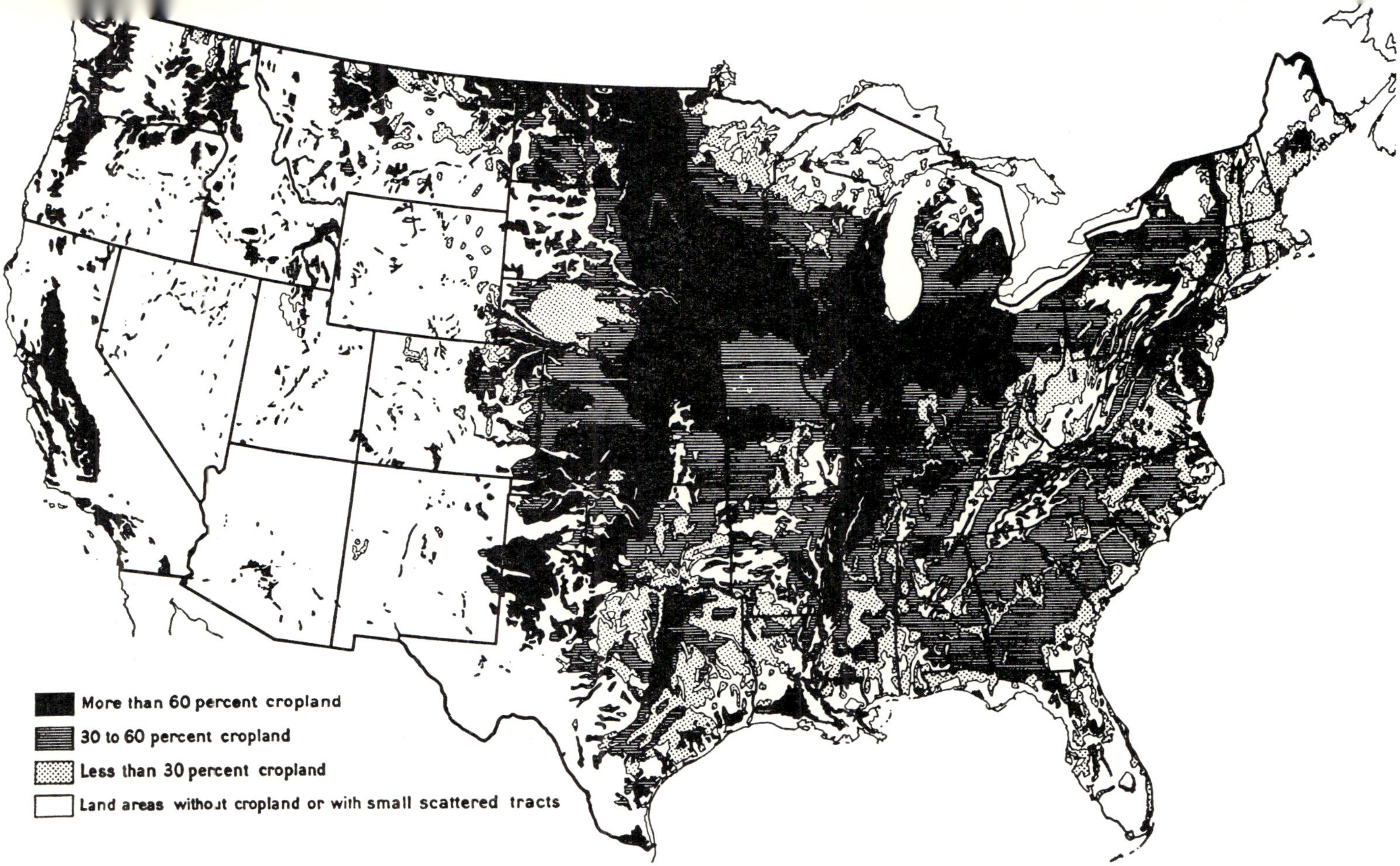

More than 60 percent cropland
30 to 60 percent cropland
Less than 30 percent cropland
Land areas without cropland or with small scattered tracts

support operations, to this extent defeating the rationale of the policy to raise the income of farm operators.[5] The demand for land for nonagricultural uses has been and continues to be another upward force. For these reasons, land values may well overstate the contribution of land to the productive process.

American agriculture is less obviously one of the largest users of capital in the economy. Comparative study of the use of capital by the various sectors of the economy is relatively new and subject to limitations inherent in the methods of investigation, but the Harvard Economic Research Project on the capital structure of American industries provides a perspective that is otherwise unavailable. Using capital coefficients, or the dollar value of capital stock required to produce a dollar of productive capacity, it is estimated that in 1947 the agricultural sector of the economy, one of the 192 defined sectors, required 11 per cent of the economy's total stock of capital.[6]

As for labor, it was not until 1880 that the number of persons engaged in all other occupations exceeded the number of persons engaged primarily in agriculture. Although the farm labor force by 1957 had shrunk approximately 43 per cent from the peak in 1916, there is a history of growth prior to that time.[7] The nonfarm labor force,

[5] If the landowner is also the farm operator there is no problem, at least until the farm is sold. At that point it is often possible to capitalize benefits of price support operations, which in effect denies their benefits to the new farm operator. On rented land the landlord may benefit almost immediately.

[6] *Estimates of the Capital Structure of American Industries, 1947,* Harvard Economic Research Project (June 1953), quoted in John H. Davis and Ray A. Goldberg *A Concept of Agribusiness* (Cambridge: Harvard University Graduate School of Business Administration, 1957), p. 133.

[7] The definitions of the farm labor force used by the U. S. Bureau of the Census and the U. S. Department of Agriculture differ sufficiently to cause a significant difference in the two series. The Census series used for the most part in this chapter, except for the reference above, shows the lowest figures, primarily because only persons primarily employed in agricultural occupations are counted, whereas in the Agriculture series all persons doing an hour or more of farm work during the survey week for pay are included, as are family members who did fifteen or more hours of farm work during the survey week without pay. The Census series does not report foreign workers entering the country to work for limited periods. See *Farm Labor,* Agricultural Marketing Service, U. S. Department of Agriculture, January 1957, for additional discussion of this point and for the employment series for 1910 through 1956. See U. S. Bureau of the Census, *Historical Statistics of the United States, 1789-1945* and its Supplement, and the *Statistical Abstract of the United States* for 1955 and 1956 for the Census series.

however, was growing at a faster rate, even before 1870. The index of farm output kept pace with the changes in cropland from 1870 through 1935.[8] The relation between changes in output and in use of nonland physical assets is similar, but not as close. From 1920 through 1935 the value of these assets in constant prices dropped sharply while farm output increased slightly. Following 1935, however, the rates of increase in farm output and in the use of the nonland physical assets are nearly the same, while the area of cropland actually decreased. Quite striking for the entire period from 1870 to date is the diminishing role of labor in agriculture. The decline in farm labor since the 1916 peak has been at an increasingly rapid rate.

Changes in the Internal Organization of Farms

Changes in these various factors and relationships for all of agriculture have their counterpart in the organization of the average farm at the different dates. At all times there have been farms of a wide variety of sizes and types. Yet the total acreage, or total labor force, or any other parameter, can be divided by the total number of farms to get an arithmetical average. Perhaps no actual farm conforms exactly to this statistical creature, but the changes it shows over time provide a clue to actual changes in individual farms. If the average farm is getting larger, this may mean a shift in the size of the modal farm, or comparatively more of the large as compared with the small farms.

The number of persons employed on the average farm has declined rather regularly and persistently since 1870, to a number now only about 60 per cent of its former level (Table 25). One occasionally hears references to corporation farms or "factories in the field," with the implication that the family-size farm is disappearing. While some very large farms today are managed and operated entirely by employees with no equity in the business, and in general use principles of business organization often associated with factories, the average

[8] The Strauss-Bean ideal index of total farm production for calendar years covers the years 1870 through 1920 and the Agricultural Research Service Index of farm output covers the years 1910 through 1957. Frederick Strauss and Louis H. Bean, *Gross Farm Income and Indices of Farm Production and Prices in the United States, 1869-1937,* Technical Bulletin 703, U. S. Department of Agriculture, December 1940, p. 125; and *Agricultural Outlook Charts, 1958,* U. S. Department of Agriculture, November 1957, p. 71.

Table 25. Changes in the combination of production factors on farms per farm worker, 1870-1954, and man-hours of farm work and productivity per man-hour, 1910-1954

	1870	1880	1890	1900	1910	1920	1930	1940	1950	1954
			Indices, 1910 = 100							
Total man-hours of farm work	—	—	—	—	100	106	102	93	67	64
Productivity per man-hour	—	—	—	—	100	109	117	152	244	276
Persons employed per farm	126	121	119	101	100	98	90	81	76	75
Acreage of cropland per farm worker	87	86	100	102	100	115	122	122	139	151
Value of capital goods (1910-14 prices) per farm worker	96	112	100	83	100	117	112	104	150	187
Value of all farm land (1910-14 prices)	109	89	106	104	100	104	109	117	136	155
			(absolute inputs used per farm worker)							
Cropland—acres	20.7	21.3	25.1	30.2	30.0	35.1	40.8	45.2	54.6	60.5
Value capital goods at 1910-14 prices ($1,000)	1.0	1.2	1.1	1.1	1.3	1.6	1.7	1.7	2.6	3.3
Value of all land at 1910-14 prices ($1,000)	2.2	1.9	2.3	2.7	2.6	2.7	3.1	3.7	4.6	5.4

SOURCE: Derived from U. S. Department of Agriculture estimates of man-hours of work productivity and estimates of cropland, from Census farm labor series, and Tostlebe's capital estimates.

farm today is more nearly a one-man operation than it has ever been in the past. In an earlier day, one or more hired men were a common feature of most farms.

Because the acreage of cropland per farm has nearly doubled since 1870 the reduction in labor used per farm is even more notable. In part, this increase in crop acreage is a reflection of the appearance of the cash-grain farms of the Plains states during the period, which are typically larger farms. However, the trend toward more acreage per farm is found for almost all types of farms. The value of capital goods other than land has roughly doubled on a per farm basis. The increase in value of farm land per farm has been somewhat less. In each case, the increase was relatively small until the World War II period; a major part of the total rise over the eighty-five-year period has occurred in the past fifteen years.

The same data can be shown in terms of the resources available per worker. In these terms, the quantity of land and the value of capital goods have tripled since 1870. The increase in the value of this quantity of land as well as the value of capital goods has been at an increasing rate (Figure 29).

The typical farm today employs less labor, but includes more land and more capital, than at any time in the past. The farmer is increasingly a manager, although chiefly of his own labor time, and less of a laborer.

Capital goods play a strategic role in agricultural production that cannot be indicated by statistics alone. The index of farm output parallels both the rate of increase in the number of farms and the rate of increase in farm land from 1870 until 1900 for land, and 1910 for farms. A good part of the increase in farm output for the period up through 1900 was brought about largely by the addition of new farm land and new farm units but with no appreciable increase in the productive capacity of the average farm. Total capital investment in agriculture grew steadily, but at a decreasing rate throughout the period of 1870 to 1920.

This does not necessarily mean that established farms in the older sections of the country were using less capital. Very likely, this is but another indication of how the rapid expansion and incorporation of new land into farms on the frontier, where capital inputs were scarce, was changing and coloring the nature of the nation's agricultural economy. Another factor was that western farms and ranches were much larger and used less capital per acre of cropland than the farm-

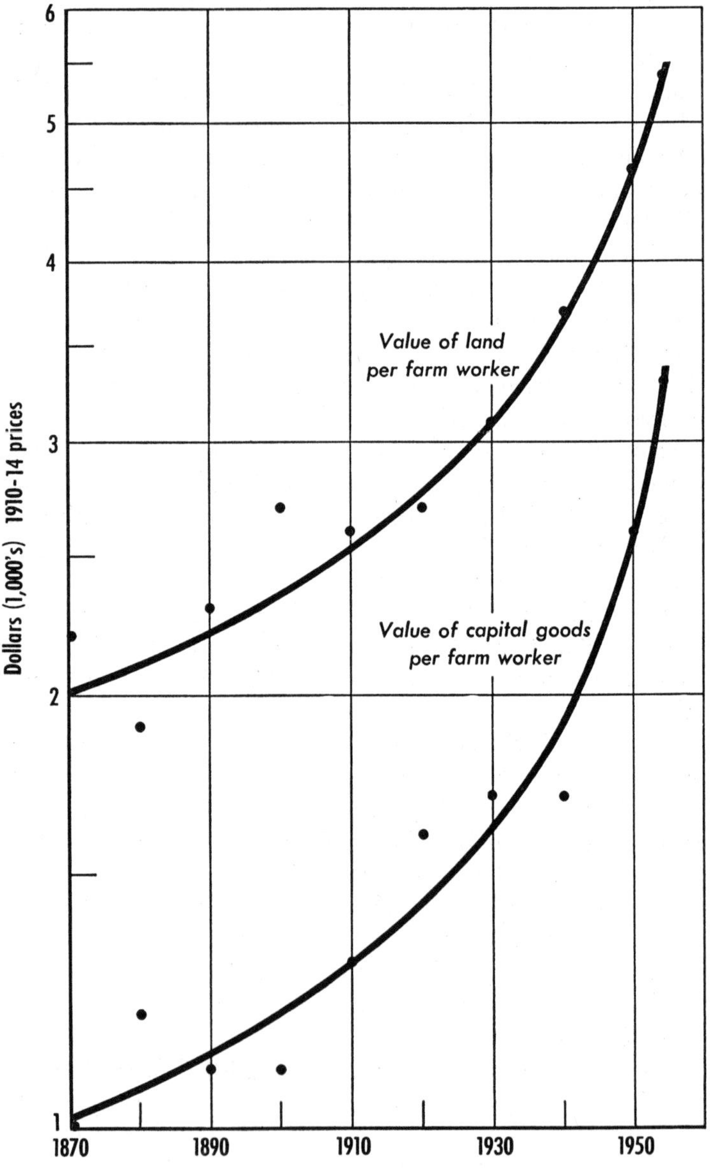

Figure 29. Trend in value of land and physical assets per farm worker, 1870–1954.

ing enterprises in the east. The relative decline was stemmed by 1900 and capital inputs per crop-acre increased until 1920. The decline in capital inputs per crop-acre, beginning after 1920 and continuing until 1940, represents the impact of the long agricultural depression. Since 1940 capital per acre of cropland has risen to a peak. In these later years, farm output increased at a faster rate than the increase in land in farms (although at the same rate of increase in cropland). Capital goods of increasing efficiency were being substituted for farm labor without decreasing farm output.

Two Revolutions

The increasing use and importance of capital goods in agriculture is but one aspect of the spectacular transformation in the manner of working and living which began in this country in the nineteenth century and does not appear to have spent itself even now. The industrial revolution is ordinarily considered to have begun in England in the latter eighteenth century, while the agricultural revolution began nearly a century earlier with the enclosure of the common lands and the introduction of new crops, particularly in England. But later changes in agriculture in the United States have made what may be called a second revolution in agriculture. While basically the industrial and the agricultural revolutions are merely different facets of the over-all phenomenon—the application of the findings of science to each of these areas—they were not strictly concurrent. However, the great interdependence of the two and the large body of causal factors common to them were enough to assure that neither lagged significantly behind the other.

The changes in agriculture, understandably, did not reach all sections of the country simultaneously. "The desire to utilize the findings of modern science in order to make farming profitable was not without influence among the leaders of the United States in the years immediately following the American Revolution," Edwards has observed. "While the forces of the agricultural revolution had long been at work, it remained for the Civil War to hasten their fruition."[9]

9 Everett E. Edwards, "American Agriculture, the First 300 Years," *Land, Yearbook of Agriculture, 1940* (Washington: U. S. Government Printing Office, 1940), p. 221.

He cited the following as forces underlying the American agricultural revolution: (1) the passing of the public domain into private ownership by means of liberal land policies; (2) the completion of the westward movement of settlement; (3) the invention and popularization of improved farm implements and machinery; (4) the extension and development of transportation facilities; (5) the migration of industries from the farm to the factory; (6) the expansion of domestic and foreign markets; and (7) the establishment of agencies for the promotion of scientific knowledge relating to agriculture.[10] In the latter he included agricultural societies and fairs, farm magazines and the establishment of the federal Department of Agriculture and the agricultural colleges and experiment stations.

The new agriculture is often characterized as a *commercial* in contrast to a *subsistence* or *peasant* agriculture. The more cumbersome phrase *tending toward a commercial agriculture* would more accurately describe the situation and the idea of an evolving but uneven process, still in operation. The kinds of processes carried on within what we now call agriculture are very different from those found in the agriculture of an earlier time; many activities have been shifted off the farm.

The decline in the farm labor force came about with the division of labor between farm and nonfarm production, with specialization within agriculture itself, and with increasing substitution of capital goods for labor. It is perhaps most dramatically underscored by the contrasting ratios of farm workers to total population for the years 1820 and 1956. In 1820 one farm worker supported 4.12 persons. In 1956 one farm worker supported 20.85 persons. Some of the manpower released from agriculture since 1820 has been employed to produce both consumer and producer goods which the farm worker in 1820 produced in addition to his output of farm products. Other manpower has been used to produce the goods which have been substituted for inputs of farm labor and land. In this way it has contributed more to farm production than it would had it continued as farm labor.

Society has also gained from this transfer of manpower. Figure 30 indicates the proportion of the total labor force over the years that would have been required in agriculture had the state of the agricultural arts remained as it was in 1820, and if total population had somehow increased as it has.

[10] *Ibid.*, p. 221.

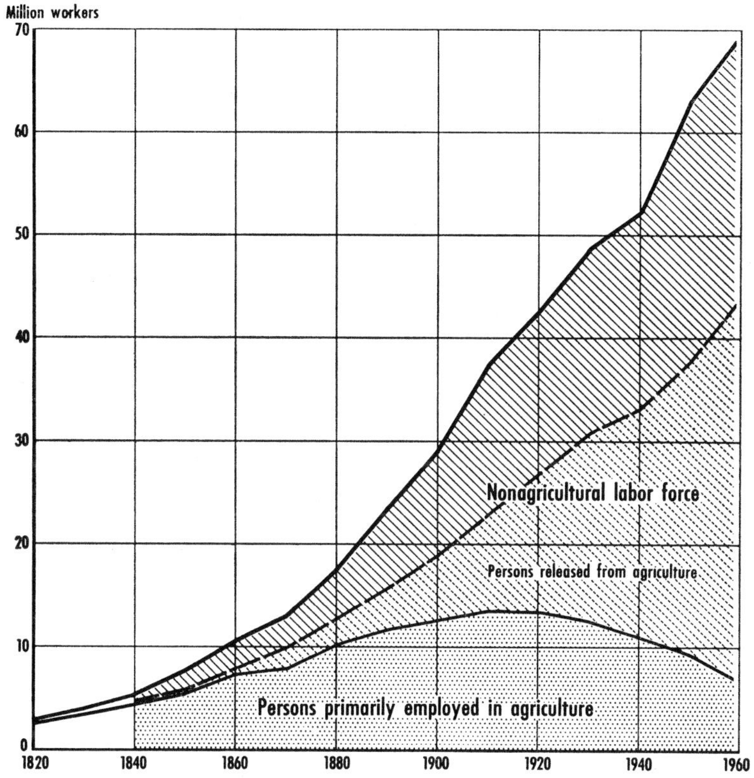

Figure 30. Agricultural and nonagricultural labor force, 1840–1954.

Agricultural technology has advanced more or less steadily throughout, although innovations have often outrun their general use. The period of advance coincided with the great westward movement. A steel plow had been patented in 1808, but the implement did not become commonplace until just before the Civil War. It was available to assist in opening the new lands on the frontier, but farmers there were often without the necessary capital to invest in much machinery immediately, particularly while they were becoming established and operating on a subsistence rather than a commercial basis. Even in more settled areas farmers tended to keep to the old ways. There too, lack of working capital was a serious hindrance. "Interest rates were high because losses were frequent. Where the farmer realized a surplus, he preferred to invest in lands, in larger homes, or in outside enterprises, rather than in labor-saving machinery."[11]

The transition from hand tools to horse-drawn machinery had been largely accomplished at the close of the nineteenth century. The next long step ahead was the substitution of mechanical power for horse power. The factory system had moved ahead rapidly under the impetus of mechanical power, but the self-propelled steam engine was expensive and too heavy and unwieldy for field work except under special conditions. Mobile mechanical power for the farm became practical only with the development of the internal combustion engine.

To appreciate what these advances meant, contrast the labor required for the production of wheat, little changed in the early nineteenth century from the methods and tools used by history's earliest farmers, with the situation at the close of the nineteenth century. The ground was plowed with a crude wooden plow and harrowed with a bundle of brush. The grain was broadcast by hand and the crop was harvested with a sickle, threshed with a flail and winnowed by hand. To produce 100 bushels of wheat in the earlier period required about 288 man-hours of labor. The same quantity of wheat could be produced in 1896 with 44 man-hours of labor with the horse-drawn equipment then commonly available.[12] The entire 1956 wheat crop in the United States required only 22 man-hours of labor for each 100 bushels of wheat produced.

The beginning of the twentieth century saw the development of a workable gasoline tractor. It was large and relatively powerful, but its use was limited to large-scale operations, particularly to the wheat-producing areas of the Great Plains. It is significant that the 1919 wheat harvest, the largest in the history of the United States until that time, was accomplished when the tractor, the combine harvester, and related machinery were spreading onto the plains. The spread of the tractor into other farming areas awaited the development of a smaller, more easily operated machine as well as the availability of accessory equipment.

The number of horses and mules reached a peak of about 26 million head in 1915; today there are fewer than 4 million. The acreage of land required to raise crops for these farm work animals, and for

[11] *Ibid.*, p. 207.

[12] C. E. Gapen and R. B. Gray, "Agricultural Power and Machinery," *Encyclopedia Britannica*, 1951, Vol. I, p. 378.

their kind in cities, has declined proportionately from roughly 100 to about 10 million acres. Farmers have substituted petroleum products and steel for oats, hay, and other farm commodities with one tractor replacing about five horses or mules.

The evolution of power in agriculture is representative of technological advance in agriculture. Like the tractor and its use of the internal combustion engine and rubber tires, much has been borrowed from advances made outside of agriculture and adapted for farm use. Not many innovations have had so immediate and direct an effect upon the available land resource as has power.

Mechanical power, mechanization of farm operations, heavier use of fertilizers, improvements in soil management methods both as to erosion control and fertility maintenance, genetic advances permitting more selective breeding of crops and livestock, more effective disease, insect, and weed control, and advances in animal nutrition have in real terms stretched the area of farm land by pushing back the economic margins of cultivation and use and by making possible greater yields on land already in use. There have been some offsets, of course, from erosion, fertility depletion, new diseases, insect pests, and other sources.

Geographic Shifts in Farm Production in the Nineteenth Century[13]

The earliest pattern of agricultural development in each newly settled area was similar in many respects. Although there were differences, each pioneering community, from the east coast to the west, was subject to a generally common economic situation and institutional environment. Survival was the foremost objective. The settler's own labor, and that of his family, an extremely limited but valuable asset, was nearly his only resource. Without the means to acquire machinery, if it had been available, and subject to competition that he certainly could not have met if the settled areas to the east had been using methods of farming beyond his means, the frontier farmer

[13] Material in this section is largely based upon Everett E. Edwards, *loc. cit.*; the series of historical studies of agricultural production, each by multiple authors, in the U. S. Department of Agriculture *Yearbooks* for 1921, 1922 and 1923; and various chapters contributed to the book: Harold F. Williamson (ed.), *The Growth of the American Economy* (New York: Prentice-Hall, Inc., 1944).

was lucky that no great gains in agricultural technology were made in much of the nineteenth century.

Climate and soils ultimately imposed limits on the use the settler made of the land. This environment might be more fully exploited after contact with the older more settled and developed areas was possible and economic conditions had changed, but these were relatively unimportant at first.

This earliest period of a subsistence agriculture continued for long periods in some areas but was relatively short in others. With its passing, involving the sometimes painful and costly process of trial and error, the agricultural land use patterns we now know began to appear. The common thread in the pattern of development was broken as each area developed a farming system, often a succession of systems, best fitting the conditions of the times.

The North and Northeast up to the Civil War. Agriculture in the Northeast at the beginning of the nineteenth century had changed but little from colonial days. Except in limited areas, subsistence farming was the rule, methods were backward, and total output was low.

The impact of the industrial revolution on the agriculture of the region was felt after the first decade. Population increased rapidly; textile mills, iron furnaces and other industries multiplied. Improved hand tools came into use, the cultivator began to replace the hoe, and the horse replaced the ox. Farming became more specialized with soils, climate and location of markets becoming the determining factors. The farmers of the area were on top of an expanding market.

The Appalachians imposed an economic as well as a physical barrier between the Ohio Valley and the East, temporarily shielding agriculture in the East from the competition of more productive lands to the west. Wagon freight and toll charges to the eastern markets were exorbitant. Shipping charges from Buffalo to New York City by wagon came to $100 a ton, or in terms of wheat, $3.00 a bushel. When the Erie Canal opened in 1825, the cost fell to about 45 cents a bushel and the trip time was cut by more than half.

Pennsylvania, New York, and Virginia were the important wheat-producing states during the period of canal use up to 1840. But even within these states the producing areas were moving west. By 1839 about half of the wheat crop was grown west of the Alleghenies although the three seaboard states accounted for 42 per cent of the production. By 1845, with the shipping charges on wheat from Chicago to New York at 25 cents a bushel, farmers in Wisconsin, Illinois, and northern Indiana as well as Ohio could compete to advantage with eastern farmers on less fertile or run-down soils. And Great Britain's repeal of the Corn Laws about this time opened an export market for grain.

Except for the valleys of the Connecticut and Shenandoah rivers, which produced sizable crops of corn, and the better pasture areas of New Hampshire and Vermont, beef cattle were produced only in limited numbers. Sheep production, although important to northern farm families as a source of wool for their homespun clothing, flourished on a large scale only after the War of 1812 and the events leading up to it cut off the supply of European woolens. Sheep numbers are estimated to have increased nearly 50 per cent between 1810 and 1814. Despite a setback when imports resumed again, the woolen industry in the East entered its period of greatest prosperity in the 1830's. Vermont, New York, Pennsylvania, and Ohio were the major sheep-producing states. But the center of sheep production had moved into Ohio by 1840 and the New England sheep grower had become a casualty.

The shift from grain production to sheep accomplished a conversion of cropland to pasture which led to the enlargement of farms and a reduction in the farm labor force. The mills of New England and lands to the west absorbed the displaced farm population. Some farm abandonment began even in this period. The growth of an urban population in the area made it possible to shift to dairying and market gardening when sheep production proved unprofitable.

The Connecticut River Valley remained one of the important centers of corn production until the early nineteenth century. But on the frontier the production of both corn and wheat followed land clearing and it was not long before the lands to the west were the important corn producers. Corn was easier to market than wheat. It could be converted to either whiskey or to beef or pork, commodities which presented less of a marketing problem where transportation costs were high. The leading corn areas in the country were south of the wheat areas, with 56 per cent of the crop being produced in Tennessee, Kentucky, and Ohio, by 1840. Virginia was still pre-eminent in the East. Hog production on a commercial basis was closely associated with the corn-producing areas.

The South until 1860. Southern agriculture had an initial advantage over northern agriculture for the climate permitted the production of items which could not be grown, or grown competitively, in England. Tobacco, rice, and indigo were all produced for an export market during the Colonial period while northern agriculture was largely one of subsistence farms. The South, however, was not without its share of such farms.

The market for southern cotton was greatly expanded by the invention of spinning and weaving machinery coupled with the establishment of the factory system which revolutionized the production of textiles in England. But cotton production was a mere 10,000 bales in 1793, the year the cotton gin was invented. In the space of seven years, cotton production increased more than seven times and doubled or nearly doubled in succeeding decades up through 1850, reaching a peak of over 4.5 million bales in 1859. At such a rate, cotton was soon the major item in American foreign trade. As early as 1809 it accounted for 23 per cent of the value of all exports and by 1860, 61 per cent.

Production centered initially in the tidewater region of South Carolina and Georgia, where it rapidly replaced rice and indigo. After 1820 an expansion north into the Piedmont of North Carolina and Virginia drove out a thriving cattle industry.

Entry into what are now the states of Alabama and Mississippi was at first effectively blocked by the Indians, but successful campaigns against them during the War of 1812 opened new areas for settlement. By 1820 the tide of immigration was in full sweep. Plantations continued to move west. Mississippi had become the leading cotton state by 1839. The development of railroads in subsequent years opened new areas for cotton, some as far west as Texas.

As the soils of some of the older producing areas of the southeast were exhausted or seriously eroded, the land was abandoned for the newer lands of the West. Had the South developed an integrated agricultural economy, the older areas might have supplied the food and feed required by the cotton plantations that were to a large degree supplied by the Middle West. Lacking industrial and commercial development on a scale comparable to that of the North, southern farmers were denied the alternatives open to northern farmers when they lost markets to the newer lands of the West.

Even though most southern farmers were still at subsistence level in 1860, they were producing half of the nation's corn and nearly 30 per cent of the wheat; cotton was, however, the only commercially important crop. Tobacco, a mainstay in the Colonial period, had lost markets because of foreign competition and the heavy tariffs imposed in Europe for revenue. Cotton outbid it for both land and labor. After 1850, however, tobacco made a recovery and by 1859 the crop was twice the size of a decade earlier. Virginia maintained supremacy in production but by 1839 the growing area had extended beyond the Appalachians.

To the Mississippi and beyond. The fifty-year period between 1840 to 1890 was marked by several major events which affected agriculture materially: (1) the United States acquired major territory in Texas, the Pacific Southwest, California, and the Pacific Northwest; (2) the frontier was pioneered in rapid fashion, first on the prairies, then, skipping over intermediate areas, in California and along the Pacific Coast, and lastly on the Plains; (3) concurrently, the railroad emerged as a major form of transportation, with lines into and across the Plains to the Pacific; and (4) the near extinction of the buffalo and the subjugation of the Indian removed some of the last major obstacles to settlement in extensive areas. These developments produced major changes in land use. At the same time, the continued growth of cities and the growth of foreign markets were of less immediate consequence in the developing patterns of land use but were important in stimulating production to meet their demands.

The effect of these changes can be seen specifically in the area and location of wheat and corn production during the same period. Wheat was often a post-pioneer crop; when the settler had satisfied his own food needs, he often turned to the production of wheat for a cash crop. The major wheat states in 1849 were Pennsylvania and Ohio. Ten years later the crop had to a great extent shifted westward, with Illinois, Indiana, and Wisconsin the major producing states. California was shortly to emerge as a major producer. Within the central part of the nation the crop by 1889 had shifted to the old Northwest, Minnesota, and the Dakotas, and into the Plains, as well as remaining important in some of the older areas and in California.

The East Central states region dominated wheat production at the beginning of the period but the situation changed with the 1890's when the Minnesota prairies and North Dakota were producing nearly a third of the crop. This was made possible largely through the ingenuity of the Minneapolis millers in developing new methods to mill the hard red spring wheat, the only wheat that could be grown in the region because of the rigorous winters. The multiple grinding method introduced in 1871 and the substitution of rollers for millstones in 1878 cut milling costs, increased the yield of flour per bushel of wheat and produced a flour far superior to that made from the soft red winter wheat of the older producing areas. The new flour was much in demand in Europe as well as the United States.

Kansas also had a hard wheat, a hard red winter wheat possessing similar characteristics which had been introduced from the Crimea in about 1872. The new milling techniques also enabled the growers of this wheat to supply the demand for high-quality flour. By 1909 Kansas was the second ranking wheat state, North Dakota the leading state.

The Corn Belt as such was a post-Civil War development. Like wheat, the production of the crop moved westward. But corn's climatic range is far more limited than that of wheat, and upon reaching the prairie states expansion halted. Other areas had somewhat similar combinations of soils and growing conditions suitable for corn production, though not on as great a scale, and the advantage of producing crops other than corn was such as to leave the prairie states unchallenged.

Soils and climatic conditions tended to limit the northern and western extent of the Corn Belt. The corn grower with a prairie soil at his disposal had an initial advantage in fertility over the grower on soils developed under forest conditions. Deficiencies in fertility, while significant, can be overcome with fertilizers, usually at a relatively low cost, but climatic conditions at variance with the minimum requirements of a crop impose rigid limits on the area in which it can be produced. Thus the crucial limit of the Corn Belt to the west was summer rainfall less than the minimum requirements of the crop, while the northern limit was established where July temperatures averaged less than 72 degrees and a shorter growing season limited the production of corn for grain.

The axis of the corn-producing area in 1860 was still in the Ohio River Valley, but in the next twenty years corn production doubled, with Illinois, Iowa, and Missouri producing half of the crop. Thus by 1880 the Corn Belt, the most distinctive and productive of American crop zones, had taken definite form.

Corn displaced wheat in the corn belt for several reasons. The susceptibility of wheat to rust and attacks by chinch bugs helped to keep yields relatively low, and although the price of wheat was higher than that of corn and production costs per acre were lower, the extremely good yields of corn reduced the per-bushel cost enough below that of wheat to more than offset the price differential. Another major factor was the rise of Chicago as a meat-packing center. When year-round slaughtering became possible and then, by 1870, commonplace in the larger packing plants, and when in 1880 refrigerated railroad cars came into regular use, the advantage of a corn-hog enterprise over wheat was even further increased. Beginning in the 1870's, shipment of corn and meat from the Corn Belt increased greatly, and when pork prices dropped during the decade, many farmers in the East abandoned hog production.

Another consequence of the land changes and economic developments at the time was the rise of specialized dairying areas. Particularly in the areas north of the Corn Belt and in the Northeast too cold for good corn, hay for winter feed and grass for summer pasturage did very well and formed the basis of the dairy industry. The Corn Belt could, and did, produce more milk per acre of farm land than these areas, but its greatest advantage was in production of corn and hogs. The early dairying was often for the production of butter and cheese, although in the Northeast the extension of the rail lines encouraged the production of fluid milk for the newly developing cities. By 1890 dairying was the dominant form of agriculture north of the major corn-producing region.

The Northeast, the Southeast and the southern Appalachian regions had in a sense stabilized by 1890 as far as new agricultural land was concerned. The total of land in farms came to a peak in the years around 1880. The southern states had recovered the drop in farm land shown by the agricultural census of 1870 and showed little change thereafter. For the Corn Belt the era of most rapid expansion of farm land occurred before 1880 although land in farms reached a peak in 1900. The Delta states and Lake states continued to show gains into the twentieth century. The spectacular increase in new farm land during the period came in the Mountain states, the Plains, and the Pacific.

Southern agriculture after the war. Forced to repair the damages of war and at the same time to reshape its entire economy, the South made a remarkable recovery in cotton production. Cotton production in 1864 was less than 10 per cent of prewar production, but by 1870 production had nearly regained the former level. High prices and the desperate condition of the southern economy stimulated this increase. Between 1875 and 1890 production doubled.

Yields gradually rose, partly in response to the use of fertilizer. The Southeast had begun to use fertilizers to offset the high losses in fertility which were the result of the farming system and the greater exposure of soil to washing and leaching. Commercial fertilizers were used to a far greater extent in the South than elsewhere during these decades. By using fertilizers on old lands and pushing cotton into the uplands and onto lands heretofore not considered for cotton, all the cotton states but Alabama and Louisiana were producing more cotton by 1879 than they had produced twenty years earlier.

The result was disastrous for the South, however. More than ever before it was dependent upon other regions for its food, as cotton farming replaced subsistence and general farming. Agricultural credit became an increasing necessity and the system which developed to provide it succeeded only in miring those who became dependent upon it deeper into debt. Cotton production of this period, and later, was partly a product of the dominant cropper form of tenure, a system that grew out of slavery and the general shortage of production capital.

When the railroads came into Texas, grazing and grain production on the black waxy prairie gave way to cotton. Between 1879 and 1899 the output of cotton in Texas trebled as the state became the foremost producer of the fiber, with Mississippi, Georgia, Alabama, South Carolina, Arkansas, Louisiana, and North Carolina following in that order.

Tobacco, unlike other agricultural products, was not as easily routed from the older areas of cultivation by new areas coming into production. In the face of mechanization in other types of agricultural production, tobacco producers were protected from low-cost tobacco from newer areas; for no matter where tobacco was grown, it still required a large amount of hand labor.

Tobacco is also a commodity with a high degree of variation in the qualitative characteristics of the leaf which determine the use made of it. Soil and climatic conditions as well as modifications in growing and curing tend to accentuate the differences. With the accumulated experience of several generations of growers to provide knowledge of the variety of seed and the special cultural and curing methods most effective for the locale, a producing area can enjoy the advantage of being the only place producing the given type of tobacco. Attesting to this great amount of variation are the twenty-eight officially recognized types of tobacco grown. For some, the producing area is relatively broad.

This sort of situation has its perils, too. Consumers' tastes and habits in their use of tobacco have changed from time to time, benefiting producers in one area at the expense of producers in another area.

With the Civil War came a shift of the center of tobacco production to Kentucky, with Virginia falling behind. Kentucky's lead in production was so firmly established that it was never seriously challenged. North Carolina, however, increased its tobacco output twelvefold from 1869 to 1899 and moved from the seventh to the second ranking state in 1899. This came about largely as a demand for the bright flue-cured tobacco, which the two Carolinas produced, developed in Europe, where it was used to manufacture cigarettes.

Adjusting to a National Agricultural Economy, 1900-1939

The first four decades of the twentieth century exposed agriculture to massive economic forces originating for the most part outside of agriculture. The war from 1914 through 1918 brought a boom, short-lived for agriculture, followed by a decade or more of painful and rather slow readjustment. Just when agriculture seemed to have regained its poise, the Great Depression of the 1930's brought farm prices lower than for a generation, wholesale mortgage foreclosures, and other symptoms of acute distress. These in turn led to governmental programs for agriculture. Less dramatic than these sharp and sudden changes in direction was the relatively steady sweep toward urbanization and the consequent growth of the domestic market for agricultural commodities. The export market, while experiencing major shifts, on the whole trended downward.

To these general economic and social changes, agriculture brought special developments of its own. There was, first of all, a great growth in the application of science to agriculture. The agricultural colleges, with their experiment stations and extension services, and the federal Department of Agriculture developed a vast fund of new knowledge. The extension services carried it to the farmer. Farmers in turn increasingly put into practice the best ideas they could get.

This was also a period during which farm mechanization moved rapidly forward, at least by the standards of that time. Irrigation and the development of summer fallowing made crop production feasible on large areas where it formerly had been impossible. The frontier of good cropland was largely gone, but considerable shifts of land from public to private ownership, mostly for grazing but to some extent for farming, continued.

The commercialization of agriculture developed rapidly. In the early 1900's farmers were still producing both "fuel" and the source of power on the farm and raising much of their own food. These activities absorbed a substantial part of their productive efforts. By the close of the 1930's the typical farmer bought much more of what he ate from the grocery store, and sold nearly everything he raised, although some of it was bought by other farmers. Cash sales were an increasing part of total farm output, and cash farm operating expenses an increasing share of total farm costs. The farmer necessarily became increasingly concerned with the market place. Trends and prices at markets, sometimes distant ones, became as much a part of his concern and thinking as the weather and his soil.

For the three major crops—wheat, corn, and cotton—readjustments in the area of production in this forty-year period ranged from large to moderate. By 1919 wheat had moved heavily into the Great Plains. The borders of the present Corn Belt, well defined by 1880, extended a bit more to the west and north by 1939. The extension was made possible by the development of earlier maturing and drought-resistant hybrids. There was a tendency to increase corn acreage in the expanding Corn Belt and to decrease it in areas outside this specialized area. No other shifts of importance in the production of these crops occurred.

The location of cotton production, on the other hand, shifted considerably during these decades. The boll weevil greatly reduced yields; the areas hardest hit were the Southeast and the hilly parts of the South generally. The crop shifted westward, to the delta area along the Mississippi, to the plains areas of Texas and Oklahoma, and increasingly to the irrigated areas of California, Arizona, and New Mexico.

The shifts in cotton acreage in each decade are particularly marked. The decreases in acreage from 1909 to 1919 were heavily concentrated across south central Alabama and adjacent areas of west central Georgia and east central Mississippi, with one other major area in central Texas. The increases in that decade were heavily in the Mississippi delta, southeastern Texas along the Gulf, northeastern Texas and adjacent southeastern Oklahoma, and a few other localized spots. In the 1919-29 decade, a major decrease was in the Piedmont areas of Georgia and South Carolina, with a smaller area in south central Oklahoma. The increases were in several areas, particularly the Mississippi delta, but somewhat further north than the earlier increase, and the High Plains of Texas, as well as in some of the irrigated areas. Arizona and California were the only states during the 1929-39 decade to maintain or increase their acreage of cotton but their output even then was small in comparison to the longer established states. Regional or even state totals fail to reveal the extent of the shift in acreage within localities or on individual farms.

Those years are remembered not so much for shifts in acres and crop zones as for the problems they raised. Declining markets, particularly the foreign market, a tendency toward excessive supplies and low prices—these were omens of a chronic situation that the agricultural economy had never had to face before.

The 1930's saw great shifts in the acreage of other important crops, such as tobacco, rice, flax, vegetables, and potatoes. Tobacco acreage nearly doubled in the first two decades, with a major shift into North Carolina and Kentucky, and substantial increases elsewhere. In the next two decades, there was an even greater concentration within North Carolina, and some decrease elsewhere. Rice is an old crop in the United States, but never a major one in point of acreage. Although once grown in the Southeast, the irrigated areas along the Texas coast, the Arkansas delta country, and the Sacramento Valley of California became the centers of production in a series of shifts that were well under way at the beginning of this century.

The relocation of vegetable production during the first forty years of the 1900's was one of the largest shifts on record, relative to the total area grown. In 1900 vegetable production was primarily a local matter. Home gardens were common, market gardening was practiced around most cities, and consumption was largely limited to local production, fresh in season and canned for the remainder of the year. By 1919 some specialization of vegetable growing in particularly favored spots, often in the West, had begun; by 1939 it was nearly complete. While improved refrigerated transportation had made this development possible, it was largely a matter of discovering methods of handling the crops on a large scale. The advantages of particular soils and climates were sufficient to bring about major specialization, with production shipped to distant markets and often for sharply limited seasons. Timing and other requirements for producing vegetables on a large scale for the fresh market are most exacting. Less exacting, however, is production for canning or freezing, which can be carried on where production conditions are favorable, with little regard to the seasonality of the output.

Potato production has undergone similar changes. Although production in 1899 was more concentrated than that for vegetables, the crop was widely grown. Even greater concentration had taken place by 1939. Specialized areas had developed in the Red River Valley, Idaho, Colorado, California, Washington, Oregon, and at southern locations in addition to older producing areas such as Maine and New Jersey. Toward the end of this period special potato production was often for a particular season, and sometimes for a particular market.

The situation with respect to fruits has been nearly the same as that with vegetables. Apples are a prime example. Apple production had been important in 1900 on a large number of farms along the Ohio River, in western Pennsylvania, central Michigan, southern Illinois, and sections of Arkansas, Missouri, Kansas, Nebraska, and Iowa, New York, Virginia, and New England. While most of these areas still produced apples in 1939, production was highly concentrated in more limited areas and also included important new areas in Washington, Oregon, and California.

Concentrations in peach orchards had changed in like manner. Areas in Colorado, Idaho, Washington and California had come into production on a large scale. Citrus production was highly localized in California, Florida, and Texas. Most other fruits had also become highly localized in special areas.

Several things had happened to bring about these changes. Improved handling and shipping methods were important in concentrating production in those acres best adapted for the crop, even though perhaps at a great distance from the market. The spectacular rise in production and popularity of the citrus fruits hastened by the discovery of vitamin C and the efforts of the producer in advertising citrus fruits as an important source of the vitamin, caused the market for apples to suffer greatly. Apple consumption per capita in 1939 was half of what it had been in 1909 but the per capita consumption of citrus fruits had nearly quadrupled and was as great as apple consumption had been.

The competition of crops other than fruit for the time and skill of the farm operator, and for capital and land, sealed the fate of many an orchard. Control of disease and insect damage was essential if satisfactory yields of the high-quality fruit consumers desired were to be produced. The demands of the type of management necessary to do this were often too much for a general farmer or someone with only a small acreage of fruit.

While these various crops were shifting in location, changes were taking place in their total area and in their yield per acre (Figure 31). The acreage of the three major crops—wheat, corn, cotton—each rose from 1900 until about 1920, and more or less regularly so. Corn acreage had begun to level off sooner; cotton acreage continued to rise to a peak in 1926. Yields per acre showed great variation from year to year, but no pronounced trends. Wheat yields were slightly upward, at least during the early years. Corn and wheat yields each declined sharply in the great droughts of the early 1930's.

A high-water mark in cotton acreage was established in the decade of the twenties while the average annual yield for the nation dropped to the lowest level since 1867. The recovery which followed was slow and yields did not return to their former level until the thirties.

While the low yields in part reflected the expansion of production into drier areas and onto relatively poor land elsewhere, they indicated in the main the seriousness of the boll weevil infestation. The first recovery in yields came as ways were found to reduce the ravages of the pest. The drought of the thirties momentarily obscured the trend to even higher yields which came as fertilizer practices were improved and larger quantities were used. Acreage restrictions imposed by the federal farm program tended to force the less productive land on each farm out of production while greater efforts were made to increase the production on each remaining acre.

Cotton yields were high enough by 1939 that although acreage had been reduced to 74 per cent of the 1929-38 average acreage, production was 87 per cent of the earlier period. The situation was much the same with corn and wheat. Corn acreage, 89 per cent as large as the average acreage for the previous ten years, produced a crop 113 per cent the size of the average for the earlier period. Similarly, although wheat acreage in 1939 was 93 per cent of the previous period, production was 98 per cent of that ten-year average.

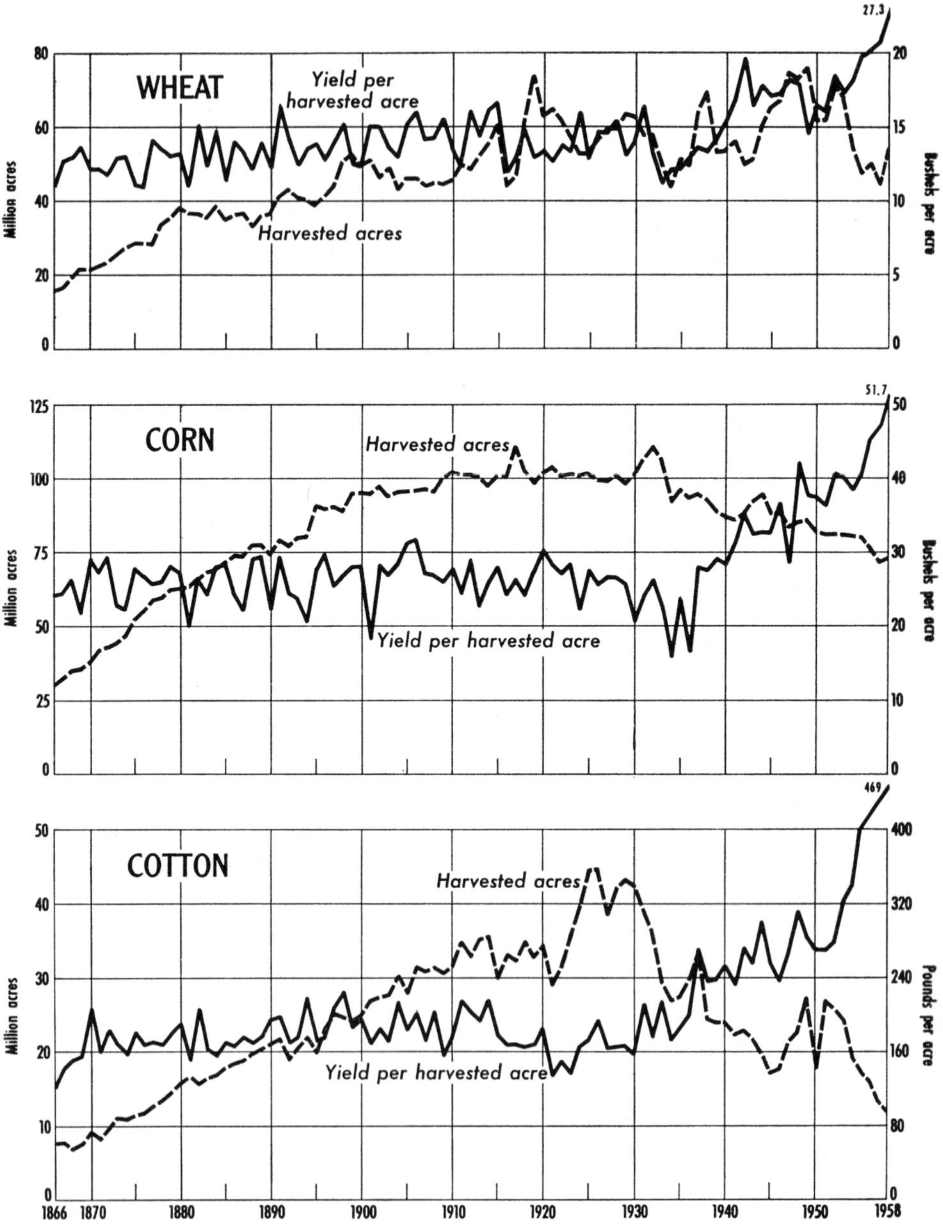

Figure 31. Harvested acreage and yields of wheat, corn, and cotton, 1866–1958.

The situation for cotton was more aggravated than that for wheat or corn because of the size of inventory of raw cotton that had accumulated as of the end of the 1938 marketing year. Enough cotton was on hand to supply the mills for two years without producing another pound. Corn carryover was equal to a quarter of the previous year's requirements while wheat carryover equalled a third of domestic consumption for the previous year.

Two essentially new crops became important during these decades, and an old one assumed new importance (Figure 32). Grain sorghums expanded in area greatly during the first two decades, as cropping expanded onto the Plains. Their ability to withstand drought and to grow again when moisture is available make them especially well adapted to this region. During the latter two decades, their area increased more slowly. Soybeans were scarcely known until the 1920's, but they rapidly became a major crop as markets developed for the oil, and as the oil meal found an important role in livestock rations as a relatively inexpensive protein concentrate. Much of their increase has been in the Corn Belt, where they seem to have replaced both corn and oats. Peanuts, to some extent, took up the acreage that went out of cotton in the southeast.

During these same decades the acreage of hay, especially of improved kinds, such as alfalfa and lespedeza, increased materially also. As the acreage of fruits and vegetables was becoming concentrated in specialized production areas, and sometimes decreasing in the process, total output was generally rising, because of higher yields. The production of nearly all these crops became much more specialized during these decades, with difficulties of control over insects and diseases often making small-scale production as a sideline nearly impossible.

While the first ten to twenty years of the century were years of increase in numbers of farms, in farm labor supply, in total area within farms, in cropland area, and in total output—and all at more or less the same rate—the second half, in contrast, was a period of relative stability in cropland area, numbers of farms, and, to some extent, in total output. The number of workers declined and total farm area increased. The end of expansion in cropland area and farm labor supply marked the beginning of increases in total output through more intensive use of about the same volume of resources. The "sorting out" process of American agriculture, under way from the earliest Colonial days, had nearly matured by the end of this period.

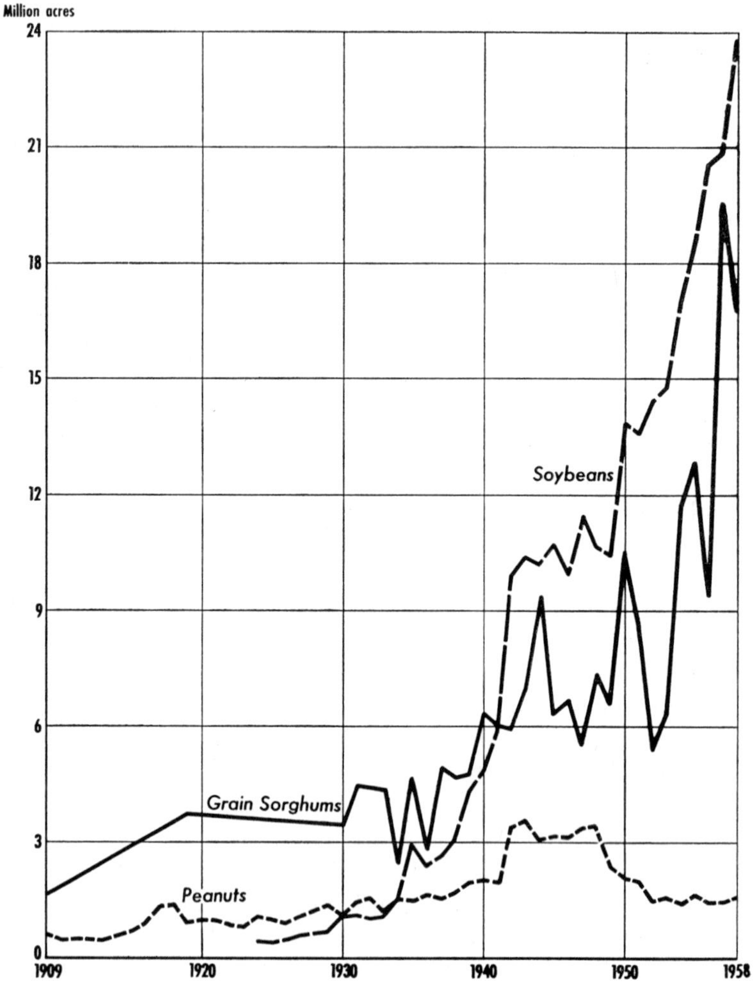

Figure 32. Harvested acreage of grain sorghums, soybeans (for beans) and peanuts, 1909–1958.

PRESENT AND RECENT PATTERNS OF FARM PRODUCTION

With no large areas of new agricultural land economically available, the area of cropland, considered in total for the United States, has remained fairly constant since 1920, although it has varied somewhat between regions. Cropland abandoned or put to other use in the East and South was largely offset by the development of new land in the West, by plowing of grassland, drainage, irrigation, clearing and flood protection. Changes of this sort tended to concentrate the production of particular crops even more within the relatively fixed boundaries of certain types of farming areas.

Production patterns for the period 1939 through 1956 changed little in general outline. It was initially a period of increased use of cropland to meet the extraordinary demands of the Second World War. The ten years ending with 1949 saw a net increase in wheat acreage of 40.8 per cent, occurring largely in the Great Plains and the Pacific Northwest. A notable decrease in wheat acreage occurred in southwestern Minnesota prairies where earlier maturing varieties of hybrid corn were gaining ground, and particularly where soybean acreage expanded rapidly.

Corn acreage decreased from southeastern Texas across the South to the Atlantic Coast. Although the Corn Belt, particularly the western portion, registered large gains in acreage, the total acreage for the nation was 4 per cent less in 1949 than in 1939.

Cotton acreage increased 17 per cent in the same period. This accounts for a small part of the drop in corn acreage in the South. But except for the Mississippi delta area, the heaviest gains in cotton acreage come elsewhere—on the High Plains of Texas, and the southern tip of the state and to a lesser extent in California, Arizona, and New Mexico. There was a scattering of acreage withdrawn from cotton throughout the Southeast.

Sorghum acreage decreased 28 per cent between 1939 and 1949, largely in eastern South Dakota and Nebraska, where it gave way to corn, and on the southern Great Plains, where cotton and wheat replaced sorghums. Soybeans, on the other hand, made a small gain nationally, as acreage increases occurred largely in the Missouri bootheel, southern Illinois and south central Minnesota.

Farm output for the 1940's averaged nearly 21 per cent greater than the 1939 output but the area of cropland increased less than 2 per cent. The wartime food and fiber requirements were satisfied by an increase in the intensity of use of a given area of land rather than by a great expansion of cropland, as had been the case some twenty odd years before. And this of course involved a greater use of capital goods but less use of manpower.

Until the 1930's crop yields on the average had shown no real improvement. This changed with the 1940's, but because the drought years had depressed yields below the usual levels and extremely favorable weather for the years 1942-1944 pushed them up; to be realistic, yield comparisons must take this into account. Thus, Sherman Johnson has estimated that normal weather would have produced an output 120 per cent above the 1935-39 average instead of the actual 128 per cent level.[14] The increase is significant, even discounting the weather.

When the extraordinary demands of the war and postwar years disappeared, the nation had an agricultural plant geared to a higher level of production than it really wanted, an output that continued to expand under the impact of further advances in technology. Agriculture was in itself, unable to do much about it. Some of the increased output came from the release of lands formerly used to produce feed for horses and mules. Some came from new cropland made available for the most part by irrigation. Together these were the equivalent of 38 million crop acres. Presumably a population increase of 16 per cent between 1939 and 1950 would have absorbed this increase in cropland, but the improvements in technology *alone* between 1939 and 1950 were sufficient in themselves to provide the additional population with a diet at the level of the 1935-39 diet.[15]

Dramatic changes have taken place in the yield indices of different crops since 1935-39 (Table 26). Although diets improved at the same time that the population was increasing, both food grain and feed concentrate carryovers reappeared and began to mount rapidly in 1949. Marketing quotas and acreage allotments on tobacco and peanuts were proclaimed and approved by growers as early as 1948. Though lifted on the 1948 peanut crop, they were used on both crops

[14] Sherman E. Johnson, *Changes in American Farming,* Miscellaneous Publication No. 707, U. S. Department of Agriculture, December 1949.

[15] Byron T. Shaw, "Land Resources for Increased Agricultural Output," *Journal of Farm Economics,* Vol. 34, No. 5 (December 1952), p. 675.

in 1949 and 1950, and on cotton in 1950. Acreage allotments for wheat, corn, rice and potatoes were also in effect for 1950. Controls on all crops other than tobacco and peanuts, which had been announced for 1951, were terminated after American armed forces were committed to action in Korea.[16]

Table 26. Index of average yields per acre for designated crops, 1910-1957

(1935-39 = 100)

Crop	1910-14	1925-29	1935-39	1940-44	1945-49	1950-54	1955-57
Corn, all	104.0	105.6	100	128.0	142.2	154.2	176.5
Cotton	88.6	75.7	100	114.6	119.0	131.0	179.0
Wheat	109.0	106.9	100	129.6	128.9	130.8	156.0
Tobacco	92.2	87.1	100	115.8	133.1	145.9	171.5
Soybeans	—	68.1	100	99.0	101.8	109.8	117.1
Oats	100.7	101.0	100	109.0	117.4	116.0	125.5
Hay	92.8	98.5	100	108.9	108.9	115.2	124.1
Potatoes	85.0	97.3	100	116.3	170.9	214.1	240.9
Rice	72.0	86.3	100	91.5	93.7	107.9	140.1

SOURCE: Computed from U. S. Department of Agriculture data.

Except for the increases in output which were ascribed to the additional cropland, the gains were largely nonreversible. The new or improved production methods were here to stay and would be displaced only by those methods which offered additional improvement. Once a farm has been organized around power equipment, for instance, it is extremely unlikely to revert to horse power or hand methods, even in the event of extreme economic depression.

How significant and how permanent are the changes in acreage devoted to various crops which occurred between 1949 and 1957? Price support, acreage control activities, and the existence of diametrically opposed national objectives for the Korean War period and for the years immediately preceding and following it, tend to confound the

[16] Wayne D. Rasmussen and Gladys L. Baker, *A Chronology of the Department of Agriculture's Food Policies and Related Programs, January 1947 to December 1951,* Bureau of Agricultural Economics, U. S. Department of Agriculture (mimeo), 1952, pp. 17, 19, 27, 30, 45, 47-49, 52, 64, 74, 77.

situation and make a satisfactory answer difficult. In wheat, for instance, some of the shift to the production of barley and flax in North Dakota was induced by other factors, such as the improved competitive position of those crops as a consequence of rust damage to wheat. But, to the extent that the rust problem can be overcome, wheat will continue as the dominant crop in the area. Harvested wheat acreage in the Pacific Northwest was down nearly 28 per cent in 1957 from the 1946-55 average although wheat production was reduced less than 3 per cent. The yield response of wheat following the use of fertilizers and sweet clover green manure has been responsible for an increased acreage of that legume to an extent that is revolutionizing previous systems of cropping. But the weather, and acreage allotments on corn, wheat, and cotton—although much less effective in the case of corn than with the other crops—were probably responsible for the greatest changes. The climate of the Plains, producing periods of drought, dust, depression and despair alternating with the periods of optimism that adequate moisture and bumper crops can create, has ever caused a fluctuation in cropland harvested, although the area of available cropland varies less.

Nineteen forty-nine was anything but a normal year in the annals of wheat production. Never before had so large an acreage been planted and harvested, although higher yields for the four previous years and again in 1952 and 1953 gave a greater total output of wheat. The 1957 harvested acreage of wheat, on the other hand, was the smallest acreage of the crop that has been harvested since 1897 with only two exceptions—1934 and 1904.

Wheat acreage seeded in the Lake states and the Corn Belt, from Iowa east, where it had expanded most rapidly from 1944 to 1949, did drop back sharply in 1954, under the impact of acreage allotments and marketing quotas. There was almost an equal reduction in the acreage of wheat seeded on the Plains, but with the severe drought conditions on the central and southern Plains the acreage harvested on the Plains was even more drastically reduced. In the same years nearly normal crops were harvested in Montana and the Dakotas.

The reduction in wheat acreage which has taken place in the Middle and South Atlantic states is largely unrelated to the acreage control program. On most of the farms there, wheat acreage has never been great, and no farmer with 15 acres or less of wheat would be subject to penalty or fine for exceeding his allotment as long as he stayed below this limit.[17] While the wheat acreage lost there is not likely to be recovered, as moisture conditions improve in the drier areas of the Great Plains, some increase in the *planted* acreage is probable.

While the Plains states lost ground in their share of the national wheat crop during recent years, even the areas with lowest yields are likely to continue in wheat production, barring any change in public policy in this respect. That is effectively demonstrated in a study of returns from wheat as against grass on the semi-arid Plains.[18]

Wheat is a more profitable use of Plains land, even with yields as low as 5 to 8 bushels per acre, than is seeded pasture for grazing. All income from the latter may be deferred for as much as five years. For the shift to be profitable, the land must be left in grass for a number of years and stock watering facilities, not included as a cost in the study, would be required on some farms. Wheat yields for the ten county areas studied in eastern Colorado averaged 10.7 bushels per seeded acre, compared with a 1945-54 average yield for the ten Plains states of 12.1 bushels per seeded acre.

Acreage allotments have been particularly effective in reducing cotton acreage, but much less so in reducing cotton output. The 99.4 per cent compliance with allotments of cotton producers in 1954 is

[17] The wheat history used in establishing acreage allotments for the 1957 crop year indicated the following average wheat acreages per farm by regions (processed material prepared by the Commodity Stabilization Service, U. S. Department of Agriculture): Middle Atlantic, 12.4 acres; South Atlantic, 9.5 acres; North Central, 17.5 acres; South Central, 12.6 acres; Plains, 119.8 acres; Intermountain, 53.4 acres; Pacific, 135.0 acres.

[18] Using average costs of production and average gains of steers grazing seeded pasture, both over a twelve-year period and assuming the price of wheat to be $1.70 per bushel, and valuing each pound of beef gained at 20 cents per pound, Sitler found that when wheat yields averaged 9.2 bushels per acre and the seeded pastures produced 21.8 pounds of beef per acre, net returns were the same. With the same assumptions and with stock water available and the conservation reserve payment of the soil bank program included (which precludes grazing for the first five years), it is most profitable to seed all land yielding 8 bushels of wheat or less to grass. Without the government payment, seedings would not be undertaken on land yielding more than 5 bushels of wheat per acre but with wheat priced at only $1.40, and no payments, land yielding 6 bushels or less would have to be seeded and grazed to yield the highest net return from the use of land. Harry G. Sitler, *Economic Possibilities of Seeding Wheatland to Grass in Eastern Colorado,* Agricultural Research Service 43-64, U. S. Department of Agriculture, 1958.

in marked contrast to the 40 per cent compliance of corn producers. While there was only 76 per cent compliance among wheat producers, these operators planted 91 per cent of the wheat crop.

Cotton acreage dropped a third under the impact of acreage allotments in 1950. With their subsequent removal, cotton acreage climbed back to its former level until a similar cut of one-third was made again in 1954. In the meantime, however, some changes had occurred (Table 27). Although the order of the top four states did not change, the different rates at which cotton acreage in the various states was reduced widened the gaps in some instances, as it narrowed them in others.

Table 27. Cotton acreage in 1957 as a per cent of the 1939-48 average acreage in the major producing states, and the rank and change in rank of the states

State	1957 acreage as per cent of 1939-48 acreage	1957 rank	Change in rank
Texas	78	1	0
Mississippi	56	2	0
Arkansas	59	3	0
Alabama	43	4	0
Georgia	37	6	−1
Oklahoma	38	7	−1
South Carolina	45	8	−1
Louisiana	47	10	−2
North Carolina	47	12	−3
Tennessee	71	9	+1
Missouri	77	13	−2
California	181	5	+7
Arizona	173	11	+2
New Mexico	144	14	0

SOURCE: Computed from U. S. Department of Agriculture crop estimates.

In 1909 practically no cotton was grown west of the 101st meridian, and nearly two-fifths of the cotton acreage was in North Carolina, South Carolina, Georgia, and Alabama. As recently as the 1939-

48 period, 54 per cent of the nation's cotton acreage lay east of Texas. In 1954 the four southeastern states previously mentioned accounted for less than a fifth of the cotton acreage. Production had increased so tremendously in the Southwest, that for the period 1946-55, 51 per cent of the acreage of cotton harvested was harvested west of Louisiana. California's spectacular rise during the period from twelfth to fifth ranking cotton state was largely responsible for this.

Sharp reductions in the acreage of upland cotton of all the major producing states except California brought the 1957 cotton acreage in the United States to its lowest ebb since 1878. A crop from an estimated 13.6 million acres was harvested and the reduction of 2.1 million acres below 1956 was achieved through the acreage reserve feature of the soil bank program.

A study of the effects of acreage allotment programs, conducted by the United States Department of Agriculture, indicates that the programs affect differently farms in each of five sample cotton-producing areas—the Mississippi Delta of Mississippi, Arkansas and Louisiana, the Clay Hills area of Mississippi and Tennessee, the Southern Piedmont area of South Carolina and Georgia, the southern High Plains area of Texas and the Upper and Western San Joaquin Valley of California.[19] In each area cotton is the major enterprise on most farms, but farm size, tenure of operators, other crops grown, and production alternatives vary widely.

The Mississippi Delta is "characterized by productive soil, generally adequate rainfall and level topography which make it well suited to cotton production. Mechanization has made rapid progress in the area." In the Clay Hills "soils are less productive and the topography is less favorable for crop production. In general, farms are smaller and less mechanized and yields of cotton are lower. . . . Farms in the Southern Piedmont have undergone drastic changes in the last decade. Industrial expansion has provided employment opportunities for both operators and croppers. Cotton, although still a major crop, is less important. . . . than was the case 20 years ago. Livestock production is increasing and pasture and hay crops are becoming more important." On the High Plains a high proportion of the farms are specialized. "Topography is level, and mechanization of cotton production has increased rapidly." Diversified farming predominates in

[19] *Effects of Acreage-Allotment Programs, 1954-1955,* Agricultural Research Service 43-47, U. S. Department of Agriculture, 1957.

the San Joaquin Valley although "individual farms tend to be highly specialized. Crop production depends upon irrigation."[20]

Changes in the use of land, labor, and capital goods varied. "Farmers in the High Plains and Delta planted other cash crops on most of the acreage diverted from cotton," but in the Southern Piedmont and Clay Hills, where alternatives were more limited, 40 and 24 per cent respectively of the diverted acreage remained idle in 1955. While there was an increase in the proportion of the cotton acreage upon which fertilizer was used, there was an even greater increase in the amount applied per acre. Producers of irrigated cotton on the High Plains, who had been using very little fertilizer, stepped up their use the most, but still used less than half the amount used in the Delta, where an 18 per cent increase occurred. Fertilizer use per acre was increased about 30 per cent in the San Joaquin Valley. There was also a greater use of insecticides in most of the areas.

One of the most significant changes took place in the Delta, Clay Hills and Southern Piedmont areas, where cropper and share tenants and their families figure importantly in the labor force. The study reports that in the three-year period of 1953 through 1955, 17 per cent of these families left farms in the Southern Piedmont, 21 per cent left from the Clay Hills area, and 34 per cent left from the Delta. This movement from the farms was brought about largely because of the reduction in cotton acreage.[21]

The Southeast has lost ground in cotton production in the past, and there is a strong probability of it continuing to do so in the future unless it accomplishes a rather drastic revolution in the organization of the production of the crop. To be sure, some changes are already under way as is evidenced by the out-movement of tenant and cropper families; yet this is only a beginning. The situation in cotton now is such that the lower yielding areas of the Southwest are able to compete effectively with the Southeast. Mechanization has made the difference. But mechanization in the Southeast will make relatively little progress on the hill lands where small farms still prevail. The number of farms with small acreages of cotton is a measure of the lack of mechanization. Harvesting is the major bottleneck. Although all operations from the preparation of the seedbed up to the picking can be mechanized readily, a family with only its own labor can pick

20 *Ibid.*, pp. 33-36.
21 *Ibid.*, pp. 41-43.

no more than 10 to 20 acres of cotton and even with primitive equipment one man with a horse or mule can plant and cultivate more cotton than that. Census of Agriculture figures show that in 1954 for the cotton states as a whole, 82 per cent of the farms producing cotton had less than 25 acres and 49 per cent had less than 10 acres. For a machine-picking operation to be economic, however, at least 100 acres of cotton should be available for harvest.

The years since 1949 have seen a strengthening of the Corn Belt as the center of corn production in contrast to the situation in the Plains states with wheat. While the 1950 corn acreage allotments reduced the acreage of the crop in the North Central region by 8 per cent in contrast to the net reduction for the nation of 4.5 per cent, the region had clearly gained ground as against the rest of the nation by 1953. When acreage allotments were imposed the following year, the area was harvesting nearly 71 per cent of the nation's corn acreage in contrast to the average of 62 per cent from 1939 to 1948 or the 64.5 per cent in 1950.

Except in the cash grain areas, the acreage allotment program has been much less effective for corn than for wheat or cotton. Unlike those other two great cash crops, as much as 70 per cent of the corn yield never enters commercial channels but is fed on the farm where it is produced; and that makes the administration of truly effective production controls extremely difficult. Thus, when controls were reimposed in 1954, corn producers harvested an acreage of corn that was 99 per cent of the previous acreage whereas the program goal called for an acreage only 88 per cent as large as that in 1953. The 1957 harvested acreage, 9 per cent below that of 1953 and 17.4 per cent below the 1939-48 average, was much nearer the desired goal. The reduction, probably only temporary, has been attributed to the 5.2 million acres of corn allotment placed in the acreage reserve of the soil bank program although the net reduction in acreage from 1956 was only 3 million acres.

This further concentration of corn production in the North Central states has been marked by continued northward expansion of corn production. Minnesota's increase is in sharp contrast to the declining acreage in Nebraska, which decline has given Minnesota third rank in acreage. South Dakota now outranks Missouri as a corn state. Wisconsin, Michigan, and North Dakota have likewise shown absolute increases in recent years. The sharpest reduction in cornfields has come in the South Central states. In the North Central states the acreage taken out of corn was more than offset by increases in other field grains and in soybeans. Much the same holds true elsewhere in the nation.

Higher yields came about as other inputs were substituted for land. As an over-all result, the net reduction in field crops harvested between 1953 and 1955 was 7.8 million acres. Yet farm output in 1955 was 112 per cent of the 1947-49 average, a continuing rise from 108 per cent of the output above that same base period in 1953.

Economic Organization of Agriculture Today

Before attempting to estimate the future use of land for agriculture, let us briefly examine its present economic structure, to which the history of the past decades has brought us.

Agriculture in the United States at midcentury uses about 523 million acres of land, or 27 per cent of the total land area. This excludes unimproved grazing and forested lands, which we shall consider in later chapters. Of the total agricultural area, about 400 million acres are cropland, of which 80 to 85 per cent produces harvested crops. Fallow, idle cropland and crop failure, together with some 69 million acres of improved pasture which is cropped in rotation, account for the remaining cropland.

Since World War I, the area used for crops has remained essentially unchanged, varying by not more than 3 per cent at any census period. Feed grains and hay and forage crops are by far the most important farm crops both in terms of acres used and in terms of their farm value (Table 28). Corn acreage exceeds that of any other single crop and its farm value is also the greatest. The feed grains and forage crops appear less favorably as a component of cash receipts because of the extent to which they are marketed through livestock produced on the farm. The value of these crops is largely reflected in the $17.3 billion worth of livestock and livestock products marketed in 1957.

Continuing heavy production of feed grains and wheat in excess of annual requirements for most of this decade has resulted in the availability of grain stocks equal to 43 per cent of the current average crop of feed grain and more than twice annual requirements of wheat for both domestic use and export. Cotton supplies from current production and carryover are now nearly 65 per cent greater than requirements. Increasing yields continue to offset attempts to reduce surpluses by restriction of acreage planted to the crops.

The value of assets used in agriculture is roughly $150 billion. Of this total, more than half but less than two-thirds is for land, and the remainder for all other items. That is an average of $18,000 worth of assets for each agricultural worker, an investment much higher than the average for all workers in the American economy.

Table 28. Acreage, production, and value of crops by groups, recent years

Crop	Acres harvested (millions)			Index of production (1947–49 = 100)		Value in 1957 (millions of dollars)	
	1947-56	1957	1958	1957	1958	Farm	Receipts from marketing
Feed grains	138.5	141.7	136.9	122	134	5,557	2,328
Hay and forage	78.5	79.9	77.0	126	125	2,337	[1]
Food grains	67.7	47.1	57.1	79	117	2,097	1,877
Oil crops	21.3	27.2	29.1	148	181	1,228	1,154
Cotton	21.8	13.6	11.9	77	81	[2]1,837	[2]1,784
Vegetables[3]	7.5	7.0	7.2	97	101	1,774	1,540
Fruits and nuts	3.2	2.8	2.8	108	110	1,333	1,415
Tobacco	1.6	1.1	1.1	83	87	936	967
Sugar crops[4]	1.2	1.2	1.2	124	124	230	[1]
Grass and legume seed crops[5]	.4	.7	.8	[6]	[6]	116	[1]
All other	.6	.5	.4	[6]	[6]	54	1,315
Totals	342.3	322.8	325.5	106	118	17,499	12,380

[1] Included in "All other."
[2] Includes the value of cottonseed.
[3] Includes dry peas and beans, potatoes, and sweet potatoes.
[4] Excludes products of maple.
[5] Acreage shown is that which is not duplicated in hay and forage acreage.
[6] Included in total but not estimated separately.

SOURCE: Compiled from U. S. Department of Agriculture Crop Reporting Board data.

While cropland area has undergone little change, the change in the number of farms in recent years is a different story. Farm numbers, remarkably steady from 1910 through 1930 at about 6½ million, have decreased since 1939 at a phenomenal rate. With the count at 4.8 million in 1957, this means that one out of every four farms in existence in 1939 has disappeared. The remaining farms differ greatly in economic size (Table 29). Twelve per cent of them produce 58 per cent of the economic output of the industry; many small ones produce very little.

Original farm sizes were determined as much by land survey systems and public land disposal methods as by economic requirements. While major farm consolidations occurred in earlier decades, they have lagged greatly behind economic forces pushing toward larger farms. The situation today, although improved, is much the same.

Table 29. Number of farms according to economic class, 1954

Class of farm	Sales per farm	Number of farms (1,000)	Percentage of all farms	Percentage of value of products sold
Commercial farms				
Class I	$25,000 and over	134	2.8	31.3
Class II	$10,000 to $24,999	449	9.4	26.9
Class III	$ 5,000 to $ 9,999	707	14.8	20.5
Class IV	$ 2,500 to $ 4,999	812	17.0	12.1
Subtotal		2,102	44.0	90.8
Class V	$ 1,200 to $ 2,499	763	16.0	5.7
Class VI	$ 250 to $ 1,999[1]	462	9.7	1.4
Subtotal		1,225	25.7	7.1
All commercial		3,327	69.7	97.9
All noncommercial farms		1,455	30.4	2.0
All farms		4,782	100.0	100.0

[1] Operator worked off farm less than 100 days and nonfarm income less than value of products sold.

SOURCE: U. S. Bureau of the Census, *U. S. Census of Agriculture, 1954,* Vol. II, Ch. 2.

Adjustments in farm boundaries and farm sizes are often hard to make. The farmer may lack capital or credit, and there may be no nearby tract of land for sale or lease.

Labor requirements have declined from 23.1 billion man-hours in 1910-14 to about 14.6 billion man-hours at present, or by more than a third. The farm operator himself and his unpaid family help have always been the mainstay of the farm labor force although from a fourth to a fifth of the labor force has been composed of hired workers from 1910 to date. The reduction in the labor force has reduced both types of labor by similar proportions.

The 1954 Census of Agriculture found hired workers being used on only 21 per cent of the commercial farms. Of the 2.7 million hired workers found on all farms at that time, only about a fourth were considered regular workers, employed for 150 days or more. Farm operators and the unpaid members of their family who did farm work numbered 6.9 million. In addition, 0.7 million seasonal workers were also employed.

This breakdown of the farm labor force must be further modified, for of the 4.1 million farm operators in 1954, 26 per cent worked on their farm less than 15 hours during the survey week. Even on the commercial farms, one out of every six operators had spent no more than 14 hours on farm work. Unpaid family labor, hired labor, or the use of the services of custom operators make such operations possible.

On many farms an excess supply of labor still exists. The possibility of working at a full-time job off the farm the year around or seasonally, or of taking less than full-time employment off the farm, often in contracting to do work on other farms with their own, perhaps specialized, equipment, provides a means of translating this time into income. The rise of part-time farming in recent years, and in off-farm work by full-time farmers has been striking. The most reasonable expectation is that this is a transition stage. Another generation is unlikely to try to work on and off the farm at the same time, although it may live on the farm and work in town.

While major changes in labor and capital were occurring in the input mix into agriculture, total output has nearly doubled. Moreover, it has increased as fast as markets were available; if larger markets and more attractive prices had prevailed, output could have risen even faster. It is to these changes in relationship between total inputs and total outputs that the name "revolution" can most appropriately be applied.

The federal Department of Agriculture and the colleges of agriculture in each state with their research arms as well as their on-campus and off-campus education activities, have been important in generating and channeling research findings to farm people. They have not been alone in this effort, however. But more important than the identity of the agents conveying the knowledge and the channels through which it has passed, is the fact that knowledge has been applied in large measure. This has been a basic factor in the great increases in the efficiency of agriculture.

A major difficulty with which agriculture must contend today is the relatively inelastic demand for its products. Per capita consumption has increased for some foods and decreased for others, but the overall quantity of food consumed per person has changed very little. If prices fall, consumption rises but little; and if average incomes rise, food consumption rises only a little; most Americans now have about as much to eat as they want. Estimates of the average elasticity of all food consumption indicate that a decrease in price of 10 per cent will bring about an increase in food consumption of from only 1.5 to 3 per cent, while a 10 per cent increase in available income will increase consumption by only 2 per cent. Thus, farmers as a group are faced with a slowly expanding but, at any given date, rather rigidly fixed total market. In this respect, agriculture is in a situation basically different from much of industry.

On the other hand, the individual farmer is faced with an almost infinitely elastic market. In most cases, his individual output is so small that no conceivable increase in his output can affect the price. Faced with a difficult cost-price situation, many a farmer finds his individual answer in expanding output. Yet the increases which he and many others in similar situations make, may bring about an even worse price-cost relationship. The nature of competition is such that the individual farmer would be acting against his own best interests to reduce his output if alternative crops provide nothing better. But, on the other hand, it may even be possible for him to apply previously unused production techniques which increase his total costs but also increase his output enough so that his unit cost of output is less than or equal to the price.

In agriculture, where differences in the relative efficiency of the bulk of the commercial farms is small and where alternative employment for the operator's own labor, his land and equipment are often no better elsewhere, prices must fall to very low levels before an adjustment in production takes place. Producers will continue to produce, making an effort to minimize their losses, until a large number quit, perhaps more than enough to effect an equilibrium, with the result that prices become exceptionally attractive again, encouraging many to re-enter with the result that no permanent adjustment is attained.[22]

The supply curve for agriculture as a whole is also highly inelastic. A major productive factor on the individual farm is the labor supply of the farmer and his family. Its full productive employment is often a major objective of farm organization, for although unused labor time may bring leisure, it does not produce income. Also, technological developments for the most part require increased output from the individual farm in order to justify their cost. Eventually such increases in output bring both lower prices and lower incomes to producers of the commodity involved, but the farmer who adopts the innovation before it has been accepted widely enough to increase production significantly and to send prices down will benefit handsomely for a time. If a new method promises to reduce costs a farmer cannot reject it; to do so would put him in a comparatively weaker competitive position.

Agricultural technology has been dynamic in the past. There seems

[22] A classic example is the hog cycle. Kenneth Boulding treated the situation in general terms, using it to make a case for government intervention. He also argued that the least "desirable" are not necessarily driven from an industry by low prices. Consequences similar to Gresham's Law may be involved. Producers who eke out a living may be willing to keep going, even with the barest return, because they use little or none of the improved methods and equipment and thus keep their out-of-pocket costs low; whereas those using the costlier methods, though more efficient in the use of all resources concerned, will soon look for other opportunities. It should be noted that it would require the persistence of a number of such producers to displace the higher cost producer because of the differences in output. Such conditions are most likely to occur in a rather severe depression.

A somewhat similar analysis might be made of successes and failures in establishing new farm businesses. The farm operator who enters farming at a time of general deflation and becomes established in years of rising prices is not necessarily a more efficient manager than the operator who enters farming at a peak of the business cycle but is forced out by a series of years in which his costs rise or remain fixed while the prices on the commodities he produces fall. Kenneth Boulding, *Economic Analysis* (New York: Harper and Brothers, 1941), pp. 450-51.

good reason to believe that new techniques will be found and even more rapidly used in the future. With the internal economics of the farm pressing for their rapid use, we may well expect a constantly shifting supply curve toward greater output. While some of this may be achieved with past levels of inputs, in general the new technologies are likely to mean greater inputs into agriculture, at least by the farmers who continue to produce.

A combination of inelastic supply and inelastic demand obviously means widely fluctuating prices, if prices are left free to move in response to the usual supply and demand forces. It has been this situation, plus the chronic tendency to overproduction, which has led agriculture to seek relief outside the competitive market.

Only by acting in concert, an achievement difficult to attain with so many producers, unless the powers of government can be used, can producers reduce surpluses and raise prices. But to accomplish this means that the equivalent of entire farm units consisting of land, labor, and capital goods must be taken out of production instead of land alone. Otherwise, inputs other than land will simply be substituted, with little or no change in the resulting production.

Quoting again from the acreage allotment program study:

> "Allotment programs that control acreage do not always control production. With acreage restricted, producers tend to step up the use of fertilizer and other yield increasing practices. Yields of wheat increased by 15 per cent and yields of cotton by 28 per cent between 1953 and 1955. Yields of rice increased by 16 per cent between 1954 and 1955. Much of this increase in yield probably would have been accomplished without acreage allotments, but the allotment programs undoubtedly accelerated the use of yield-increasing practices. The smaller acreage of cotton was concentrated on the best land and a higher proportion of the wheat acreage was planted on fallow."[23]

Unless these excess resources are shifted to nonagricultural uses, if possible, the surplus problem is simply transferred from one commodity to the next. This is indicated by these further excerpts from the study.

> "Despite important shifts in acreage of individual crops, allotment programs have affected major uses of land very little. The total

23 *Effects of Acreage-Allotment Programs, 1954-1955,* cited above, p. 1.

planted acreage of all field crops decreased by only 1 per cent from 1953 to 1955; relatively little land was shifted from harvested crops to pasture. . . .

"Diversion of land and other production resources from allotment crops to feed grains other than corn resulted in a 10 per cent increase in production of feed grains and a record accumulation of total stocks of feed grains. . . .

"In most of the [cotton] areas studied, farmers were in position to make adjustments, and land diverted from cotton was used to produce other crops, chiefly feed grains, plus soybeans in the Mississippi Delta, and specialty crops in the San Joaquin Valley of California. In the Clay Hills of Mississippi and Tennessee and the southern Piedmont of South Carolina and Georgia, where alternatives were more limited, from one-fourth to two-fifths of the diverted acreage was idle in 1955. . . . In the Piedmont, Clay Hills and Delta areas, many cropper and share-tenant families left the farms as a result of the reduction in cotton acreage. . . .

"In [most of] the wheat areas studied farm production and farm returns . . . were more affected by the weather than by acreage allotments and marketing quotas. . . . Alternative uses for land are more restricted in the drier wheat areas than in most farming areas. Diverted acreage . . . was used mainly for production of feed grains and for summer fallow. Some acreage was used for flaxseed in North Dakota and some for dry peas in the Northwest. These uses generally are less attractive than wheat. . . .

"The effect of reductions in acreages of corn on compliance farms was more than offset by other adjustments, which included: Increases in the acreage of corn on noncompliance farms; increases in the use of fertilizer and other improved practices on both compliance and noncompliance farms; and increases in acreages and production of feed grains other than corn. . . . Only small acreages went into hay or rotation pasture.

"Effective use of the land, labor, and equipment released depended on opportunities for (1) shifting the acreage diverted to other income-producing uses, (2) more intensive use of both allotment and diverted acreages, by applying more fertilizer and other improved practices, (3) rental or purchase of additional land, and (4) employment off the farm."[24]

24 *Ibid.*, pp. 2-4.

FUTURE USE OF AGRICULTURAL LAND

So far we have traced the evolution and the present organization of American agriculture in its use of land. What, in the light of history and of the present trends, seem the most likely uses of American farmland in the future, specifically up to the year 2000?

There are two lessons of significance to guide our estimates of probable future changes:

1. Agricultural output has increased as fast as, and sometimes faster than, the demand for it. The only exceptions to this statement have been during wartime, when temporary shortages of a few commodities have appeared. These shortages were not only temporary, but also brief in duration and limited to a few commodities. They were due to the rapidity of the increase in demand, not to the absolute level of the demand—that is, they were timing phenomena, largely. At all other times in our history, there has been an ample supply of agricultural commodities. For much of the time, there were over-ample supplies, seriously price-depressing volumes of production, or actual surpluses. The latter have been in large measure due to government action to hold agricultural prices to levels deemed just and fair. Government programs to meliorate the effects of over-ample supplies have existed at many times, but on a major scale during the past twenty-five years. This increase in output to fully meet the increases in demand is truly remarkable when one considers the very great and long-sustained increases in population and in income per capita. One must conclude that agricultural output in the United States is a function primarily of agricultural demand, not of natural supply factors primarily or in a limiting sense. If the demand for agricultural commodities had been higher in the past, or were to be higher today —and were to find expression in higher real prices and in assured markets for larger volumes—there is little doubt that agricultural output would have been or would now be higher. There is a large reserve productive capacity in American agriculture, which may be larger today than throughout most of the past. This demand-oriented supply, this large reserve capacity in agriculture, is surely a far cry from the dismal outlook of Malthus!

2. In the agricultural production history and present situation, land is only *one* of the productive factors. The proportions between land and labor and capital, as productive factors, have changed over the

decades. Technology has been a major variable and a basic cause of change. With the rise of the modern agricultural technology, land as such plays a much less limiting role in agricultural output than it did when technology was less advanced and less dynamic. It is possible to devise models of future agricultural output with widely differing amounts of the various productive factors, yet to reach the same total output in each case.[25] This increased interchangeability of land with other productive factors greatly complicates the estimation of future agricultural land use and of land "requirements"; it also greatly reduces the importance of land and of errors in estimating land needs.

Probable Demands for Agricultural Products

If agricultural output and agricultural land use have been determined more by demand for agricultural commodities than by any other single factor, then the obvious place to start in estimating future use of agricultural land is with demand for agricultural commodities.

If total population increases to 240 million in 1980 and to 310 million in 2000, as we have estimated, this alone will increase the demand for agricultural commodities. Total demand is also affected by consumer incomes. We assume that average disposal income per capita will rise from $1,630 in 1955 to $2,525 in 1980 and to $3,660 in 2000, in the same general prices. Even after allowing for the marketing services which will absorb a large part of the increased expenditures for food at retail, increases of 55 per cent and 124 per cent in personal disposable incomes in 1980 and 2000, respectively, could mean increases in demand for agricultural commodities of the order of 11 per cent and 25 per cent, respectively, above 1955, again assuming a constant general price level.[26]

For the most part these increases in demand will represent, not an increase in the weight of the aggregate of food consumed by the average person, but the use of foods of higher cost. This is merely a continuation of a trend that has long been under way. The retail weight equivalents of the food consumed per capita in the United States from 1909 (the earliest year for which reliable estimates are

25 Vernon W. Ruttan, "The Contribution of Technological Progress to Farm Output: 1950-75," *Review of Economics and Statistics,* Vol. 38, No. 1 (February 1956).

26 This assumes an income elasticity of 0.2.

available) to 1957 have worked down from an average of 1,592 pounds per capita for the earliest ten-year period to 1,520 pounds for the ten years ending with 1957. Consumption has equalled or exceeded that of the earlier period only for the years 1941 through 1947, when consumption averaged 1,627 pounds.

Consumption has increased for dairy products other than butter; for eggs; fats, oils, and butter as a group; meat, fish, and poultry; citrus fruits and tomatoes; and coffee, tea, and cocoa. The consumption of dry beans, peas, nuts, and soya products as a group has changed little over the period, while vegetables of the leafy, green and yellow sorts and sugar and sirups were consumed in smaller quantities in 1956 than during the whole period, but in greater quantity than in 1909.

Per capita consumption of potatoes and sweet potatoes and flour and cereal products is about half the level it was in 1910. Consumption of other vegetables and of fruits has decreased to a lesser extent.[27] The increases in the value of food consumed per person in the future will also mean a decline in the per capita consumption of less-valued foods.

Changes in the eating habits of Americans are a reflection of a number of factors, some of which can be expected to bring about additional dietary changes. For one, human energy requirements may no longer be as great as they once were. Diets need provide fewer calories; the activities of the population became less strenuous with machines replacing human muscle, with both the work day and week shortened, with warmer interiors in which to live and work, and with a greater proportion of the population living in the milder areas of the nation. Increasing proportions of youngsters and of people in their later and less active years also makes a difference. A growing awareness of the health hazards involved in a diet which provides calories greatly in excess of normal requirements has been a factor in limiting the caloric intake, but overeating remains a problem for a significant number of people. A further reduction in the caloric level of American diets can be expected.

Estimates of what these trends, together with population increases, could mean in terms of increases in the demand for agricultural commodities appear as indexes in Table 30. The product of increased population and increased per capita demand leads to an estimate of in-

[27] For details in changing food habits for the United States as a whole, see *Consumption of Food in the United States, 1909-52* and the *Supplement for 1957*, Agricultural Handbook No. 62, U. S. Department of Agriculture, 1953.

creased total domestic demand, from 100 in the 1954-56 period to 162 in 1980 and 235 in 2000. However, it by no means follows that agricultural output need step up to this extent. In the 1954-56 period, about 4 per cent of total agricultural production was surplus, in an economic sense, at prices as supported in those years. Part of this surplus went into a net accumulation of storage under government loan or purchase, part was government-subsidized export more or less outside of the commercial channels of trade, and part was diverted into noncommercial channels in this country.

Table 30. Approximate estimate of total demand for agricultural production, 1980 and 2000

Item	Average 1954-56	1980	2000
Total population (millions)	165	240	310
Domestic demand for domestic agricultural products (index)			
per capita	100	111	125
total	100	162	235
Use of domestic agricultural output (index)			
for domestic consumption	89	144	209
for export	7	7	7
net surplus	4	0	0
total	100	151	216

Net exports took about 7 per cent of total agricultural production in the years 1954-56. An attempt to project or forecast the demand for American agricultural products for export is hazardous indeed; it is primarily a political, not an economic, projection; and one which necessarily involves many considerations in the countries to which the exports might go, as well as conditions in the United States. It would appear that most agricultural economists and others in the United States are pessimistic about the long-run future foreign demand for American farm products on an unsubsidized basis.[28] Lacking any bet-

[28] For a somewhat different view see Raymond A. Ioanes, "Projections of Foreign Demand for Selected United States Agricultural Products, 1965 and 1975," *Policy for Commercial Agriculture*, Joint Committee Print, Joint Economic Committee, 85th Congress, 1st Session.

ter basis, we assume therefore that the total volume of exports in 1980 and 2000 will be the same as in 1954-56. This does not preclude some shifts in the make-up of the exports; American wheat, cotton, and tobacco will probably run into increased competition from foreign sources in the decades ahead.

The combination of these factors leads to an increase in total demand for agricultural commodities of about 50 per cent in 1980 and about 116 per cent in 2000, each compared with 1954-56. This projection assumes no surpluses and a constant level of exports, and thus measures only the increase in domestic demand. Some of this increase in demand could be met by increased imports, but we think it unlikely that there will be a significant increase in imports of competing commodities. Most of the "agricultural" imports into the United States now are not strictly competing with domestic production; sugar and wool are perhaps the two major exceptions. Imports of those commodities not produced in the United States in adequate volume for domestic needs are assumed to increase in the future but this will not replace the domestic production of commodities now produced primarily in this country.

Increases in total output of agricultural commodities of these magnitudes are equivalent to annual increases in output of the order of about 1.7 per cent annually.

This matter of probable future domestic demand for agricultural commodities may also be approached by a consideration of consumption trends of different foods or groups of foods. As incomes rise in the future, consumption patterns will change, as they have changed in the past. To the extent that incomes average higher, and perhaps the range between high and low incomes narrows, there will be less *economic* excuse for inadequate diets. But level of income alone does not guarantee an adequate diet (see Appendix D, Table D-10); knowledge of what constitutes an adequate diet and the careful management of the food budget, particularly if it is small, are also required.

The 1955 household food consumption survey of the United States Department of Agriculture indicated that households with incomes under $2,000 after taxes, compared with all households, use per person less livestock products of all kinds with the exception of bacon and salt pork and nearly the same quantity of eggs. Their use of potatoes and sweet potatoes is also, surprisingly, below the average, as is their use of all other vegetables and fruits as well. On the other hand they consume greater quantities of dry legumes, flour and cereal products, fats and oils, and sugar and sweets.[29] Low-income families tend to cut their consumption of meat and milk and depend heavily upon starches.

The 1955 survey, in contrast to a similar survey in 1936, which disclosed "poor" diets in as many as a third of the households, found such diets in only about a tenth of the households surveyed. Calcium deficiencies were most commonly found in all income groups. Vitamin C, or ascorbic acid, as well as riboflavin, thiamine, and Vitamin A were also found lacking, but less frequently. The incidence of such deficiencies among households with higher incomes roughly indicates the level at which the problem ceases to be one of economics and becomes one of education in nutrition (Appendix D, Table D-11). On balance, diets have improved in the past twenty years because families at the lower end of the income distribution have enjoyed higher real income and more people are aware of the requirements for an adequate diet.

The American diet includes considerably more sugar than required for a diet that is palatable though not elaborate and that is not dangerously lacking in any elements necessary to good health. Diets on the average lack those nutrients which an increase in the consumption of milk, of citrus fruits and tomatoes, of the leafy green and yellow vegetables, and a smaller increase in cereals and potatoes would remedy.

As the general level of income rises in the future, the greatest changes in eating habits are most likely to be those made by the lower income groups. A crude estimate of changes that might be expected in the over-all pattern of food consumption if the diets of different income classes were upgraded can be gained from using data derived from the 1955 survey of food consumption (Table 31). Considerable error is introduced by expanding data for one week to estimates for a year's consumption without making various adjustments for seasonal changes, particularly in prices. However, there is probably less danger of error in comparing the relationship of the average consumption pattern for each income group with that for the average of all households and applying such relationships to the apparent yearly consumption rates as indicated in other statistical series.

29 Based upon data from tables in *Dietary Levels of Households in the United States*, Report No. 6, Household Food Consumption Survey 1955, U. S. Department of Agriculture.

Table 31. Average quantity of various foods consumed by different income groups and its relationship to that consumed in all households, spring 1955

(Quantities in pounds per person per week)

Commodity	Under $2,000 Quantity	Under $2,000 Per cent	$4,000-4,999 Quantity	$4,000-4,999 Per cent	$6,000-7,999 Quantity	$6,000-7,999 Per cent	$10,000 and more Quantity	$10,000 and more Per cent	All households Quantity
Dairy products, excluding butter (milk equivalent)	8.50	89	10.07	105	10.09	105	10.81	113	9.57
Meat, poultry and fish	2.92	76	3.93	102	4.35	113	4.61	120	3.84
Bacon and salt pork	.39	126	.27	87	.29	94	.26	84	.31
Eggs	.91	99	.89	97	.92	100	1.00	109	.92
Dry legumes	.29	193	.11	73	.09	60	.05	33	.15
Shelled nuts and nut butter	.05	55	.10	111	.10	111	.11	122	.09
Potatoes and sweet potatoes	1.76	94	1.93	103	1.92	102	1.68	89	1.88
Green leafy and yellow vegetables	1.80	92	1.87	96	1.97	101	2.39	123	1.95
Other vegetables	.95	77	1.24	100	1.36	110	1.41	114	1.24
Tomatoes	.56	71	.80	101	.88	111	1.07	135	.79
Citrus (juice equivalent)	.70	56	1.26	102	1.50	121	2.15	173	1.24
Dried fruit	.05	100	.05	100	.05	100	.06	120	.05
Other fruit	1.67	74	2.36	104	2.78	122	2.95	130	2.27
Flour equivalent	3.86	136	2.58	91	2.46	87	2.21	78	2.84
Fats and oils	.98	110	.86	97	.86	97	.90	101	.89
Sugars and sweets	1.52	110	1.33	96	1.33	96	1.26	91	1.38

SOURCE: Computed from *Food Consumption of Households in the United States,* Report No. 1, Household Food Consumption Survey 1955, U. S. Department of Agriculture, 1956.

If per capita disposable income for 1980 is in the neighborhood of $2,500 and households average 3.2 persons per unit, then the patterns of the $6,000-$7,999 income group may be expected to approximate the "all households" pattern for 1980. This group may be compared with estimates for 1975 made by Rex F. Daly in the United States Department of Agriculture (Table 32). Both Daly's estimate and ours, on this basis, contemplate more meat and poultry, more dairy products,

Table 32. Per capita consumption of major livestock products and food crops for 1955 with projections for 1975 and 1980

(Pounds per person per year)

Commodity	1955[1]	1975[1]	1980[2]
Meat (carcass weight)			
Beef	81.2	85.0	—
Veal	9.4	9.0	—
Lamb and mutton	4.6	4.0	—
Pork	66.0	75.0	—
Total	161.2	173.0	182.1
Poultry and eggs			
Chicken (eviscerated)	20.9	27.0	—
Turkey (eviscerated)	5.0	5.2	—
Total	25.9	32.2	29.3
Eggs (number)	366	403	366
Dairy products			
Total milk (fat solids basis)	700	720	735
Fats and oils: Food (fat content)	45.0	45.5	44
Fruits (farm weight)			
Citrus	88.6	115.0	107.0
Other	110.5	132.0	135.0
Vegetables (farm weight)			
Tomatoes	54.3	65.0	60.2
Leafy green and yellow	80.7	95.0	82.5
Other	72.1	80.0	79.2
Total	207.1	240.0	221.9
Potatoes and sweet potatoes			
Potatoes	101.0	85.0	—
Sweet potatoes	9.0	9.0	—
Total	110.0	94.0	112.1
Dry legumes (clean basis)	8.2	7.0	4.9
Grain products (grain equivalent)			
Wheat	172.0	160.0	150.0
Rye	1.7	1.5	—
Rice	5.3	5.5	—
Corn	47.3	45.0	—
Oats	6.8	6.5	—
Barley	1.8	1.8	—
Total	234.9	220.3	204.0
Sugar, cane and beet	96.3	93.0	92.4

[1] Taken from Table 4 of Rex F. Daly, "The Long-Run Demand for Farm Products," *Agricultural Economics Research,* Vol. 8, No. 3, July 1956.

[2] Based upon consumption pattern of families with $6,000–$7,999 income, as shown in Table 31.

more fruits, and more vegetables per capita in 1975 or 1980 than now, but not to exactly the same extent in each case. Each contemplates a considerable reduction in grain products and legume consumption per capita. Daly estimates a considerable reduction in potato consumption per capita; our method of estimate gives indication of a slight increase.

Using the consumer food survey as a basis for projection has this obvious merit: It is based on almost a photograph of actual consumer behavior. But this approach has limitations, and some of them are not peculiar to the method. Consider, for instance, the assumption of no change in relative prices. Price relationships have been assumed unchanged from 1955—a convenient although not an entirely realistic assumption.[30] Price elasticity of both demand and supply must also be considered, as well as trends in taste. Further, there is always the possibility that either the sample or the behavior pattern of the sample at the time of the survey was not typical of the population it represented. The survey did not obtain a measure of the quantities of food consumed away from home. Nevertheless, it is of interest to note the similarities of these projections and those by Daly, derived by using income elasticities for individual commodities and taking into account certain trends as well as changes in relative prices. Both are built upon basically similar income assumptions.

For the nonfood commodities such as cotton, wool, and tobacco, and for the industrial market for fats and oils, Daly makes projections on a per capita basis for 1975 (Table 33).

The level of prices assumed is most critical in estimating nonfood uses because of the wide range of substitution that is possible for these commodities. Tobacco is the only exception. Synthetics complicate the picture. Although research can be expected which will make the natural fibers more attractive for particular uses, it is sometimes possible to "build" a synthetic fiber which will meet the specifications desired to a degree superior to the natural fibers. The natural fibers must then compete on the basis of cost. Since 1943 the prices of cotton and rayon,

[30] Rex Daly has revised his estimates of per capita consumption for 1975 using two levels of prices. Level I averages approximately 90 per cent of the 1951-55 average for the major crop and livestock items while level II assumes the general level of farm prices to be about 70 per cent of the level of prices in that period. While not identical, these levels approximate the prices paid—prices received relationship of 90 per cent and 70 per cent of parity respectively. The estimates above correspond to a parity ratio of 92 per cent. For the results of Daly's revised estimates, see his contribution to the papers in *Policy for Commercial Agriculture,* cited above, pp. 108-118.

Table 33. Per capita nonfood use of major farm products, selected periods 1925 to 1955 and projections for 1975

(in pounds)

Commodity	1925-29	1947-49	1951-53	1955	Projection (1975)
Nonfood fats and oils					
Soap	n.a.	13.6	8.8	6.7	4.0
Drying oil	n.a.	6.6	6.3	6.3	5.0
Other industrial	n.a.	4.9	6.8	7.1	11.5
Total	n.a.	25.1	21.9	20.1	20.5
Cotton	27.7	29.5	29.3	26.5	32.0
Wool, apparel	2.1	3.1	2.3	1.7	1.8
Tobacco[1]	9.0	12.0	12.8	12.2	15.4

[1] Unstemmed processing weight per person 15 years and over including Armed Forces overseas.

SOURCE: Rex F. Daly, "The Long-Run Demand for Farm Products," *Agricultural Economics Research,* Vol. 8, No. 3 (July 1956), Table 5.

in pounds of usable fiber, have been extremely close, with rayon generally underselling cotton. Daly's per capita cotton consumption figure is based upon a price assumed to be approximately 22 per cent below the present level, assuming no change in the price of rayon.

The per capita consumption rates are a critical item in an estimate of future requirements but are only one dimension of the problem of deriving a demand for agricultural land. Various patterns of consumption might be used. It would be extremely difficult, if not impossible, to reduce the uncertainty to an appreciable degree in choosing between them. Recognizing that there are limitations to all projections of this sort, and except for cotton, we have chosen to use the Daly estimates.[31] We shall also assume that eating habits will not be greatly

[31] Other comparatively recent published estimates, the first three of which predate the Daly estimate, are those of W. W. Cochrane and H. C. Lampe, "The Nature of the Race Between Food Supplies and Demand in the United States, 1951-1975," *Journal of Farm Economics,* Vol. 35, No. 2 (May 1953), pp. 203-222; Harlowe W. Halvorson, "Long Range Domestic Demand Prospects for Food and Fiber," *Journal of Farm Economics,* Vol. 35, No. 5 (December 1953), pp. 754-765; Nathan Koffsky, "The Long-Term Price Outlook and Its Impact on

different in the year 2000 from those of 1980, recognizing however that there will be some changes.

On the basis of these per capita consumption estimates and our own population estimates, the total domestic requirements of agricultural products in terms of 1956 are set forth in Table 34.

To the estimates of domestic utilization in Table 34 must be added an allowance for export and some consideration of the present surplus. There seems no reason to treat these differently here than in Table 30. Accordingly, total output would not need to rise as much as the increase in domestic utilization because we assume that export demand will not rise proportionately and, further, that present net surpluses will disappear.

These index data in Table 34 can be translated into actual quantities for domestic use, as in Table 35. A substantial increase in tonnage of each type of feed grain and forage is called for because of the projected major increases in consumption of animal products of all kinds. These requirements assume an improvement in the feeding efficiency of livestock which may prove to be even greater than the 10 per cent improvement implied here. A larger tonnage of oil crops is also needed. On the other hand, the domestic market will be able to absorb considerably less food grains in the future than were produced in 1956; some of this difference may go into export. However, even if wheat production remained just the same as in 1956, the volume that would have to be exported by 1980 is much larger than any reasonable estimate of available export outlets on an unsubsidized basis. By the year 2000, however, a much larger total population would require about the 1956 output of wheat, even though per capita consumption was less. At the price assumptions used, domestic requirements for cotton by 1980 would exceed 1956 production, and the even higher requirements for 2000 would also exceed this production.

American Agriculture," *Journal of Farm Economics,* Vol. 36, No. 5 (December 1954), pp. 790-98; and John D. Black and James T. Bonnen, *A Balanced United States Agriculture in 1965,* National Planning Association, Special Report No. 42, 1956.

Daly's indices of per capita consumption of farm commodities, as they were presented in his contribution to the papers in *Policy for Commercial Agriculture,* cited above, use 1956 equal to 100. The per capita indices used hereafter are not exactly in accord with the actual figures used in Tables 32 and 33 because two separate price assumptions are made. Daly's 1975 figures for the two price levels have been increased proportionately to the population increase for 1980 and 2000.

Table 34. Indices of estimated domestic utilization of agricultural products, 1980 and 2000, at two price levels

1956 = 100

Commodity	1980		2000	
	I[1]	II[2]	I[1]	II[2]
Meat animals	154	163	200	210
Dairy products	150	154	194	200
Poultry	163	169	211	218
Eggs	142	142	183	183
Food grains, potatoes, sweet potatoes, and dry beans and peas	132	132	170	170
Oil crops				
Food use	153	154	198	200
Industrial use	159	173	205	224
Fruits and vegetables	165	167	212	216
Sugar	142	143	183	185
Feed crops and seed	136	143	177	185
Cotton[3]	106	131	149	212
Tobacco	166	166	214	214

[1] 1956 price relationships.

[2] Price level 20% to 25% below 1956.

[3] Assumes cotton's share of the total fiber market, after all fibers are converted to a cotton equivalent basis, as follows: 1980 (I) 39%, (II) 48%; 2000 (I) 33%, (II) 47%. On the same basis, cotton had 59% of the total market in 1955.

NOTE: Based upon Tables 4, 6, and 7 of Rex F. Daly, "Prospective Domestic Demands for Food and Fiber," in *Policy for Commercial Agriculture,* U. S. Congress Joint Economic Committee, 1957. Estimates for cotton are those of the authors.

Probable Supply of Agricultural Commodities

If supply invariably and exactly expanded to the extent that demand increased, and not more, then obviously there would be no problem. But, in suggesting that the chief determinant of agricultural output in the past had been the demand for agricultural commodities, we certainly did not mean to suggest that such an exact and invariable relationship exists. In the past, there have been differences in the rate at which the two expanded, and the crux of the problem for the future may be the rate of increase in supply. As careful a look at supply possibilities as we can make therefore seems warranted.

Table 35. 1956 domestic production and estimated domestic requirements of agricultural crops, 1980 and 2000, at two price levels

Item	Unit of production	1956[1]	Projections for domestic utilization			
			1980		2000	
			I[2]	II[3]	I[2]	II[3]
Feed grains						
Corn, for grain	Mil. bu.	3,090.0	4,211	4,404	5,477	5,706
Oats	Mil. bu.	1,163.2	1,558	1,635	2,025	2,119
Barley	Mil. bu.	376.9	387	408	501	531
Grain sorghum	Mil. bu.	206.2	234	246	304	318
Forage						
Hay, all	Mil. tons	108.7	145	155	178	185
Corn and sorghum for silage	Mil. tons	63.1	84	90	103	108
Sorghum for forage	Mil. tons, dry weight	4.6	6.1	6.6	7.5	7.8
Oil crops						
Soybeans for beans	Mil. bu.	449.4	524	533	678	691
Peanuts, picked and threshed	Mil. cwt.	16.1	19.3	19.4	26.5	26.6
Flaxseed	Mil. bu.	48.0	60.5	66.0	78.0	85.0
Food grains						
Wheat, all	Mil. bu.	1,004.3	740	740	950	950
Rice (rough)	Mil. cwt.	49.5	18.2	18.2	23.4	23.4
Rye	Mil. bu.	21.2	28.1	29.4	36.5	38.0
Vegetables						
Truck crops	Thousand tons	19,222	31,800	32,100	40,750	41,500
Potatoes and sweet potatoes	Mil. cwt.	260.6	316.3	316.3	400.3	400.3
Dry beans and peas	Mil. cwt.	21.8	18.5	18.5	28.0	28.0
Fruit	Thousand tons	17,914	29,540	29,920	37,970	38,500
Cotton	Mil. bales	13.3	8.9	11.0	12.5	17.9
Tobacco	Mil. lb.	2,179.0	1,124	1,124	1,451	1,451

[1] Total production, some of which was exported or went into surplus.
[2] 1956 price relationships.
[3] Price levels 20% to 25% below 1956.

Agricultural output has increased greatly over the decades, but unevenly as one considers particular commodities, different years, or various regions. The increase, more obvious and spectacular for the past two decades than for earlier periods, has at times been masked by fortuitous events—the ravages of insect infestations, the outbreak of disease, and the vagaries of weather. Weather, however, when more favorable than normal, has also tended to overstate the actual gains.

In contrast to these unplanned variations in output are those consciously planned by farmers in response to more favorable prices, or to changes in production methods resulting in reduced unit costs, or both. Usually, the increase in output has been made possible by a change in the "mix" or proportions of labor, land, and capital goods used in production.

A relatively simple method has been employed to eliminate most of the effect of year-to-year climatic variations.[32] Instead of examining the yields of each major crop in each major region, considering all the factors responsible for variations in yield, and attempting to eliminate yield variations due to weather alone, we have taken the yields of three major crops—cotton, wheat, and corn—each of which contributes much, directly or indirectly, to total farm output, and which between them reflect climatic conditions in different parts of the country. The yield of these crops has been upward, especially in recent years; this reflects new varieties, new methods of production, more fertilizer, and other factors. Some of the variations around these yield trends may be due to year-to-year variations in fertilizer or other factors, but we have assumed that all the variation is due to weather.

Having calculated the percentage deviations from trend yield each year for each of the three crops, these deviations are averaged and total agricultural output corrected to the same number of percentage points, but in the opposite direction. This, of course, makes no direct allowance for deviations in yields of other crops; but some, at least, would be affected similarly by the same factors that affected the yields of these crops. The result is to convert the strongly upward but irregular trend of farm output into a much more regular series. The adverse effect of the severe droughts of 1934 and 1936 is eliminated. (Figure 33). Since the upward trend in yields is accepted, the trend of the corrected series must equal that of the original series; the significant accomplishment of the correction has been a substantial improvement in the regularity in the upward movement. After adjustment, the index of farm output moved upward or downward by two points or less (or stayed constant) in twenty-five of the forty-seven year-to-year changes; in only fifteen years did it change as much as

[32] For the basic data and a discussion of the methods used, see Marion Clawson, *Journal of Farm Economics*, Vol. 41, No. 2 (May 1959).

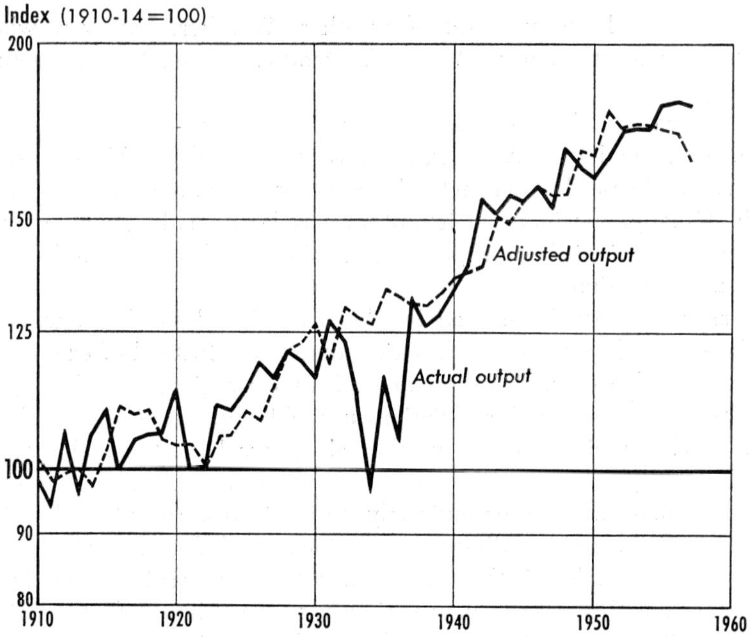

Figure 33. Farm output, unadjusted and adjusted to normal crop yields, 1910–57.

five points from the preceding year. In contrast, the unadjusted output index showed the relatively small movement of two points or less in only thirteen years, and the relatively large movement from year to year of five points or more in twenty-six years. Moreover, the size of the comparatively large changes from year to year in the unadjusted data was far larger than in the adjusted data. The relatively smooth series of the adjusted farm output index seems to suggest that the influence of factors other than annual climatic variations has been relatively steady and constant from year to year.

The fact that farm output, adjusted for normal weather, has increased so regularly at a rate of nearly 1.7 per cent annually in the past is not proof that it will do so in the future, but it is at least suggestive. The rate has continued regularly for many years; perhaps it can continue equally fast over the next forty years. Should this be so, total farm output will rise at least as fast as demand. Our earlier estimate of future demand, on an over-all basis, was also for an annual increase of about 1.7 per cent annually.

The area of agricultural land required in 1980 and in 2000 can be approached by way of the yields of the different crops. Crop yields per acre have increased in the past; this, indeed, is one of the chief ways in which new technology and new mixes of inputs have found expression. Yield increases have been greater in the past twenty years than previously; the massive effects of the major droughts during the 1930's obscured the tendency to higher yields then. It seems highly probable that acre yields will rise further in the future. In 1951 a Department of Agriculture estimate of maximum attainable yields, through full adoption of then known improved practices, was for a substantial increase;[33] and another estimate made during the same year, for maximum yields in 1975, indicated still higher yields.[34] While each of these assumed new crop varieties, neither was based upon major new genetical improvement that might result from the application of radiation to force plant mutations.

In light of these two earlier studies and of past trends in crop yields, we have made assumptions as to average yields in 1980 and in 2000. In general, our estimates for 1980 are for much higher yields than now, roughly in the same order as those of the 1951 studies. Our projections for 2000 are still higher. This is obviously an estimate as to future technological developments, and hence uncertain at the best. However, it might be pointed out that most such projections in the past have underestimated the rate of increase in crop yields.

Taking the estimate of total domestic agricultural commodity requirements, as given in Table 35, and the crop yield estimates of Table 36, it is possible to estimate the crop areas required (Table 37). With the exception of fruit, these acreages are for crops of domestic utilization only. They assume no net accumulations of crop surpluses in their respective years. The areas required in the future, at both levels of demand, are substantially below present crop acreages. However, as will be seen, these figures are not strictly comparable. Much of the apparent surplus of present area over future need is in food grains, particularly wheat. In part, this reflects some reduction due to the elimination of the present surplus. But mostly it means that projected increases in yield are as great as, or greater than, projected increased demand for the livestock products to be produced by

33 *Agriculture's Capacity to Produce,* Agriculture Information Bulletin No. 88, U. S. Department of Agriculture, June 1952, p. 59.

34 President's Materials Policy Commission, *Resources for Freedom,* Vol. 5, *Selected Reports* (Washington: U. S. Government Printing Office, 1952).

Table 36. Average crop yield, 1947-56 and 1958, yield estimates and projected yields for 1980 and 2000

Commodity	Unit	1947-56 average yield	Yield estimates A[1]	Yield estimates B[2]	1958	1980[3]	2000[3]
Feed grains							
Corn, for grain	Bu.	38.8	61.1	78.4	51.7	70	90
Oats	Bu.	34.3	51.7	60.0	44.7	52	60
Barley	Bu.	27.2	32.2	35.0	31.6	38	43
Grain sorghum	Bu.	19.2	22.5	35.0	36.7	44	58
Forage							
Hay, all tame	Ton	1.55	2.38	—	1.82	2.00	2.40
Corn silage	Ton	[4]8.27	—	—	8.85	12	14
Sorghum dry forage	Ton	1.20	—	—	2.00	1.45	1.70
Oil crops							
Soybeans for beans	Bu.	20.3	29.8	32.8	24.2	25.0	33.0
Peanuts, picked and threshed	Lb.	870	1,517	1,400	1,213	1,500	1,800
Flax seed	Bu.	9.0	10.7	—	10.3	10.7	12.0
Food grains							
Wheat, all	Bu.	17.7	22.3	24.4	27.3	24	28
Rice (rough)	Lb.	2,465	2,674	—	3,309	3,800	4,500
Rye	Bu.	12.8	16.5	—	18.2	21	25
Vegetables							
Truck crops	Tons	4.2	—	4.9	5.0	5.3	6.6
Potatoes	Cwt.	148.2	319.3	379.1	180.0	320	385
Sweet potatoes	Cwt.	53.9	—	—	65.0	85	110
Dry beans	Lb.	1,088	1,224	—	1,186	1,500	1,700
Dry peas	Lb.	1,136	1,359	—	1,219	1,590	1,800
Fruit							
Citrus	Ton	9.41	—	13.34	9.77	13	14
Noncitrus	Ton	4.46	—	4.54	5.53	8	10.5
Tobacco	Lb.	1,315	1,619	1,566	1,626	1,700	1,900
Cotton lint	Lb.	317	524	501	469	550	700
Sugar cane	Ton	21.6	30.6	—	24.7	30.6	35
Sugar beets	Ton	15.3	17.9	17.9	17.2	20.0	24.0

[1] Yields adjudged to be the maximum attainable with full adoption of improved practices known in 1951; no year specified. *Agriculture's Capacity to Produce,* Agriculture Information Bulletin No. 88, U. S. Department of Agriculture, June 1952, p. 59.

[2] Implicitly and explicitly indicated as maximum yields for 1975 based upon data available in 1951. See President's Materials Policy Commission, *Resources for Freedom,* Vol. V, *Selected Reports* (Washington: U. S. Government Printing Office, June 1952), pp. 65-69.

[3] Estimates by the authors based upon projections of 1939-57 yield data considered in light of A and B.

[4] 1949-58 average.

Table 37. Land required for major agricultural crops for domestic utilization, 1980 and 2000, assuming two separate price relationships and total acreage devoted to those crops in 1956.[1]

(Acres harvested in millions)

	Actual[2]	Projected for domestic utilization only			
		1980		2000	
Crop[3]	1956	I[4]	II[5]	I[4]	II[5]
Feed grains					
Corn, other than silage	69.1	60.2	62.9	60.7	63.4
Oats	33.7	30.0	31.4	33.8	35.3
Barley	12.9	10.2	10.7	11.7	12.3
Grain sorghum	9.3	5.3	5.6	5.2	5.5
Forage					
Hay, all	73.3	72.5	77.5	74.2	77.1
Corn and sorghum for silage	8.0	7.0	7.5	7.3	7.7
Sorghum for forage	6.3	4.2	4.6	4.4	4.6
Oil crops					
Soybeans	20.6	21.0	21.3	20.5	20.9
Peanuts	1.4	1.3	1.3	1.5	1.5
Flaxseed	5.5	5.7	6.2	6.5	7.0
Food grains					
Wheat	49.8	30.8	30.8	34.0	34.0
Rice	1.6	.5	.5	.5	.5
Rye	1.6	1.3	1.4	1.5	1.6
Vegetables					
Truck crops	3.8	6.0	6.1	6.1	6.3
Potatoes and sweet potatoes	1.7	1.3	1.3	1.3	1.3
Dry beans and peas	1.8	1.2	1.2	1.6	1.6
Fruit[6]	2.6	—	—	—	—
Citrus	—	1.1	1.1	1.3	1.3
Other	—	1.9	1.9	1.9	1.9
Tobacco	1.4	.7	.7	.8	.8
Cotton	15.6	8.1	10.0	8.9	12.8
Total	320.0	270.3	284.0	283.7	297.4

[1] Yields assumed are those estimated in Table 36.

[2] These are total acreages—domestic utilization, surplus, and export. Other columns are domestic utilization only.

[3] The following minor crops, which together totaled 4.4 million acres harvested, have not been included: buckwheat, popcorn, legume and grass seeds, cowpeas, velvetbeans, broomcorn, hops, tung nuts and nuts. Also omitted are the sugar crops of which 1.1 million acres were harvested in 1956.

[4] 1956 price relationships.

[5] Price level 20 to 25 per cent below 1956.

[6] Bearing acreage only.

these feed grains. Increases in the feeding efficiency of livestock are also involved. About the present acreage of forage will be needed.

In order to get a full and accurate comparison of projected crop areas and present acreages, it is necessary to include the area required for export crops. Earlier, we estimated that exports in 1980 and in 2000 would have the same total value or volume as in 1954-56, but not necessarily exactly the same composition. Exports have recently taken about 7 per cent of all agricultural output, but about 12 per cent of the total area of cropland was used to produce these crops. About half of the acreage devoted to export crops was in wheat, and this was the major reason for the discrepancy between value and acreage. Wheat takes a large acreage of cropland, but average yields per acre are low. Even if the value of exports should remain about the same in the future, the acreage required to produce these crops could well be less. Higher yields for wheat and other export crops would mean less acreage, but some reduction in wheat exports is also assumed. Exports in the future might require 30 million acres of cropland, an estimate that may well prove to be too high.

A summary of agricultural land use in 1980 and in 2000 is given in Table 38. The crop acreage required for domestic utilization ranges from about the present area to moderately higher, depending upon the date and the level of demand. The requirements for export crops are less. The total harvested crop area varies from somewhat less than at present to slightly more, depending upon date and demand level. We assume that the area idle, in crop fallow, or on which crops have failed will decline. For one thing, if the area of wheat should be reduced in the dry farming area, this would take some land area out of fallow. One effect of modern agricultural technology is to make crop failure less common. Extension of supplemental irrigation to humid regions may prove a further major factor in this direction. Cropland pastured has also been held slightly above the current level and it is anticipated that this land will be capable of carrying a much greater number of livestock.

On the basis of the data in Table 38, it would appear that, at the most, the needs of the future can be met with about the present area of cropland. This estimate assumes the lowest prices for farm commodities, relative to other prices. Other price and population assumptions indicate the need will be modestly less than at present. The apparent surplus of farm land is greater for 1980 than for 2000 and

Table 38. Summary of agricultural land use in 1954-56, and projections for 1980 and 2000

(Millions of acres)

Item	1954-56 approximate average	1980[1] I[2]	1980[1] II[3]	2000[1] I[2]	2000[1] II[3]
Harvested crop acreage for:					
Domestic utilization and surplus...	285	270	284	284	297
Export....................	48	30	30	30	30
Subtotal...............	333	300	314	314	327
Fallow, idle, crop failure..........	64	50	50	50	50
Cropland pastured..............	66	70	70	70	70
Total..................	463	420	434	434	447

[1] For origin and meaning of these figures, see Table 37.

[2] 1956 price relationships.

[3] Price level 20% to 25% below 1956.

greater at relative prices that are comparable to the 1956 price level than at a substantially lower price level.

What factors are likely to affect the area of land actually available for crops in the years 1980 and 2000? Some cropland will be lost to urban growth if, as we have assumed, urban building takes an additional 13 million acres by 1980, and a further 11 million acres by 2000. Based upon past experience and the present use of land within standard metropolitan areas, roughly half of this growth would come from farm land and still less would come from cropland. Transportation, especially highways, will take some cropland, but only a part of the total increased land for this use we have estimated—3 million acres by 1980, and 2 million acres more by 2000. Some of the poorer cropland might well be converted into recreation areas, perhaps into state parks. We have estimated that the area of intermediate-type recreation areas, alone, might increase as much as 40 million acres by the year 2000; but cropland's loss to this use would presumably be minor. Some cropland might go into forestry; but usually this would only happen to land, low in productivity, that ceased to be profitable to continue in agriculture. Thus, this type of shift in land use is more likely to be a result than a cause of reductions in cropland use.

On the other side of the ledger, cropland is likely to gain some areas in various ways. The Soil Conservation Service has estimated that there are 105 million acres of land suited to cultivation now in woodland, and another 110 million acres in grassland.[35] Tracts now growing trees, particularly those in commercial farming areas of the Corn Belt, individually may be so small as not to be worth the cost of clearing for some years to come; but there and elsewhere, woodland, together with grassland, stands as a reserve upon which cropland may draw in the future. Some land, mostly now in farms, is likely to be drained and added to cropland. Some new irrigated areas are almost sure to come in.

It is impossible to place an accurate quantitative estimate on the balance between these additions and subtractions. It seems that the net change in cropland would not be great—not great, that is, compared with the present area of cropland. The most likely outlook appears to be for an ample, even over-generous, supply of cropland for the period up to 2000. *Surpluses of cropland are much more probable than deficiencies* in this country during the next forty years.

Regional Patterns of Agricultural Change

The transformation of American agricultural production, in geographic terms, presents a spectacle of kaleidescopic change in response to the nation's growth and development. Population increase and movement, the development of the transportation and communication facilities necessary for an effective national market, the variability of soils and climate and the economic impact of these differences as new lands were opened to cultivation, the uneven acceptance of new techniques, the changing tastes of consumers—these and other factors have shaped the regional patterns that now appear.

Plainly, however, the nation's farmers have *not* yet fully adjusted their methods and output, even to today's technology, costs, and markets. As to future adjustments, growth in population can be counted

[35] H. H. Wooten and James R. Anderson, *Major Uses of Land in the United States—Summary for 1954,* Agriculture Information Bulletin No. 168, U. S. Department of Agriculture, January 1957.

upon to increase the domestic market. But the human increase in this country is unlikely to offset the increasing productivity of the agricultural plant. The direct demand for space that more and more people exert will be of considerable importance in some regions. Preemption of agricultural land for other purposes may serve somewhat to dampen farm output. But farmers in other regions may find their competitive position improved thereby.

Something of this can already be seen in the Northeast and on the West Coast. New communities have sprung up almost overnight in the heart of productive farming areas. Levittown, Pennsylvania, mushroomed from fields which had been producing commercial vegetables. Prune orchards and orange groves in California, which once produced for the breakfast table, have been uprooted to provide space for more breakfast tables. In Arizona homes are now built on land which once produced irrigated crops, and water formerly available only for irrigation is directly used in those homes.

Changes such as these have not created shortages of fruit and vegetables, nor are they likely to do so. Pressure from urban developers for land now in citrus groves has become especially evident, because the areas best suited to citrus are some of the most congenial for winter or year-round residence. Should the greater part of such locations in California and Florida, for example, become permanently given over to split-level and ranch houses, and if the enlarging demand for citrus products continues to mount, it is conceivable that the considerable potentialities for citrus production in Central America may be developed.

A crucial difficulty for agriculture in the forty years ahead will be to find alternative employment for its people and land. The human problem centers around people who are now farming. Many of them, because of lack of training or other experience, their age or their health, are almost certain to keep on farming, regardless of income. They simply do not have the means or equipment to change their way of life and to grasp new opportunities. For some of them, a combination of farming and off-farm work will provide enough money to live upon for the rest of their days.

As for younger people who have grown up on farms, or who are part of farm families now, a transfer to other occupations presents less formidable difficulties; and, under the impact of economic forces and pressures, it is already in process. It becomes harder each year for young people to get into farming with prospect of ownership, or even as renters. Farming may once have been a "poor man's business," but that is no longer so. The capital required to obtain enough land and equipment has risen to a point that discourages would-be farmers or keeps them out.

When it comes to changing the systems and patterns of farming and ranching to accord with the realities imposed by larger units of operation and fewer operators, the regions that present the gravest difficulties in adjustment are those that have most depended upon an export market. The Great Plains, long dependent on its output of hard wheats, offers an outstanding example on this score; and climatic conditions at the heart of this vast region impose limited alternatives of production.

However, a continuing rapid expansion of population, all but certain west of the Plains, offers bright prospects for the further growth and expansion of livestock feeding enterprises on the Plains. Farmers there can capture a considerable part of a saving in freight costs. They can eliminate the costs of transshipment of beef on the hoof to midwestern points for feeding, with the meat shipped back west after slaughter. Through developing present potentialities in the production of drought-resistant grain sorghums, as well as barley, the Plains may well develop mixed systems of farming and be less dependent on shipped-in feeds.

Wheat will undoubtedly remain a mainstay in Plains farming. The natural advantages of the region, in this particular, will be accentuated as trends toward farm enlargement and mechanization further reduce production costs. But this, necessarily, will further depopulate the region and add to the number of wheat farms operated by nonresidents. The problem of offsetting such depopulation by developing greater opportunities for nonfarm employment is already evident, and is certain to become more urgent, year by year.[36]

With fewer and larger farms, greater unit yields, and a consequent reduction in unit costs, not all of the Plains land now in wheat can stand up to the competition. To retire large areas from wheat to grass would require a price relationship between wheat and cattle much more favorable to cattle than now exists. Such a shift is not impossible, however; and it would be given impetus by the discovery and devel-

36 George Montgomery, "Adjustment Problems Faced by Commercial Wheat Farmers in the Great Plains," in *Policy for Commercial Agriculture,* cited above, p. 217.

opment of seeding methods that reduce the risks and time required to establish a usable stand of grass.

The Southeast, where the sharpest reductions in cotton acreage have occurred in recent years, has brighter prospects for adjustment than the Plains. Individual incomes can be expected to rise and in part to create an increased demand for livestock products. The process of urbanization would be accompanied by a substantial decrease in farm numbers but an increase in farm size.

That this evolution is already under way may be seen in such signs as the 50 per cent decrease in number of cotton growers since 1930, and the 20 per cent decrease in cotton acreage that occurred between 1943 and 1950, a period in which acreage allotments were not in operation. The smaller cotton acreage is more intensely handled, with yield increases often offsetting the decreased acreage.

Emigration of the farm population to urban centers will facilitate the trend to larger cotton farms made possible by machine harvesting, and made necessary, because of the cost of the machine, if harvesting is to be mechanized. Soybeans and corn are expected to make important gains where cotton yields are low and machine harvesting is not practicable. The increase of livestock production will be an additional stimulant. The shift from cultivated crops to pine trees will undoubtedly be accelerated. Cotton production will tend to become even more concentrated in those areas throughout the South and Southwest where large farm units exist on nearly level land. Supplemental irrigation in the humid area; the use of fertilizer on all or nearly all the crops, in contrast to only 58 per cent of the acreage in 1954; heavier applications of fertilizer per acre; and increased attention to boll weevil control will increase cotton yields greatly. This will spell the end for the grower who may be several notches above the fast disappearing "forty acres and a mule" operation.

The growing emphasis on livestock in contrast to cotton is dramatically illustrated by the $130 million of cash income which Georgia farmers received from broilers in 1957—twice their cash return from cotton. The phenomenal rise of the broiler industry was facilitated by the introduction of contractural arrangements wherein costs, risks, and management skills have been shared by the broiler grower and feed suppliers, hatcheries, or processors. Whether integrated operations of this type are desirable or not, they have overcome problems of credit and have provided management skills and technical assistance without which the enterprise would not have been possible. The development of other livestock enterprises in the area may require a similar contribution of capital and management skills from off-farm sources, at least for farmers whose only experience has been with cash crops.

The shift to livestock enterprises is likely to be especially difficult on tenant-operated farms, in the absence of rental arrangements which recognize the longer term investments required for pasture improvement and perennial crops and livestock housing that the landowner would be called upon to contribute.

The Northeast will become less and less significant as a farming region. Its trend toward greater urbanization will put increasing pressures on the farm owner to sell. Such pressures, first in the form of increases in property taxes, and then in offers to buy at extremely attractive prices, can be resisted only where the farm operation is making as intensive use of the available land as possible. For a dairy operation, this means a great expansion of the dairy herd, and more land as well if it can be obtained, with greater attention to the production of forage than of feed grains. Fortunately, the climate of the region favors pasture and forage crops.

New technology, high costs, and narrow profit margins have tended to increase the optimum size of most farm enterprises. There is little reason to believe that this will change in years ahead. Increased size is usually impossible without a high degree of mechanization which, when accomplished, commits the operator to the enterprise and to the most efficient operation he can give it. With each enterprise increasingly calling for special attention, for special equipment, and for high managerial skill, it becomes next to impossible to keep abreast of everything, and this in itself increases the trend toward specialization.

Specialization has already gone far in other farm enterprises—broiler, egg, and fruit and vegetable production. These enterprises together with nurseries, green houses, and similar intensive and highly specialized agricultural activities, will fight the rearguard action as agriculture in the region retreats in the path of urban advance.

Urban development poses no real threat to the prairie region, although even here cropland will be lost to other uses. So Corn Belt farmers, too, must anticipate changes ahead. Farm sizes here must increase, which again means fewer farms. Investment per farm worker will also increase. There will be somewhat less pressure to specialization, in part because of the degree of specialization already attained in this area, and because there will probably be much less urban pressure for the use of the land in the western Corn Belt. More important as a stimulant to greater efficiency in farm operations is the potential competition from southern farmers if, in their shift from cotton to feed grain and livestock, they find themselves able to exploit advances in technology in the manner of the Corn Belt.

The Corn Belt enjoys climatic and soil advantages which have made it the outstanding area in the production of feed grain and finished livestock. More than this, Corn Belt farmers have benefited from a combination of topography admirably suited to mechanization, and from farm units of a size nearer the optimum for the use of such equipment. They have developed, moreover, a background of experience and skills in a combined grain-livestock enterprise, and are generally willing and able to make the investments necessary to reduce unit costs and maintain their competitive position. Thus far, all these advantages lie with the Corn Belt; and while no major changes in farming are anticipated there for the future, the region's present margin of advantage may be greatly reduced.

The dairy region of the upper Midwest, already under pressure to expand output per farm, will undoubtedly see fewer but larger farms with dairy herds of much greater size than the average farm today and producing much more milk per cow. Intensification and specialization of the dairy enterprise will be somewhat less than in the Northeast, except for areas closest to the large urban concentrations. Fruit and vegetable enterprises should continue to hold their own, perhaps making some gains, especially where the Great Lakes modify the climate.

On the West Coast, in the Southwest, and in irrigated areas in the Mountain states, with water so much the limiting factor, a great deal will depend upon the efficiency by which water sources are used, the extent to which their volume is increased or maintained, and the extent to which the water supply is made available to agriculture. On irrigated and nonirrigated land alike, the Coast and the Mountain states have every reason to expect an increase in livestock feeding enterprises, and an expansion of dairying and poultry enterprises seems equally certain. Such more intensive pursuits are likely to edge into the acreages now given over more expansively to the growth of wheat, rice, sugar beets, and cotton.

Transportation costs, now another factor limiting production alternatives, will be somewhat lessened with the fast growth of population that is anticipated in the Far West. In time, it is not unlikely that the clustered production of special commodities around western urbanized centers and areas will come more and more to resemble the production pattern of specialized agricultural products adjacent to the longest established metropolises and metropolitan areas of the East Coast.

Possibilities for Adjustment

In considering the future use of land for agricultural purposes, careful distinction should be made between what is possible, that which would be economically sound if all adjustments could be made smoothly, and those changes which seem most probable. This nation could, if necessary, obtain sufficient food and fiber for its needs from a much smaller area than at present; the physical possibilities for increased output from land now in crops are so great that a significantly smaller area could be made to produce the needed supplies. Our analysis suggests that a moderately smaller area of land than now in crops will be adequate in 1980 and 2000. If agricultural technology moves forward as we estimate, there will be powerful forces tending toward greatly increased output from the present lands. On the face of it, some reduction in crop acreage appears economic. But the hindrances to ready adjustment between the various productive factors, on the one hand, and the market demand, on the other, are many and powerful, and the lags between need for adjustment and its completion are often long. Accordingly, one cannot be sure that the changes actually made in crop area will be those which economic analysis—on the basis of freely adjustable production factors—would seem to indicate. This is a matter to which we return in Chapter VIII, where the demands for land for different purposes are reconciled.

Two major questions seem pertinent: (1) How dependable are our estimates of the future? and (2) What problems does the nation face if our estimates are wrong?

At the best, the demand and supply estimates are approximations, and rather rough approximations at that. Several factors affect both demand and supply, but one is dominant for each. On the side of demand, the major variable—major both as to its size and as to its unpredictability—is birth rate; other population factors are smaller and more predictable, as are differences in per capita rates of consumption. On the supply side, the major variable is the rate at which new agricultural technology will be developed and adopted for general use. We do not know how either of these variables will change in the future; no one could possibly really know. Both variables have changed in the past in ways or in degrees then unanticipated; we cannot be sure that either or both will not so change in the future.

Altogether too little is known about *why* birth rates have changed or rates of technological change have occurred. We have only a general basis for estimating why or how much they may change in the future. The agricultural demand-supply balance of the future may be represented as a race between the stork and the agricultural scientists; and, at any moment, the outcome is in doubt.

Since we cannot profess a high degree of confidence in the results of this type of analysis, interest naturally shifts to the possibilities of adjustment if it should turn out that we have either grossly overestimated or underestimated supply-demand balances. If we have overestimated the demand for agricultural commodities, underestimated the supply, or both, then the country would be faced with an excess capacity in its agricultural plant. This might take the form of an actual physical surplus; or it might be only a potential one, held in check to some degree by low farm prices or government control programs, or both. The situation would be similar, except in quantitative terms, to that which the country now faces.

The United States has had little success in its attempts to use the powers of government to control agricultural output. Part of the difficulty has been the assumption that agricultural surpluses were temporary and short-term phenomena. As a result, programs have been designed to deal only with temporary situations. In addition, the efforts have centered almost exclusively on land, ignoring the man on the land and the other resources at his command. Programs simply have not been devised and put into operation which would adjust agricultural output to effective demand in a time of rapid and continued technological advance. The formulation and implementation of policies to accomplish this remains a major challenge.

If errors in our estimates of supply and demand are such that we have underestimated the amount of agricultural land that will be required, several adjustments are possible:

1) The deficiency could be made up by imports. Even today we import large quantities of sugar and wool rather than attempting to depend upon domestic production, and the number of such imports could well increase in the future. Moreover, our industrial exports

would surely provide the foreign exchange to pay for needed imports.

2) More cropland could be made available by clearing land of forest or of stones, while additional land could be irrigated and more could be drained if the operations were subsidized or if prices increased sufficiently to make them profitable. The previously mentioned acreages suitable for crops, but now woodland or grassland, constitute an important reserve. Some 35 million acres of this land lie in the southeastern coastal plain. Most of it requires either clearing or draining, or both.[37] Other lands, often with some deficiencies but still usable for crop production, in nearly all parts of the country could be developed were the need great enough. Increasing the area of cropland is not the only nor perhaps the best way to increase agricultural output; but it is a way still open to this country.

3) More intensive farming could be stimulated in many ways. Our publicly supported research programs are certainly a stimulant to more intensive cropland use, and they might be stepped up. We subsidize the use of limestone and we have subsidized the building of fertilizer plants or the use of fertilizer or both, and we could again. If prices of farm products rose relative to costs, this would provide a powerful stimulus toward more intensive farm operation on the present crop acreage.

4) The pattern of food consumption could be adjusted to place more reliance on foods of plant origin and less on foods of animal origin, and thus make a major reduction in the area of cropland needed.

In all probability, some adjustment would take place in each of these directions, rather than in any one of them alone. In total, we have ample capacity for adjustment to any conceivable increase in demand for agricultural commodities up to the year 2000.

[37] James R. Anderson, *Land Use and Development, Southeastern Coastal Plain*, Agriculture Information Bulletin No. 154, U. S. Department of Agriculture, May 1956.

V

Forestry as a land use

In the United States 484 million acres, or 25 per cent of the total land area, is used for commercial forestry. Additional millions of acres have some forest cover but are used primarily for other purposes. The use that forestry makes of land and the relation of this land use to other land uses is the concern of this chapter. The chief focus therefore is on forest growth and the production of forest materials rather than on the processing and marketing of forest products or on the other values of the forests—for recreation, grazing, wildlife, and as watersheds.

First to be considered are some of the special characteristics of forestry as an economic activity. Following this we review briefly the history of forest land use in the United States, after which the present organization of forestry—area of land, volume of tree growing stock, annual output, condition of the forests, their ownership and other relevant factors are examined. Lastly, in the section on future forest land use, our projections of the probable acreage required, the demand for forest products, and the intensity of forest operations show the kind of problems that will require solution in the course of the next twenty to forty years.

SOME SPECIAL CHARACTERISTICS OF FORESTRY

Wood is a renewable resource comparable to wheat or corn, but its primary competition is with nonrenewable resources. As a building material, for instance, particularly in dwelling construction, lumber competes with brick, stone, gravel and sand, cement, and various other minerals. And in packaging, where wood fiber is extensively used both as lumber, in boxes and barrels, and now primarily as paper, it competes in the latter case with various synthetic wrappings. Until 1850 wood was the mainstay of home heating, but it has long since been displaced almost entirely by coal, oil, and gas. To some degree in many instances, then, the renewable resource of wood fiber has lost out to nonrenewable resources. If the competing mineral resources were in short supply, which they are not, it would be possible to draw more heavily on wood fibers than we now do; but it is clear that wood can never regain the important place it held 150 years and more ago. The forest resources necessary for such a reversal do not exist.

Ideally, forestry should be considered as a land use according to the degree to which it is a conscious growing of trees for specific purposes. The extensive forest cover the white man found in America did not constitute forestry. Neither did his abuse of the trees. Intensive forestry requires conscious decisions about what kinds of trees to grow, with what practices, when to harvest them and how, coupled with purposeful action. In this sense, forestry in the United States is comparatively recent—hardly more than fifty years old anywhere, and widespread only for the past twenty years or less. However, there are different degrees of forestry. Extensive forms of forestry have been followed for many decades. Forest harvest has proceeded on a commercial basis for the past century and a half, even since early Colonial times. Without care, forests will grow and produce some wood fiber, even though fire protection will permit greater growth and various cultural practices more and better growth. Whatever grows can be harvested some day, if its value warrants. Thus, commercial forestry practiced at all levels of intensity is covered in this chapter.

To the forest owner, the growing tree is both capital and product. If allowed to grow, it is the capital to which the sun, soil, and water are adding volume and sometimes value per unit of volume; when harvested, it is the result of several years past output. In this respect, a tree is like a growing calf. Once a tree has reached a merchantable size, the owner has the choice of cutting it or of allowing it to grow for another year or longer. One of the major costs, if he chooses the alternative of growing, is that of the income he could secure from use of the capital value of the tree in some alternative investment. He could cut the tree and invest the revenue from it in government bonds or in the stock market, or elsewhere.

This capital-product duality of trees has an interesting and highly significant economic consequence: a rise in the price of the product increases many of the costs of production proportionately. When stumpage values rise several fold, as they have in recent years, one might expect this to be an incentive to greater forest product output through more intensive forestry; for almost all other products a great rise in price serves as incentive for their increased output. But the rise in stumpage price has increased the capital value of the growing trees proportionately, and thus some of the costs of further growth have risen to the same degree. To the extent that other costs, such as taxes, labor and management, may not rise or do not rise proportionately, a higher price for stumpage is an incentive for more intensive forestry and greater output from the same area. But a rise in stumpage prices does not have nearly the same effect upon forestry output that it does upon, say, agricultural output.[1] It may increase the pressure for an increasing rate of harvesting.

When prices of agricultural commodities rise, some of the greater profitability from their production is capitalized into higher land values. The farmer receives his greater total income largely in the form of higher rent on land or as interest on the higher value of land. When stumpage prices rise, this same process works to increase the value of forest land. The greater income from land in either case is realizable only by continued product of the particular crop. Land being more valuable, there is an incentive to use it more intensively. But the increase in value of standing stumpage is realizable without further intensification of output, by harvesting the marketable trees. The rise in stumpage prices provides much less incentive for preserving the trees for further growth than it does for keeping the forest land actively in forest production.

As a result of this capital-product relationship, the interest rate has more relevance for forestry than for any other major land use. Planting trees for harvest twenty years or longer in the future obviously

[1] This relationship is described in some detail in a book by William A. Duerr, *Fundamentals of Forestry Economics* (New York: McGraw-Hill Book Co., 1960).

involves considerations of probable returns at the end of that time and probable costs during this period; one of the major items of the latter is interest upon the original capital and the accumulated tax and holding costs up to harvest time. But, in the more common case of a growing forest which is at least partially harvestable today but which will have much greater volume at some future date if allowed to grow, the rate of interest on the capital realizable from liquidation of the forest is likely to be the greatest single item of cost in holding for further growth. A reduction in the interest rate will considerably extend the time at which the most profitable harvest should be made; its effect is direct and unmistakable, compared with the slight and somewhat uncertain effect of increased stumpage prices.

Forestry still relies primarily upon "wild" trees. That is, most forestry uses the trees that grow naturally on the site, either for cutting or for reproduction from their seed. Where there is more than one species the harvesting practices can influence the species and even the quality of the individual trees which grow again after harvest. In some areas and for some species, artificial planting is gaining favor. Some forest tree species have been imported, and by selection or crossbreeding forest genetics research is developing new strains. But while selection of genetically superior stock is rapidly developing, the choice is still primarily from among various strains or species of "wild" trees. In man's history he has gradually developed farm crops very different from their wild ancestors—different in ways which immensely improve their usefulness when cultivated, but sometimes destroy them if they are left in the wild state. Nothing comparable to this can be said of tree development. In part, this is because the growing cycle for trees is so long, but in larger part it is because the supply of wild trees has been ample to meet our needs.

Even at its best, forestry is practical applied ecology; in this it resembles grazing. Man does not usually grossly alter the environment for trees as he does for farm crops, by plowing, seeding, cultivating, and annual harvesting. Instead he tries to understand and take advantage of natural processes, not to thwart or pervert their operation. The kinds of trees, their individual size and form, their rate of growth, and many other factors of the forest can be altered, and the alteration may be in the direction of either less or more output. Careless cutting and indiscriminate use of fire have greatly reduced the output of many forest areas; but careful and intensive forestry may result in much higher values than natural seeding, thinning, and decay.

FORESTS IN AMERICAN HISTORY

Some idea of the history of forest use and exploitation in the United States is necessary to an understanding of today's use of the forest lands and their expected use in the future. Nearly all of the United States east of the Mississippi, additional areas in what are now Minnesota, Missouri, Arkansas, Louisiana, and east Texas, and much of the mountainous area farther west, were originally forested. The first explorers and later the settlers, were enormously impressed by the denseness of the growth and the variety and size of the trees. "When explorers landed, America was trees."[2]

The North American continent is endowed with a far greater and richer variety of forest species and types than any other part of the temperate world. In contrast to Europe, where less than a dozen conifers and hardwoods are the only native timber species of commercial value, the continental United States has over a hundred. And the original forests of the western states are entirely different from those in the East; they are larger in size and contain few broad-leaved trees. Except at the northern and southern edges, and very occasionally elsewhere, hardwoods are dominant everywhere in the eastern forests.

Within the broad definition of a forest may be found a wide variety of tree and other plant species which change with variations in soil and climatic conditions. The lower limit of tree growth in the eastern United States follows closely along the 25-inch annual precipitation line, except in river bottoms where soil moisture conditions may be exceptionally favorable. In the West, mountainous areas which catch this amount of precipitation or more, and Pacific coastal areas such as the Douglas fir and redwood forests, where winter rainfall and summer fog are adequate to hold down transpiration rates, are also highly favorable for tree growth.

The United States has been divided into six major forest regions: the Northern, the Central Hardwood, the Southern, and the Tropical in the eastern United States; and the Rocky Mountain and Pacific Coast in the West (Figure 34). In each of these forest regions common characteristics and tree associations are found which blend away at the edges into the adjoining forest regions. The Southern region extends up along the Atlantic Coast, while the Northern forest runs

[2] Richard G. Lillard, *The Great Forest* (New York: Alfred A. Knopf, 1948).

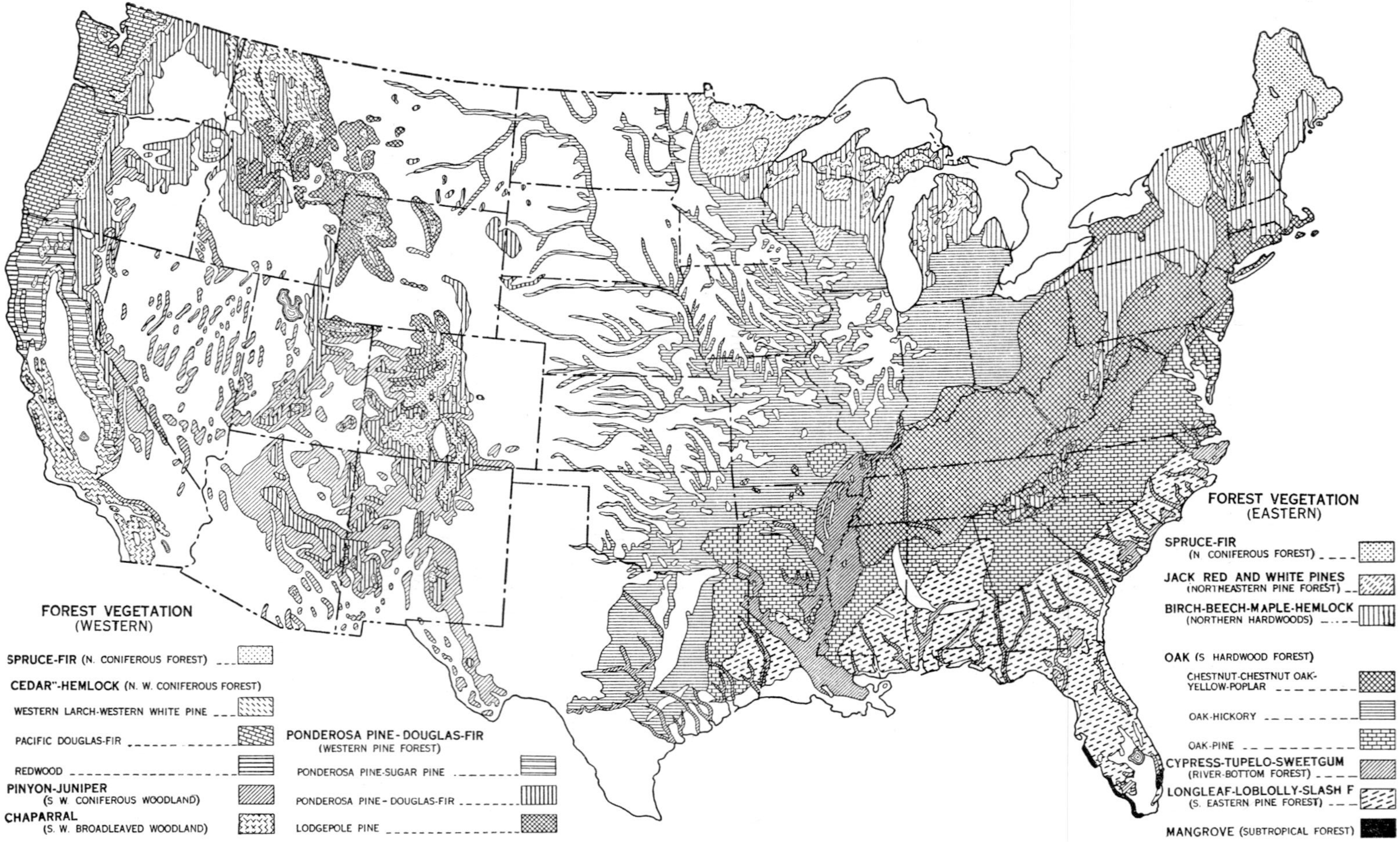

Figure 34. Forest vegetation of the United States. (Map from U. S. Forest Service.)

south along the upper slopes of the Appalachian Mountains. Whereas the combination of tree species found in each forest region is largely determined by annual precipitation, relative humidity, and temperature, the tree associations *within* each region, which are called *forest types,* are influenced by soil, soil moisture conditions, and slope. Pine forest types are usually found on the lighter and sandier soils; broad-leaved hardwoods on heavier soils. Other northern conifers, such as white and red spruce, balsam fir, and hemlock, prefer moist cool locations and frequently are found only on northern slopes in the southern parts of their regions. While many species, such as oaks, pines, aspen, and red maple, do very well on poor soils, tulip poplar, black walnut, basswood, and yellow birch make their best growth on rich, moist, but well-drained soils. Many species are found in swamps and bottomlands which, if well drained, grow satisfactory timber. These include black spruce, tamarack, white cedar, and black ash in the North, and bald cypress, black gum, swamp white oak, pond pine, Southern white cedar, and others in the South.

While timber trees may be found on a variety of poor-quality soils (measured on the basis of value for agriculture), they do have minimum requirements of fertility and moisture to yield satisfactory crops of wood products. Poorly drained swamps produce only stagnant tree growth of little or no value. Gravelly, droughty, south-facing slopes, as in the Ozarks of Missouri, are such poor sites that only short-boled timber ever develops.

The virgin forests as the white man found them were mostly mature. Fire, insects, and diseases had taken their toll, but man had not cut them. As a result, most of the trees were as large as their variety and site qualities would permit. When trees grew over-mature, in time they fell and decayed. From time to time natural or man-caused fire killed or destroyed trees, and hurricanes blew them over. The forests were thus composed of standing trees of varying ages, largely mature, with fallen trees and vines and undergrowth common. The size of the trees and their quality for ship timbers or lumber were as impressive to the pioneers as the vast extent of the forests.

It has been estimated that the virgin forests contained 7,500 billion board feet of timber, compared with about one-fourth of that volume now. But the economic value of this forest was low. In part, its very volume made it nearly valueless; when any commodity is as superabundant as good timber was in those early days, its value must be low.

Were these virgin forests, then, "capital" in the economic sense? Forest products had never been income up to the time the white man came; hence there had been no saving of current income in order to form capital. The forest capital, small as it was, thus represented no saving in the usual sense of the term. Certainly, the forests white men found did not represent forestry, in the sense of a conscious decision to establish or maintain them. The primeval forests were closer in character to the gold deposits later found in the West than to the traditional forests established and managed by private or public landowners in European countries.

The American forests acquired value only because and as the total economy developed. As total population grew, but especially as urban population increased, commercial forest products acquired value. At first, the value was only in their harvest, and forested land was not sought for ownership unless this was necessary for harvest of the virgin growth. Only very much later did forest land and timber come to have value for continued production.

This initial stock of forest products, immense in volume, high in quality, and widely dispersed over the continent, dominated the forest economy for many decades. It is difficult to single out a date when the virgin forest ceased to dominate the forestry scene, and besides, the dates differ in different regions; but on a national scale one may say that the inherited forests were dominant almost to the outbreak of World War II. They are still important though reduced in area; they are the source of a major part of the lumber cut today, and particularly of certain qualities of lumber and of trees for plywood.

Early Uses of the Forest[3]

Early colonists found the forests both a curse and a blessing. A source of raw material for home building, fuel, and food, the forests also impeded cultivation and had to be cleared with primitive tools and backbreaking labor (occasionally it took a month to clear one acre for the plow). For nearly two hundred years white settlement

[3] Jenks Cameron, *Development of Governmental Forest Control in the United States* (Baltimore: Johns Hopkins Press, 1928); John Ise, *The United States Forest Policy* (New Haven: Yale University Press, 1920); S. T. Dana, *Forest and Range Policy* (New York: McGraw-Hill Book Co., 1956).

was limited to the narrow strip of coastal lands hemmed in by the formidable Appalachian Mountain range immediately to the west. The forested mountains harbored the enemies of the English colonists—the Indians and the French; and so the westward migration was stopped for nearly six generations.

Timber from the forests, especially boards and ship timbers, was an immediately exportable product which could be sold to pay for much-needed imports. But even before the American Revolution, with the clearing of agricultural lands, the timber trade was forced to go back farther into the hinterlands. As the forest receded it became more difficult to move logs and timber except by large rivers suitable for log driving. White pine, New England's famed exportable product, became so scarce that the King in 1711 issued a decree restraining the colonists from cutting tall, straight white pines needed for shipmasts. This "Broad Arrow" policy (reserved trees were blazed with an arrow mark) met with immediate resistance from settlers desiring this timber for local and export trade. It was evidence of local timber scarcity and was one cause of Colonial opposition to central authority which eventually led to the Revolution. Much of the early free and easy attitude toward public resources continues—one sees it to this day in attitudes toward water, fish, and wildlife—although the development of the conservation movement early in the present century with evolving legal restraints has brought about changes.

Many other forest products and by-products formed integral parts of the Colonial economy. Fuelwood was the only source of heat energy. Lye made from wood ashes, syrup from maple trees, pine resins and tars for ship caulking could be converted into cash from a resource which was free for the taking. Waterfowl, deer, fish of all variety, passenger pigeons, and small upland game supplied important parts of rural diets and were often sold in quantity in city markets. Fur-bearing animals yielded an important export item for New England and French Canada. Early clearings in the forest surrounded by split rail fences overgrown with vines and shrubs at first encouraged a greater abundance of game than existed in the unbroken forest. But gradually, as cultivation became more intensive and hunting pressure increased, wildlife diminished in the more populous areas. But a man could always go to the frontier and find timber to build a cabin, locate land to clear for a farm, trap furs for cash needs, and shoot game enough to eat.

Forest Utilization Moves Across the Nation

From the earliest settlements shortly after 1600, until roughly 1800, forest utilization was similar throughout the country. Land was cleared for farming and the forest cover destroyed. Houses were heated primarily by wood burned in fireplaces—surely inefficient converters of fuel. Fuelwood shortages around Philadelphia prompted Benjamin Franklin to invent the more efficient Franklin stove. Lumber or logs were used for building materials, almost to the exclusion of other materials. But there was no large urban population—even in 1800 only 6 per cent of the people lived in towns of 2,500 or more. Hence the use of timber was primarily local. The export of shipmasts and ship timbers was economically important, but in volume it was merely a chip from the immensities of the forests. In a number of ways this situation began to change after about 1800. Slowly at first, and then with increasing tempo, the forests began to be cut in larger and larger volume, and the centers of commercial forest operation gradually moved west across the country.

Although a significant area of the original forest along the Atlantic Seaboard had been removed to make way for agriculture prior to the American Revolution, it was still but a small fraction of the total. The opening up of the original Northwest Territory allowed new settlers to fill the mountain valleys, to move into the unsettled portions of the South, and to clear their way through the richer soils of the states north of the Ohio. While much of this land was forested, and hence had to be cleared, the wood products were used locally, or simply burnt. Commercial lumbering was still concentrated in the Northeast, particularly New England. However, logging soon began in the Adirondacks and northern Pennsylvania, where tributaries of the Hudson and Susquehanna served as transportation arteries, and it wasn't until the 1840's and 1850's that the big onslaught on the Michigan pineries began.

Until 1860 there were only vague stirrings of concern over the local depletion of timber supplies. Live oak of the South was a prime ship timber and so essential to the United States Navy that the Congress in 1817 reserved the Santa Rosa peninsula in Florida for exclusive naval use. As cities grew, the local supply of fuelwood became a limiting factor; coal was soon discovered as an economical substitute. Other evidences of interest in the future of timber supplies came from the state legislatures of Missouri, Georgia, and Massachusetts, but it was not until the post-Civil War period that national attention began to focus on forest perpetuation.

Nearly all the farm settlement up to about 1860 was in or very close to forested areas; the urban population of the country was still less than 20 per cent of the total, and much of that in comparatively small cities. Except for the export trade, the demand for lumber was relatively small and local. But after the Civil War this changed. Settlement pushed on to the prairie areas of Illinois and Iowa, and into northern Missouri; then on to eastern Nebraska and Kansas; and still later to the Great Plains. Concurrently, the cities grew at a rapid pace. All of this took lumber, for lumber was yet the almost universal building material. The railroads were pushing over the entire continent, making it possible to ship lumber from the forested areas to places where it was needed.

The first big westward push was into the Lake states, where the chief species was white pine. The heyday of lumbering in this region was from 1870 to 1890. Vast forest acreages were cut and, partly through the widespread fires that followed, the ecological conditions required by white pine were largely destroyed. People believed that all this land, like the land to the south, was potential farm land; hence, that clearing the forest and preventing its regrowth was economically and socially sound. Time was to prove this to be utterly false, but meanwhile many people, especially immigrants, sacrificed their labor and capital in fruitless attempts at farming; and the regeneration of the forest was seriously retarded.

As the northern pineries became depleted, attention shifted to the South, and extensive liquidation of its pine forests began roughly about 1890 and continued until about 1925. This was a somewhat different type of lumber, but its uses were much the same. It was during this period that both total and per capita lumber production hit its peak.

By 1900 the southern pine region was the chief source of lumber for the national market, and it continued so for several years. On the Pacific Coast, some logging for local needs had started early, and some export market had developed, especially for those forests where streams could float the logs and where proximity to tide water made loading onto ships simple. Really large-scale Douglas fir logging for shipment of lumber to eastern markets was largely a post-World War

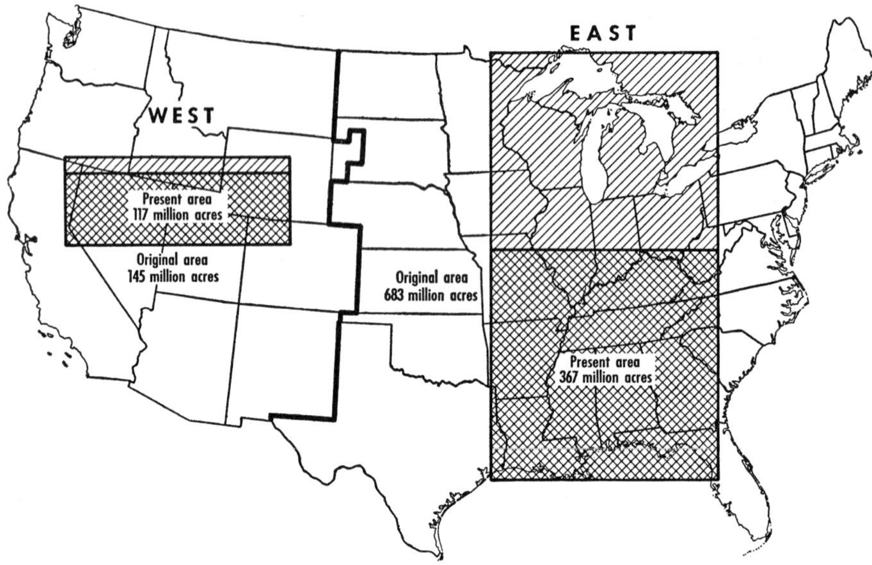

Figure 35. Original and present commercial forest area.

I phenomenon. Extensive logging had started in the Puget Sound region of Washington, and gradually spread southward. Intensive and large-scale logging reached into southern Oregon only during World War II. Major commercial logging had developed earlier in the Sierra Nevada and northern Rocky Mountain areas, and continues there today. Practically all the remaining virgin timber grows in these same general regions, and it is being cut at a relatively rapid rate.

The effect of forest exploitation as it has occurred across the nation has been to reduce the area of commercial forest land from its original more than 800 million acres to today's 484 million acres (Figure 35). The reduction has been greatest, both in terms of total area and as a percentage of the original area, in the eastern half of the country, where the forests were cut not mainly for commercial lumbering, but to clear the land for farming. In both East and West, changes in area of forest land do not tell the whole story; there have been important changes in the forest cover itself, in the volume of growing stock, and hence in the output.

Economic Changes in Forestry

Through the years, both total output of wood fibers and output per capita have changed in volume. The chief manufactured forest product in the United States has always been lumber, at least in terms of volume of wood fiber involved; though rising in importance, other products are still much smaller in volume than lumber. The consumption of lumber, and of other industrial wood users, such as railroad ties, mine timbers, and piling, has been influenced by forces operating in opposite directions: (1) growth in population and growth in total economic output, have increased the demand for these products; and (2) exhaustion of local supplies, reliance on more distant sources of

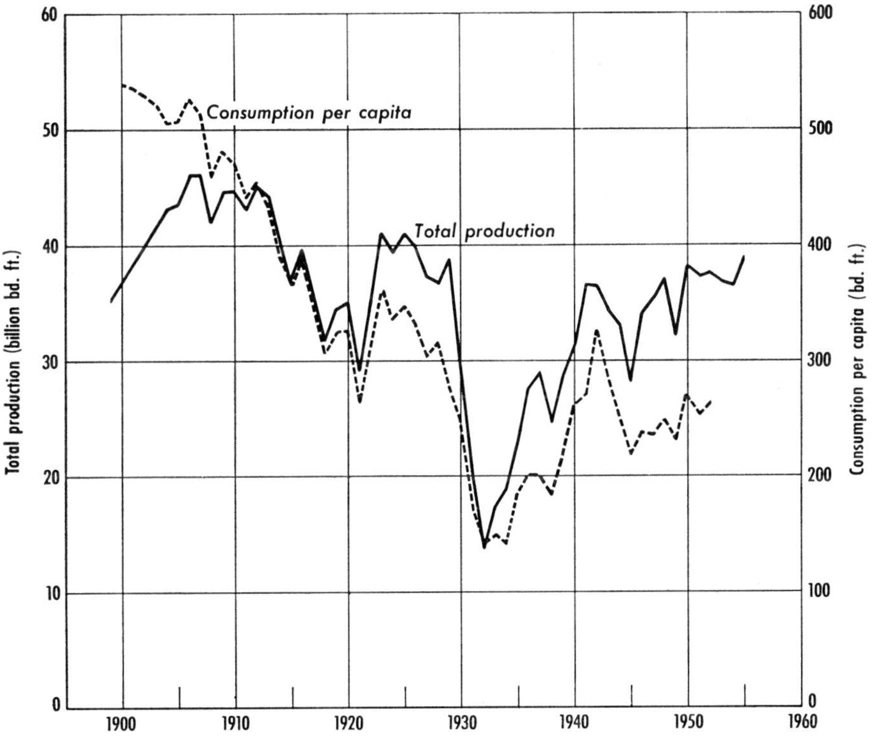

Figure 36. Total lumber production and consumption per capita in the United States, 1899–1955.

supply, greater transportation costs, general lowering of timber quality, and development of substitute materials have all operated to reduce the per capita consumption of lumber. Data on lumber production and use are limited and perhaps inaccurate before the turn of the century. As nearly as can be judged, however, total lumber output was mounting more or less steadily until about 1906, when it reached a peak. Since then, total consumption has varied upward and downward, largely in response to the general business cycle, and more particularly to the cycles in building (Figure 36). There was some decline during the early stages of World War I, a rise later in the war, and a sharp decline after the war; then lumber demand rose rapidly through the 1920's as building reached a high level, but fell during the depression to levels not known for many years. During World War II demand rose again; and since the war it has mostly been at a comparatively high level, due to the high volume of home and other building. Per capita consumption has trended downward more steeply, as total population has risen; and from year to year follows a slightly different course, as exports and imports have varied.

Before World War I the overwhelming portion of the total cut was sawlogs for lumber; paper and pulpwood were largely imported. Other roundwood was used for various purposes, especially railroad ties, mine timbers, and piling (Figure 37). In recent years, however, peeler logs for plywood manufacture have replaced much of what used to be in the miscellaneous category of roundwood. Pulpwood, now a major user of small timber, in terms of total volume is still far below sawlogs, but sawlog output has not yet regained the level it had reached before the depression. These changes in economic relationship partly reflect differences in demand for wood, but also to a large extent reflect differences in location and quality of the timber available. The southern pine forests of today produce pulpwood efficiently, but they are not virgin stands. While much excellent old growth timber remains in the West—primarily Douglas fir, but also Ponderosa and other pines—transportation across the country makes the lumber expensive in eastern markets, and the imminent scarcity of timber of this grade in increasing its value wherever it is sold.

Per capita consumption of lumber, paper, and plywood shows characteristics similar to those for total output (Figure 38). Even if the drastic declines of the depression years are ignored, the downward decline in per capita consumption of lumber has been in excess of 1 per cent but less than 2 per cent annually. In absolute terms, it has

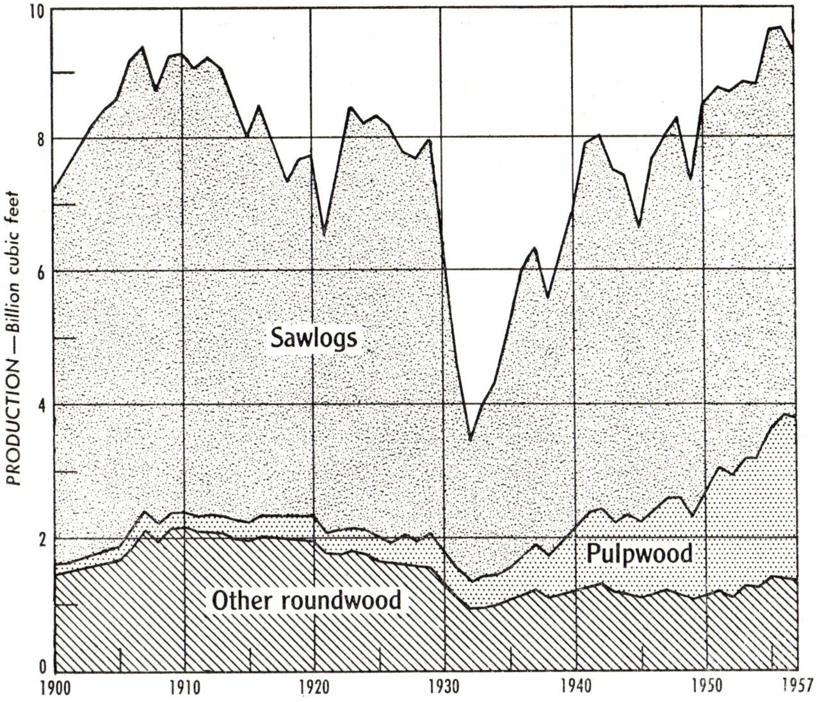

Figure 37. Annual production of commercial roundwood, other than firewood, in the United States, 1900–1957. (Chart from William A. Duerr, *Fundamentals of Forestry Economics,* McGraw-Hill Book Co., copyright 1960.)

declined to less than half its peak. In contrast, the per capita consumption of paper and paperboard has been regularly upward at a rate close to 3 per cent annually. Although newsprint imported from Canada has been important, a major trend has been toward domestic production of paper, especially from southern pine. The trend of softwood plywood consumption has been steeply and rather regularly upward, at about 10 per cent annually. Plywood has been truly the growth industry of the forest products field; temporary declines in consumption have been experienced, but the upward trend has been strong and persistent.

Underlying these changes in total and per capita consumption of the various forest products have been important shifts in the use of

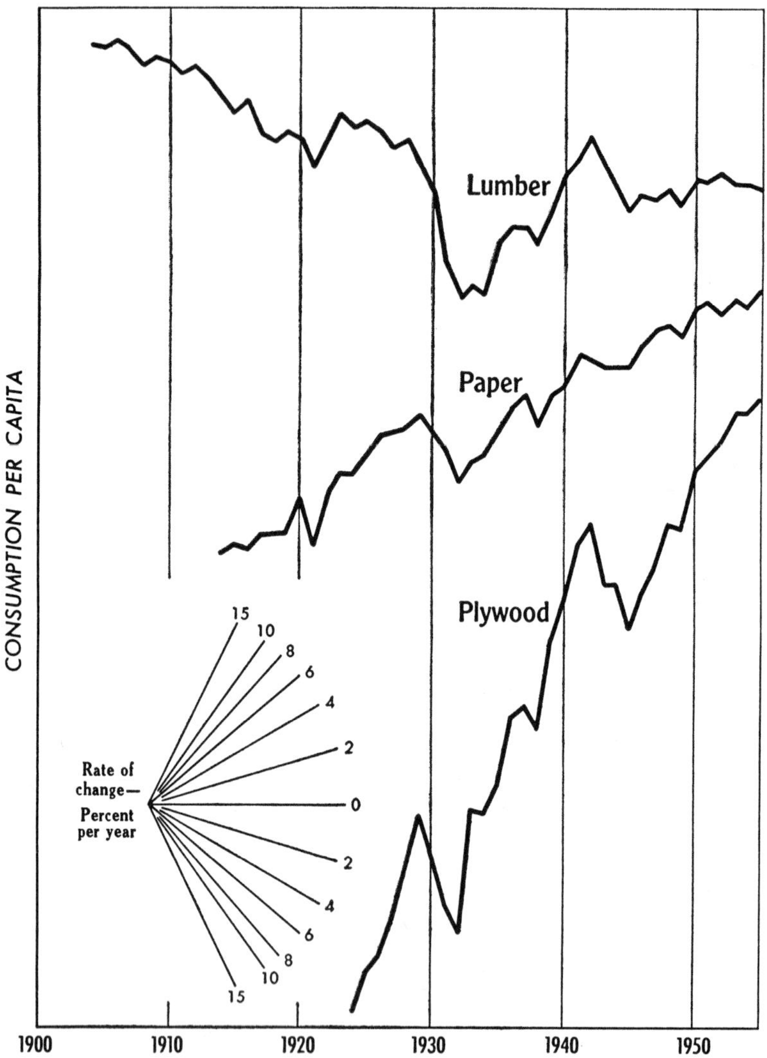

Figure 38. Rates of change in annual per capita consumption of timber products in the United States: lumber, 1904–55; paper and paperboard, 1914–55; softwood plywood, 1925–55. (Chart from William A. Duerr, *Fundamentals of Forestry Economics,* McGraw-Hill Book Co., copyright 1960.)

each and in the development of substitutes. In building, mineral products have substituted for large quantities of lumber; in packaging, paper and paperboard have replaced many of the boxes and barrels; and in both industries plywood increasingly is replacing lumber as a construction material.

Wood, once the chief source of fuel in this country, now provides only about a third of the energy it once, at its peak, provided for this purpose. The huge growth in use of other fuels has reduced wood, as a fuel, to relatively minor status. Today, much of the wood used for fuel is burned on the premises where it originates, as slabs, trimmings, and edgings, and other unmarketable by-products. Sawmills often get their power by burning wood to make steam, and the larger modern mills often have power to sell, even after salvage of material for pulp and other purposes.

Changes in relative prices of forest products have accompanied and sometimes brought about these changes in consumption patterns. Lumber is one important commodity for which price relative to the general price level has risen persistently for more than 150 years (Figure 39). The average annual rate of increase in real prices over the entire period is roughly 1½ per cent. During boom or inflationary periods, deflated prices rose faster than the trend, and during depressions or low points in the long-term building cycles they actually declined, even in relation to the general price level. But the regularity and persistence of the upward trend over so long a period is striking—and unusual for economic data. In sharp contrast are deflated prices for paper since 1890—variable from year to year, and even over a period of years, but completely lacking in trend either upward or downward. With such divergent price trends, it is not remarkable that the trend in per capita consumption of lumber has been downward, or that of paper and related products upward.

These trends in the prices of lumber and paper, "the finished products," have their counterparts in the trends in log prices and stumpage prices. Price data on the latter are seriously incomplete in terms of national averages, but the data for Douglas fir, which is confined to the Pacific Northwest may give a clue to the broader picture (Figure 40). Prior to World War I, Douglas fir lumber sold for between \$10.00 and \$15.00 per 1,000 board feet; its price rose rapidly during the war and scarcely fell afterward, remaining in the neighborhood of \$20.00 per 1,000 board feet during the 1920's. The price fell during

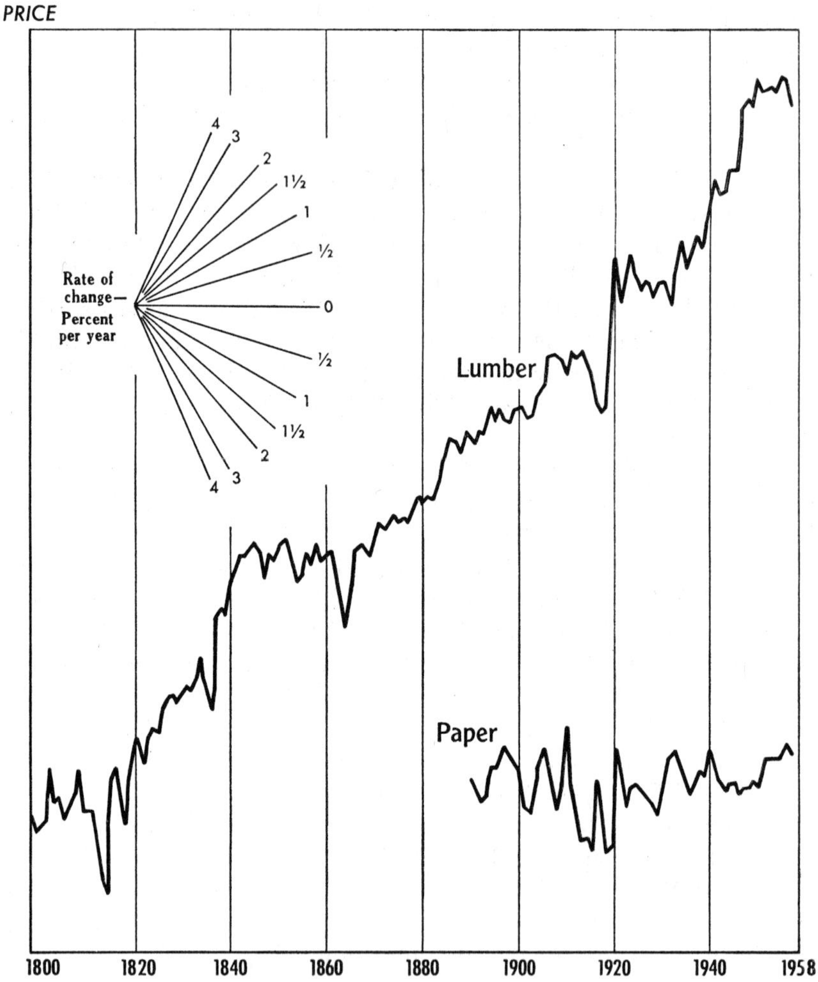

Figure 39. Rates of change in deflated wholesale prices of timber products in the United States, annual averages: lumber, 1798–1957; paper, paperboard, and pulp, 1890–1957. (Chart from William A. Duerr, *Fundamentals of Forestry Economics,* McGraw-Hill Book Co., copyright 1960.)

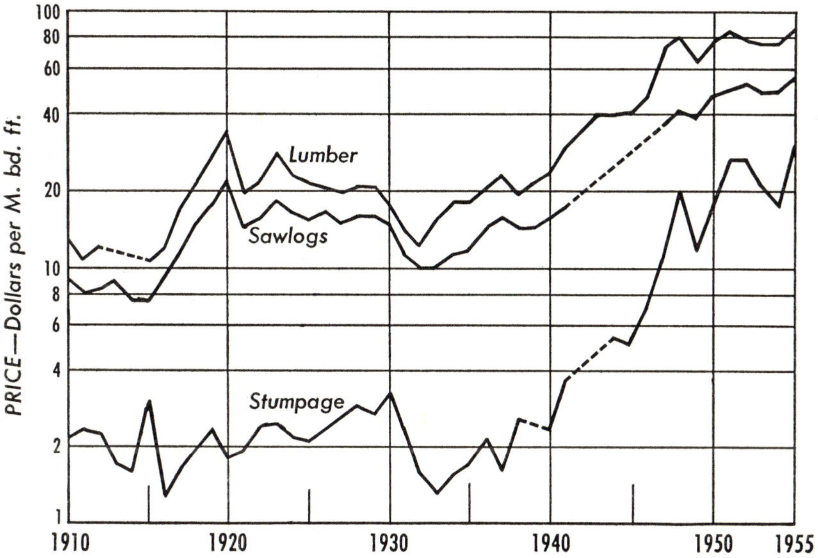

Figure 40. Rates of change in Douglas fir stumpage, sawlog, and lumber current prices, 1910–55. (Chart from William A. Duerr, *Fundamentals of Forestry Economics,* McGraw-Hill Book Co., copyright 1960.)

the depression to about the prewar level, but since has risen rather regularly and steadily to a peak of about $80.00 per 1,000 board feet in recent years. Sawlogs of the same species have varied in price in a very similar manner, at a level lower by $4.00 to $20.00 per 1,000 feet. Important divergences in price movements of the two products have occurred over short periods of time, due primarily to short-run changes in demand, but the general trends have been closely similar.

The price of Douglas fir stumpage shows a very different course. From the pre-World War I period until 1930, only a modest upward trend is shown, nothing like that of lumber and logs; it declined during the depression to a level far lower than prewar; but it has since risen rapidly and regularly, paralleling but somewhat steeper than the rise in lumber and log prices. Proportionately, however, the rise in stumpage has been far greater than the rise in lumber, because stumpage prices started from so low a level. While lumber prices were increasing by roughly four times from the depression low to the present, stumpage prices were increasing by roughly twelve to fifteen times; the absolute rise in lumber prices was about $45.00 per 1,000 board feet and that of stumpage about $18.00; thus, roughly 40 per cent of the rise in lumber prices was carried back into stumpage prices. A disproportionate rise in the price of raw materials during a period of generally rising demand for finished products is a common economic phenomenon, of course.

A major factor behind these upward trends in prices of lumber and stumpage, and to some extent also behind their divergent trends, has been the increasing role of transportation. While, for the present precise quantitative information on this matter is deficient, and for the past is almost totally lacking, there is common agreement as to the general situation and trend. In a much earlier day—the 1800 to 1850 period, for example—most of the lumber and other forest products came from local timber. Since then each shift of lumber manufacture, from the East to the Lake states to the South to the West, has resulted in a larger proportion of the consumer's dollar spent for lumber going into transportation costs. The higher quality, lower stumpage prices, greater availability of timber, and hence usually lower logging and sawmilling costs in the new areas, often meant a lower net price than would have been possible from locally sawed lumber, even with the greater transportation costs; but, more importantly, the local supplies were often simply inadequate in quantity and quality.

It has been estimated that in 1952 over $2 billion was spent for transportation of timber products.[4] About half of this went into rail transportation; the average distance moved for lumber was over 1,000 miles, and for pulp, paper, and paperboard was over 700 miles. In contrast, most common building materials move less than 400 miles from manufacturing to use sites. The average price paid for transportation of all lumber that moved by rail was over $21.00 per 1,000 board feet; for the most common lumber movement, that from the western major producing areas to the major eastern consuming areas, the average price was over $31.00 per 1,000 feet. It has been estimated that the amount spent for transportation exceeds that paid to timber

[4] James C. Rettie, "Some Factors Influencing Past Consumption of Timber Products," originally presented as Chapter V, "Timber Resource Review" (preliminary review draft, 1955). Not included in final version.

producers for stumpage by a wide margin. If in the future the sources of timber shift back nearer to the consumer, then the swing in transport costs can be shared by the timber grower and by the timber consumer.

These shifts in consumption, prices, and transportation are simply the specific manifestations of the general change that has been under way in the forest economy of the United States—a shift from harvesting mature or virgin stands to growing new timber supplies. In the first phase, the center of interest was on harvesting methods; now, more concern is given to the problems of growing trees. One major shift in emphasis, not yet fully realized, concerns the importance of the site and its productivity. When the emphasis was on timber harvest there was little concern with how long the tree had to grow to reach a certain size, as long as the resulting log was suitable for processing; but now that more emphasis is directed at tree growing, the comparative productivity of the site is a matter of the greatest importance for the economic soundness of the whole enterprise.

Institutional Factors of Forest Land Tenure

While these changes in the chief centers of forest exploitation and in various strictly economic relationships were going on, major changes in forest land ownership and tenure were also occurring. At the beginning of Colonial times all the land was the property of the Crown; it was gradually transferred to ownership of grantees and through them to individuals. The same general pattern was followed after Independence. The additions to national territory, as well as the areas outside of the original colonies, became the property of the new national government, and by it were transferred gradually to individuals. The general pattern of disposal was aimed at the establishment of farms; at no time in the history of the public land were the major disposal policies well designed to meet the needs of the lumberman or timber owner. While it was possible for those interested in forestry to buy land, at various times there were limitations on the area of land one purchaser could obtain.

While the institutional framework was thus most unfavorable to the acquisition of public lands in units large enough to provide the base for sustained yield forestry operations, the economic base was equally unfavorable. Too much virgin timber was available for cutting, just a little further away from market, to make growing timber profitable. At the same time, the land laws were generally enforced so laxly that land could be obtained through fraud or timber cut from the public lands in trespass. All the standard accounts of public land history emphasize both practices.[5] In the Lake states, in particular, much of the timber was cut in trespass; in the Northwest a large part of the forest land passed into private ownership by methods not contemplated by the sponsors of land legislation. At the same time, it should be recognized that those who bought timber, including those who did so after 1900 when the government's forestry leaders cried "timber famine," often suffered losses. The quantity of mature timber yet waiting for liquidation was too great to make even modest prices economic, when the waiting costs were counted in.

Gradually, large blocks of the best timber were built up into single-ownership units in different regions. This was done by buying large areas from the federal government or the states, by buying railroad grant lands, by use of the railroad "in lieu" selection procedures, by use of individuals as dummy selectors, and in other ways. As late as the beginning of World War II, however, there were thousands of small forest land owners in many of the commercial forest areas. During and since the war a major move toward consolidation of these small ownerships took place, extending even to large ownership units which could be combined to make still larger ones. This has been especially apparent in the Pacific Northwest and the South, particularly on the most productive lands. Nevertheless, by far the greater part of the privately owned forest land is still held in relatively small tracts.

A different direction of forest land ownership began toward the end of the nineteenth century, when the national forests (at first called forest reserves) were established. These have placed in federal ownership some 85 million acres of commercial forest land, which is managed on a sustained yield basis. The national forests were established chiefly between 1890 and 1910, and more particularly between 1905 and 1910, when President Theodore Roosevelt acted upon the urgings of Gifford Pinchot. There was widespread concern over the indiscriminate timber cutting and the lack of forest regeneration in the cut-over areas. Many people were fearful for the future timber supply of the country. For many years thereafter the national forests were

[5] Cameron, *op. cit.*

largely "storage forests," with only limited cutting. However, the volume of cut has been growing for some time now, and while still less than the sustained yield capacity under moderately intensive management, it is far higher than it was before World War II.

Physical Aspects of Forests and of Forestry

While these locational, economic, and institutional changes were under way, both the forests and forestry were undergoing physical changes. The virgin forest which the white man found had as large a volume of standing timber as the species and the site characteristics would support. The extensive cutting, both to use the forest products and to clear the land, has greatly reduced the area of commercial forest land; to an even greater extent, they have reduced the volume of standing timber (Figure 41). The original volume of timber is not accurately known, of course, but present volume in the eastern half of the United States is estimated to be less than 10 per cent of what was originally there; as Figure 35 shows, the area of commercial timber in this region is slightly more than half of the original area. Thus,

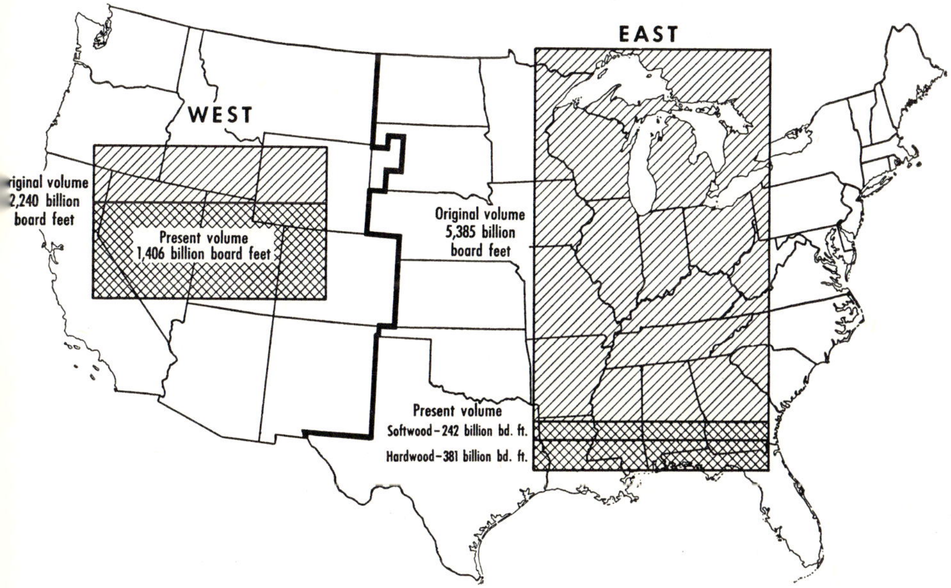

Figure 41. Original and present timber volumes.

most of the reduction in volume is due to a reduction in the volume per acre of present forest land, rather than in the clearing of land. In the West the situation is different. The present volume is nearly two-thirds of the original volume; the area is about 80 per cent of the original area.

With this great reduction in timber volume per acre, even of the lands still classified as commercial timber, the eastern forests particularly and the western ones to a lesser degree have experienced major ecological changes. The extent and character of these changes have depended not only on cutting as such, but also upon the role of fire in cutover lands, and upon the natural conditions of the land which tend toward resilience or brittleness of the area. In the earlier logging, fires were completely uncontrolled. Thousands of small fires occurred each season, and occasionally great fires swept large areas and sometimes killed hundreds of people; the Peshtigo, Wisconsin, fire of 1871 and the Hinckley, Minnesota, fire of 1893 are two well-known examples.

Various climatic, soil and vegetative combinations in the world have responded differently to man's disturbances of natural balances through the introduction of agriculture, grazing, fire, and timber cutting. In northern Europe, New England, and the upper Lake states a relative stability between man and nature has been achieved, due not so much to man's intelligent handling of resources as to favorable climate and the resiliency of these areas. Such unfavorable ecological relationships as exist are manifest in the low growth rate of the forest. On the other hand, the continental United States (except for the Pacific states) is subject to sudden thunderstorms and heavy rainfall which severely erode exposed soils. As a result, cleared lands in the central and southern climatic belts have often been extensively damaged. The northern tier of states, where snow accumulates all winter and the ground is frozen, are free from soil-eroding rainfall for nearly six months of the year. Soil erosion is not, in general, serious in these areas. The rural economy is based on dairying, and hence the land is mostly in pasture and forest so that a higher degree of equilibrium is possible.

In the South and California, where rainfall is heavy during the winter, poor land use practices often result in both serious erosion and considerable loss of water needed for infiltration. Overgrazing and poor cropping practices are the most frequent cause of rapid runoff leading to erosion.

The coniferous forests of California, which grow in an essentially Mediterranean climate, are the most delicately balanced in the United States. Dependent upon accumulated winter precipitation for summer moisture, forest conditions cannot be permanently maintained if a combination of overcutting, fire, or overgrazing takes place. The greater humidity and reduced wind velocities under forest cover are essential to seedling survival. It is fortunate that a large proportion of the California Sierra Mountain forests are under federal ownership and presumably administered in a manner to restrain some of the more destructive pressures. Fire has taken its toll, however.

Professor John Curtis, of the University of Wisconsin, has made an intensive study of changes in forest conditions in his state.[6] The most obvious biotic changes he found were in the species composition of the forests. The hardwood forests had suffered the least change. Originally made up of several species, none of which was in heavy demand by lumbermen, they had been cut lightly each time, though perhaps several times in total. Fires in these areas were also comparatively lighter than elsewhere. The major change in the hardwood forests has been an increase in the amount of sugar maple, because this species is hardier and more aggressive than others. The pine forests, on the other hand, have suffered severely. Their lumber was in heavy demand, and they were cut severely, the initial harvest sometimes removing 90 per cent or more of the dominant trees. Fires were much more prevalent and severe. Following the first fire pine seeds often sprouted, but later fires in the same areas killed the seedlings and often destroyed the most accessible sources of seed. With the absence of tree cover, the micro-environment shifted materially, in the direction of more light, more variable temperatures, more variable moisture, and much greater transpirational stress. The new environment is thus hotter and drier. As a result, large areas are now covered with relatively valueless species, or may even be barren.

While these ecological changes were occurring in the forest, important technological changes in logging and other aspects of forestry were taking place. Some of these affected transportation. The truck for road hauling and the caterpillar tractor for woods work have given a greater flexibility to logging than existed when streams, and later the logging railroad, were the only means of moving logs. Important

[6] John T. Curtis, "The Modification of Mid-Latitude Grasslands and Forests by Man," pp. 723-25, in William L. Thomas, Jr. (ed.), *Man's Role in Changing the Face of the Earth*, (Chicago: University of Chicago Press, 1956).

developments also were occurring in the mills. The band saw and later the "gang" saw yielded more lumber from a given volume of wood. Development of plywood manufacturing methods gave a higher-valued outlet for the better grades of logs. "Re-manufacture," or the making of finished products such as doors and windows at the saw mill rather than nearer the ultimate market, increased the amount of labor used locally. The development of pulp mills in the same regions has often provided a profitable outlet for the mill wastes that once were burned.

Fire fighting techniques and wood regeneration techniques have improved. Use of lookout towers to detect fires, use of telephone and radio to report them to headquarters, use of truck and specialized fire-fighting equipment to fight them on the ground, use of airplanes to parachute men to the scene of the fire, and now aerial application of chemicals to inhibit fire spread, have all greatly increased the effectiveness of forest fire fighting. Many commercial forest species of trees are now grown in nurseries and transplanted, often by mechanical planters, to areas lacking adequate forest stands. Undesirable species of trees are killed chemically or mechanically. Various methods of loading and hauling pulpwood have so lowered the costs of thinning stands of timber designed for later sawlog harvest as to make this practice economical under a much wider variety of conditions than formerly. All of these technological changes have affected the kind of forestry it is most profitable to pursue; in general, their effect has been to increase the intensity of forestry operations, especially on the more productive sites.

One by-product of these changes has been to change materially the definition of commercially merchantable timber. Confusion has resulted, because in the past different timber inventories and estimates of future timber availability used differing definitions. The merchantability of a tree, or of a log within it, depends not only on the physical characteristics of that tree or log, but also on the costs of getting it out; on the available technology by which it can be transformed into lumber, plywood, or other usable product; and on the price at which those products can sell, which in turn depends on the demand for forest products and upon alternative sources of supply. A great deal of timber is taken from the woods today which formerly would have been left there to rot, because it was an unwanted species, was too small in diameter, too short in length of sound log of adequate diameter, or too full of rot or other defect to be easily manufactured.

Attitudes of the Public and of Foresters

Possibly the greatest change in American forestry over the past hundred years or more has been in the attitude of the general public as well as of those trained in forestry. The early attitude, at its best, was that the forest was limitless and inexhaustible so that its conservation, protection, and renewal was unnecessary. Perhaps a more widespread idea was that the forest was an obstacle to progress, to be gotten rid of as rapidly and cheaply as possible. The headlong cutting of the virgin forests, the repeated fires and the destruction they left, the extensive land frauds and trespass, and the growing local shortages of timber in time led to different attitudes on the part of at least some thoughtful leaders. Establishment of Arbor Day in 1872 was one manifestation of this concern; so was passage of the Timber Culture Act in 1873. These early moves were almost entirely unsuccessful in their effect upon the forests, but they do reveal a changing attitude. This concern over future timber supply was responsible for establishment of the Forest Reserves in 1891, now called the national forests. When these were provided for, they had no popular support; they were the product of a minority of farsighted leaders.

The national forests were thus perhaps the first effective step toward forest conservation. Gifford Pinchot became head of the Forest Service in 1905 and shortly led a conservation crusade which did much to awaken the nation to the need for forest conservation. Private forest owners and operators were often denounced as greedy and shortsighted, and the forest conservation problem was presented as one of controlling their actions. Proposals for various forms of federal control over private forestry operations were made over many years until the present decade, but the opposition was at all times so great as to prevent effective action. Gradually, as contending viewpoints reached a compromise, a different approach became possible, involving research, co-operative fire control, and various other public aids to private forestry. These programs became established with the Clark-McNary Act of 1924, and have continued and expanded to the present. In this way, forestry has somewhat paralleled the development of publicly supported research and education in agriculture.

Today, "conservation" is a symbol and slogan of such emotional appeal that no public figure would dare denounce it. With often only a vague idea of what conservation actually means, the general public yet strongly supports it as a principle of resource preservation. At the same time, the economic and other changes in forests and forestry already described, have substantially changed the programs of many forest landowners until today the larger forest owners are, in general, following as good forest practices as are the public agencies. Even where, because it is not economical or for other reasons, good forestry is not practiced, there is an awareness that it should be practiced, and those responsible are likely to be apologetic about their apparent failure to comply with what is held to be sound forest management.

FORESTS IN THE UNITED STATES TODAY[7]

Nearly 650 million acres may be classed as forests in the forty-eight states today. However, of this, only 484 million acres are capable of commercial forestry (Table 39 and Figure 42). The 163 million acres of noncommercial forest chiefly grow types or species which cannot attain a size or quality of tree that has commercial value; or are in locations where climatic and other factors produce stands of too low a volume to warrant harvesting; or are in reserved areas, such as national parks. Three-fourths of such lands are in the West. In this study most of the noncommercial forest lands are classified as grazing land, or land for wildlife or other uses. Of the commercial forest, nearly three-fourths of the total area is about equally divided between the North and South, and only one-fourth is in the West.

These commercial forest lands contain approximately 2,000 billion board feet of sawtimber and about 500 billion cubic feet of growing stock, the distribution of each being quite different from that of the area of such land (Figure 43). Approximately 20 per cent of the sawtimber is hardwood, chiefly found in the North and South, and 80 per cent is softwood. Two-thirds of all sawtimber is found in the West, and nearly 85 per cent of the softwood sawtimber volume is there; only a little hardwood sawtimber grows in the West. The remaining sawtimber is about equally distributed between North and South. The

[7] Unless specifically noted to the contrary, the data in this section come from U. S. Department of Agriculture, Forest Service, *Timber Resources for America's Future* (Washington: U. S. Government Printing Office, 1958). The calculations and interpretations based on those data were made by the present authors. Except where noted, Alaska is not included in the over-all figures.

distribution of growing stock is slightly more even as between regions, with somewhat more than half in the West and slightly less than one-fourth each in the North and South. Sawtimber volume is high in the West because of the large proportion of the area that is still covered with mature or virgin timber.

The various subregions present some interesting contrasts. The Douglas fir subregion, which contains some species other than Douglas fir, includes only 4 per cent of the commercial forest area, yet today has about 30 per cent of the volume of sawtimber and over 20 per cent of the volume of growing stock. The Pacific Northwest pine region, which includes several species, also has about 4 per cent of the area and 8 per cent of the sawtimber volume. The three Pacific Coast states of Washington, Oregon, and California contain more than half of the sawtimber volume.

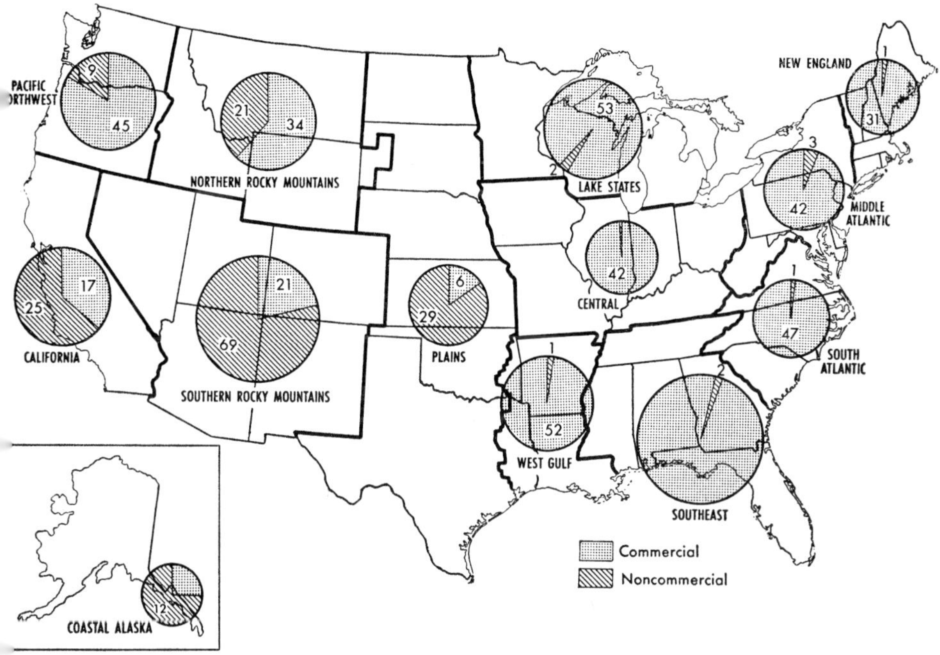

Figure 42. Forest lands of the United States and coastal Alaska, by regions. Figures indicate millions of acres. (Chart from *Timber Resources for America's Future,* U. S. Forest Service.)

Table 39. Forest land area and timber volume, by regions, 1953

Section and region	Forest land area: Total forest land, Mil. acres	%	Commercial forest land, Mil. acres	%	Non-commercial forest land, Mil. acres	%	Timber volume: Live sawtimber[1] (billion board ft.), Total	Soft-wood	Hard-wood	Growing stock (billion cubic ft.), Total	Soft-wood	Hard-wood
North:												
New England	31.4	5	30.6	6	.8	*	51	27	24	24	10	14
Middle Atlantic	44.9	7	42.2	9	2.7	2	74	13	61	34	5	29
Lake	55.2	8	53.3	11	1.9	1	50	14	36	25	7	18
Central	42.7	6	42.4	9	.3	*	83	4	79	25	1	24
Plains	34.6	5	5.5	1	29.1	17	8	1	7	3	*	3
Total, North	208.8	31	174.0	36	34.8	20	266	59	207	111	23	88
South:												
South Atlantic	47.3	7	46.1	9	1.2	1	107	51	56	34	15	19
Southeast	96.9	15	95.0	19	1.9	1	139	77	62	48	23	25
West Gulf	53.1	8	52.2	11	.9	*	111	55	56	32	13	19
Total, South	197.3	30	193.3	39	4.0	2	357	183	174	114	51	63

West:												
Pacific Northwest												
Douglas fir subregion	29.0	4	25.4	5	3.6	2	595	577	18	113	107	6
Pine subregion	25.1	4	20.0	4	5.1	3	154	154	*	33	33	*
Total, Pac. N. W.	54.1	8	45.4	9	8.7	5	749	731	18	146	140	6
California	42.6	6	17.3	4	25.3	14	360	354	6	67	64	3
Northern Rocky Mt.	55.3	8	33.8	7	21.5	12	167	166	1	43	43	*
Southern Rocky Mt.	89.6	14	20.5	4	69.1	40	69	66	3	18	16	2
Total, West	241.6	36	117.0	24	124.6	71	1,345	1,317	28	274	263	11
Continental U. S.	647.7	97	484.3	99	163.4	93	1,968	1,559	409	499	337	162
Coastal Alaska	16.5	3	4.3	1	12.2	7	89	89	*	18	18	*
All regions	664.2	100	488.6	100	175.6	100	2,057	1,648	409	517	355	162

* = Negligible.

[1] In addition to live sawtimber, there are 37 billion board feet of salvable dead trees, of which 34 billion are in the West, 2 billion in the North, and 1 billion in the South.

SOURCE: U. S. Department of Agriculture, Forest Service, *Timber Resources for America's Future* (Washington: U. S. Government Printing Office, 1958), Tables 15 and 21.

The various forest tree species also show important differences. Douglas fir, mostly found in the Pacific Northwest but also to some extent outside of that region and not comprising all forest trees within its main region, accounts for 26 per cent of the sawtimber volume and 19 per cent of the growing stock. Ponderosa and Jeffrey pines, found in the pine region of the Pacific Northwest, in California, and in the northern and southern Rocky Mountain areas, includes 11 per cent of the sawtimber and 8 per cent of the growing stock. Western hemlock and Sitka spruce, mostly found within the Douglas fir region, include 10 per cent of the sawtimber and 8 per cent of the growing stock. True firs, also found in the West, include 9 per cent of the sawtimber and 7 per cent of the growing stock. Southern yellow pine, found across the South, includes 8 per cent of the sawtimber and 9 per cent of the growing stock, and undoubtedly a far larger area. Oaks of

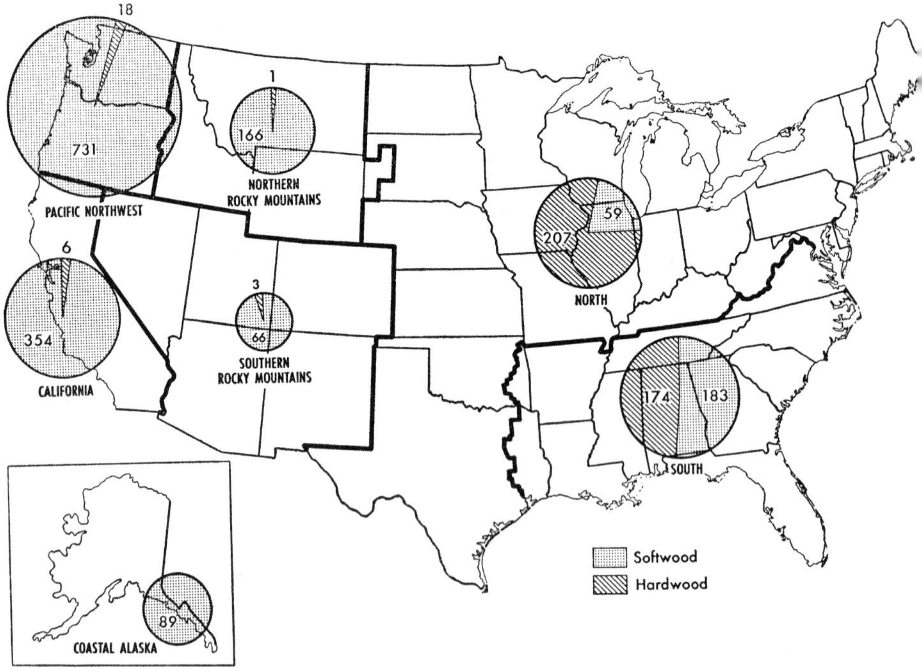

Figure 43. Sawtimber volume in the United States and coastal Alaska, 1953. Figures indicate billion board feet. (Chart from *Timber Resources for America's Future*, U. S. Forest Service.)

various subspecies include 7 per cent of the sawtimber and 10 per cent of the growing stock, and are found in a large area in the eastern half of the country. These six species or groups of species thus account for 71 per cent of the sawtimber and 61 per cent of the growing stock. While scores of other forest species are commercially used, their total volume is relatively small, although their value for special-purpose uses may be large.

Of the commercial forest land, about 10 per cent is still in old-growth timber; all of this is in the West, mostly in Douglas fir and pine stands. Another 27 per cent of the land is in young-growth sawtimber stands, with about half of it in the South, a third in the North, and least of all in the West. Slightly more than one-third of the total area is in stands of pole timber size, and the rest in smaller sizes or is unstocked. Regionally and locally, the age and size distribution is not as good as the national totals suggest.

Unfortunately, the total figures conceal some rather serious deterioration in timber quality, a subject on which information, especially of a historical nature, is sparse. Nearly 10 per cent of the sound timber volume is in cull trees; the proportion is higher in the hardwoods. Cull trees reach or exceed a third of the total in many southern hardwood stands. Much of the hardwood lumber is of comparatively low grade. In the East 20 per cent of the hardwood stands is classed as cull, and more than half as low grade. Much of the mature timber in the West is of high quality and large sizes, with half of the timber volume in trees of 32-inch or larger diameter. In the East log sizes run smaller; half of the sawtimber by volume is in trees of 14-inch or less diameter. Such information as is available suggests that larger diameter trees are becoming scarcer, and that resort is being had to smaller and smaller trees.

This picture of the present area and volume of timber may be contrasted with the estimates of original area and volume we have presented, and with other estimates of volume made in various years (Table 40). It has been noted that forested area has declined by about 42 per cent and total volume of timber by about 73 per cent from the original area and volume. The earliest estimates that are considered reasonably accurate were made shortly after 1900; they showed a total timber volume of about 2,000 billion board feet. Later estimates were about 40 per cent higher; this can scarcely represent a growth of timber of this magnitude, but rather reflects more nearly complete inventories and changing standards of log merchantability. From this highest estimate of 1909, the estimate of standing timber declined more or less steadily until an apparent low in 1943-46, which was only 57 per cent of the peak. But, again, the extent to which these estimates were fully comparable is not clear. The 1953 estimate is higher than any in the preceding thirty years; while there probably has been some increase from the low point, changing standards of log merchantability may also have affected the apparent increase.

Table 40. Estimates of standing timber volumes in the United States

Year[1]	Estimate made by	Total volume in billion board feet
1903	J. E. Deffebaugh	1,970
1908	R. S. Kellogg	2,500
1909	Bureau of Corporations—Department of Commerce	[2]2,826
1920	Capper Report (U. S. Forest Service)	[3]2,215
1930	Copeland Report (U. S. Forest Service)	[3]1,668
1938	U. S. Forest Service Forest Survey—Joint Congressional Committee on Forestry	[3]1,764
1943-46	American Forestry Association Appraisal	[3]1,621
1945	U. S. Forest Service Re-Appraisal	[3]1,601
1953	U. S. Forest Service—Timber Resource Review	[3]1,968

[1] Earlier estimates were made by C. S. Sargent for the U. S. Census of 1880, by B. E. Fernow in 1895, 1897, and 1902, and by Henry Gannett of the U. S. Geological Survey in 1900. These early efforts, which paved the way for greater accuracy in later surveys, are not comparable with them and hence are not included here.

[2] Log scale.

[3] Lumber tally or international scale.

SOURCE: Martha A. Dietz, "A Review of the Estimates of the Sawtimber Stand in the United States, 1880-1946," *Journal of Forestry,* Vol. 45, No. 12 (1947).

Ownership of Forest Lands

The history of forest land use and exploitation in this country has left its mark upon the present pattern of forest land ownership. Of the total area 27 per cent is publicly owned (Table 41). This percentage includes Indian Reservations which, technically, are owned by Indians and managed for them by the federal government. About two-thirds of the publicly owned forest land is national forest, about 14 per cent is "other federal," and the rest is state and local government land. Over 60 per cent of all this public land is in the West, where it makes up two-thirds of all the commercial forest land in that region. In the North and South, publicly owned forest land is comparatively much less common. Of the total privately owned land, about 46 per cent is in farms, 17 per cent is owned by forest industries, and the remaining 37 per cent is classed as "other." Farm ownership is particularly large in the South, and "other" ownership larger than average in the North.

Table 41. Ownership of commercial forest land by section, 1953

(Million acres)

Ownership	All sections	North	South	West and Coastal Alaska
Private:				
Farm	165.2	61.4	90.1	13.7
Forest industries	62.4	14.1	33.5	14.8
Other	130.7	66.1	53.0	11.6
Total	358.3	141.6	176.6	40.1
Public:				
National forest	84.8	10.3	10.4	64.1
Other federal	18.3	2.8	3.8	11.7
State and local	27.2	19.3	2.5	5.4
Total	130.3	32.4	16.7	81.2
All ownerships	488.6	174.0	193.3	121.3

SOURCE: U. S. Department of Agriculture Forest Service, *Timber Resources for America's Future* (Washington: U. S. Government Printing Office, 1958), Table 16.

In total in 1953, there were 4.5 million different forest landowners (Table 42); 75 per cent of the ownerships were classed as farm forests, nearly 25 per cent as "other," and less than 1 per cent as forest industry. The farm forest owners averaged only 49 acres of forest; 86 per cent of all owners of all classes had less than 100 acres, averaging

Table 42. Number and area of private commercial forest land ownerships in the United States and coastal Alaska, 1953

Ownership	Number of owners[1]	Total area	Ownership size class (acres): 50,000 and larger	5,000 to 50,000	500 to 5,000	100 to 500	Less than 100
	(1,000)	(Mil. acres)	(Mil. acres)	(Mil. acres)	(Mil. acres)	(Mil. acres)	(Mil. acres)
Farm	3,382.5	165.2	.5	4.5	23.2	59.2	77.8
Forest industry							
Lumber	21.3	34.7	18.6	10.6	3.1	1.9	.5
Pulp	.2	23.3	21.8	1.3	.2	—	—
Other	2.0	4.4	1.6	2.5	.1	.2	*
Total	23.5	62.4	42.0	14.4	3.4	2.1	.5
Other	1,104.7	130.7	15.8	15.8	19.8	36.6	42.7
Total, private area	—	358.3	58.3	34.7	46.4	97.9	121.0
Total, number of owners (thousands)	4,510.7	—	.3	2.5	46.3	586.5	3,875.1

* = less than 0.1.

[1] State basis. Owners holding commercial forest land in two or more States are counted more than once.

SOURCE: U. S. Forest Service, *Timber Resources for America's Future* (Washington: U. S. Government Printing Office, 1958), Table 17.

only 31 acres each; another 13 per cent had from 100-500 acres, averaging 167 acres each. All forest landowners averaged but 79 acres each. The forest industry ownerships were far larger than these averages, with about 2,650 acres per average holding of this type. Had it not been for these few relatively large ownership units, the average size would have been very much smaller than it was. Together, the small ownerships add up to a considerable acreage of forest land.

Why so many of these small forest land holdings continue to exist is difficult to understand. Those that are part of farms have been dominated by factors affecting size and ownership of the farm itself. But in some parts of the Midwest there are farm woodlots on land quite capable of continued crop farming; in such cases inertia seems to be the only explanation for their remaining as woodlots. For the large number of small forests of "other" ownership, it seems that prospects for profitable operation cannot be the motivation for ownership. Such tracts can at best produce only a comparatively few dollars gross income per year; the costs and bother of managing them, if indeed they were managed for profitable forestry, would almost surely be greater than the returns. It seems more probable that lands of this type are owned for their actual or supposed recreational value, or because of sentiment, or out of a misapprehension as to their income-producing capabilities.

Of the farm woodlots, a substantial share (62 per cent) are pastured. In dairy farming areas a pastured woodlot is likely to be a poorly managed one; not only is the pasturing likely to prevent the growth of young trees and to compact the soil, but it also may signify neglect of the woodlot. In the South and much of the West, forest range supports moderate numbers of cattle. If their numbers are carefully regulated and other management practices instituted, grazing can do little damage to these forests.

Productivity Status of Forest Lands

The productivity status of much forest land in the United States is not only far below its biological optimum or maximum, but also far below that found on the better managed lands of similar forest type. Productivity status can be measured in various ways; one is the degree of stocking of the land. In 1953 about 9 per cent of the commercial forest land had a stocking of less than 10 per cent, or was essentially unstocked; another 17 per cent had from 10 to 40 per cent stocking; each of these classes was distributed among the regions more or less proportionately to total forest area. Such light stocking not only means low output per acre, but probably low-quality trees as well. Even the 74 per cent that was stocked at 40 per cent or more included much land stocked at less than optimum volume per acre. If these percentages of stocking and of area are combined into a single weighted average figure, it is apparent that stocking is only slightly more than half of the practically realizable level.

Of the poorly stocked area, some 52 million acres are classified as physically and biologically plantable; of this area, only 7 million acres have been planted to date, and on only 5 million are there acceptable plantations. The trend in area planted per year has been sharply upward in recent years, yet in 1952 the area planted was only 1 per cent of the plantable area needing planting.

Another measure of productivity status of forest lands is the condition of recently cut lands. The rapidity with which a cutover area becomes reforested if clear-cutting was practiced, or the degree of stocking after selective cutting was practiced, and the quality of the stocking, are all vital to the future timber growth of an area. For the purposes of the Forest Service's 1953 Timber Resource Review, all lands on which any cutting had taken place since 1947 were defined as "operating areas," and data were obtained as to their productivity index. There were 235 million acres or 48 per cent of all commercial forest land in this category, ranging from only 32 per cent of total farm and other private forest holdings, where timber harvest ordinarily occurs infrequently but sometimes with severity, to 71 per cent for forest industry and 73 per cent for public forests.

An apparent but spurious relation existed between type of forest ownership and productivity status; the real relation was between size of ownership holding and productivity status. The condition of publicly owned and industry-owned lands was relatively good—80 per cent and 77 per cent, respectively, classed as "upper" productivity, with an index of 70 or more, and only 3 per cent and 4 per cent, respectively, classed as "lower," with an index of less than 40. In contrast, only 46 per cent of the farm and other private forests combined were classed as "upper" and 21 per cent were classed as "lower." However, when comparisons are made on a size-of-holding basis, there is uniformly better productivity on the large holdings and poorer productivity on the small ones: 78 per cent of those with 50,000 acres or more, 64 per cent of those with 5,000 to 50,000 acres, but only 40 per cent of ownerships less than 5,000 acres, were in the upper productivity class. There were no major or decisive differences between ownership classes in this regard. Small forest industry holdings were classed as "low," on the average, as were small farm or small "other" ownerships; and large units of each were about equally good. Forest industry ownerships average large, and farm forest ownerships average small, and this difference in average size accounts for most or all the difference in productivity class of their respective holdings.

Some regional differences in productivity classes appeared, but these also seem due primarily to differences in average size of holding. However, there does seem some tendency for southern forests of a particular size and ownership class to average a little lower in condition than their counterparts in the North and West.

Natural forest lands in the United States vary greatly in their annual wood output per acre, and the different forest types and species differ in their growth requirements. To a large extent, the natural conditions which determine the forest type also affect the annual growth rates. But within each type there are great differences in site quality, also dependent upon natural conditions; and these affect the volume of growth each year. A third factor affecting average annual growth is the general system of forest management—the degree of fire protection, the thinning and other stand improvement programs, the harvesting methods, and the provision for planting or otherwise getting a good new stand established.

Differences in mean annual growth rates per acre occur for the various forest types even when site quality and management practices are relatively constant (Table 43). Among the major commercial forest types, southern pines and Douglas fir are among the fastest growers. In contrast, some of the northern types are much slower. Factors other than annual growth rate may affect the profitability of forestry of a particular forest type. For one thing, the supply of a wood in relation to the demand for it may be either relatively large or relatively small, so that average prices over a period of years are relatively low or high. Total volume of timber per acre may make a considerable difference in operating costs, especially for harvesting, including road-building costs. Terrain may also affect the latter. In addition, location of the forest type with respect to markets for its products may have a major effect. Many or most of these factors, however, are applicable for large areas or all of a particular forest type.

In a somewhat different category are differences in site quality. These are measured by foresters and are usually characterized by differences in tree height of stands of given ages. Soil characteristics, particularly moisture retention, drainage and soil fertility, and micro-climatic differences, such as are associated with exposure and slope, are often dominant in site differences. For some forest types, site quality is relatively similar over extensive distances; for others, site quality is highly variable within short distances. The site index for some valley bottoms, for instance, may be very different from that for the adjacent steep slopes. These differences in site result in variations in annual wood growth and also, naturally, in forest management profitability. Annual operating costs may not vary greatly between

Table 43. Mean annual growth by regions for fully stocked, unmanaged stands of selected species on medium sites and with average breast-high diameter of eleven inches

Region	Species	Average site index	Stand[1] age (years)	Stand[2] volume (cu. ft.)	Mean annual growth (cu. ft.)
Boreal	White spruce and mixed stands[3]	—	188	4,245	23
	Spruce-poplar[4]	70	150+	5,914	39
Rocky Mountain	Ponderosa pine[5]	100	65	6,000	92
	Western white pine[6]	50	105	9,000	86
Coast	Spruce-hemlock (SE Alaska)[7]	100	105	11,250	107
	Douglas fir[8]	140	58	7,550	130
	Redwood[9]	11	32	10,600	331
South	Loblolly pine (N. Louisiana)[10]	90	45	6,750	150
	Loblolly pine[11]	90	40	4,290	107
	Shortleaf pine[11]	70	72	7,600	106

[1] Corresponding to average stand diameter of 11 inches.

[2] To approximately 4-inch top and 4-inch minimum diameter.

[3] *Preliminary Yield Tables for Alaska's Interior Forests,* Technical Note No. 14, Alaska Forest Research Center, Juneau.

[4] W. K. MacLeod and A. W. Blyth, *Yield of Even-Aged Fully Stocked Spruce-Poplar Stand in Northern Alberta,* Technical Note No. 18, Canada Department of Northern Affairs and National Resources, Forestry Branch, Forest Research Division, 1955.

[5] W. H. Meyer, *Yield of Even-Aged Stands of Ponderosa Pine,* Technical Bulletin No. 630, U. S. Forest Service, October, 1938.

[6] I. T. Haig, *Second-Growth Yield, Stand, and Volume Tables for the Western White Pine Type,* Technical Bulletin No. 323, U. S. Forest Service, October, 1932.

[7] R. F. Taylor, *Yield of Second-Growth Western Hemlock-Sitka Spruce Stands in Southeastern Alaska,* Technical Bulletin No. 413, U. S. Forest Service, March, 1934.

[8] R. E. McArdle, W. H. Meyer, and D. Bruce, *The Yield of Douglas-Fir in the Pacific Northwest,* Technical Bulletin No. 201, U. S. Department of Agriculture, Revision October, 1949.

[9] D. Bruce, *Preliminary Yield Tables for Second-Growth Redwood,* Bulletin No. 361, University of California College of Agriculture, Agriculture Experiment Station, Berkeley, May, 1923.

[10] W. H. Meyer, *Yield of Even-Aged Stands of Loblolly Pine in Northern Louisiana,* Bulletin No. 51, Yale University School of Forestry, 1942.

[11] *Volume, Yield, and Stand Tables for Second-Growth Southern Pines,* Miscellaneous Publication No. 50, U. S. Department of Agriculture, September, 1929.

SOURCE: George R. Armstrong and John A. Guthrie, "The Outlook for Western Forest Industry," unpublished ms.

sites—they may, in fact, be as high or higher for poor as for good sites. Larger amounts of growth may thus be almost wholly additions to net returns. The ability of different forest sites to absorb inputs, such as labor, fertilizer, capital in various forms, probably also differs greatly.

It was pointed out in Chapter IV that agriculture has gone through about a century and a half of a vast "sorting out" process, whereby through trial and error we have learned the productive capacities of different lands for different agricultural purposes. Such a process is never complete; yet, for agriculture, we largely know the capacities of areas, not only for present crops but for possible future ones. This process has only begun for forestry. The larger and better managed private forestry operations have made selections of forest land which certainly are based upon site characteristics, among other factors. Their management practices are related to site characteristics. But this kind of forestry is still a minor part of the whole. With few exceptions, the small private forest land owners have not made this type of differentiation among forest sites; and public forest managing agencies are required to undertake forestry on their entire holdings. Differentiation of management according to site quality has been delayed for forestry largely because the immense volume of virgin timber available for harvest made tree growing unprofitable until relatively recent times. The long growth-harvest cycle for trees, compared with that for farm crops, is also a deterring factor. One of the most useful economic developments in forestry for the next several decades is likely to be a closer differentiation according to site characteristics, with more intensive management on the better sites and more extensive management on the poorer sites.

Forest Growth-Drain Relationships

A number of forces operate to restrain trees from making their maximum growth, or even from surviving. Men may cut them for their purposes but, further, fires, insects, diseases of various kinds, and unfavorable weather all take their toll. The influence of these natural forces may be either intensified or reduced by men's activities. Most forest fires undoubtedly are caused by men, for instance, but on the other hand men may seek to control fires and limit their destruction.

Through cutting, men may open some forests to loss from other causes, or may so change the ecological balance as to materially affect survival. The actual death of trees is evident even to casual observers. Well-managed and timely forest operations may salvage much of the value of the newly killed trees. Less evident is the loss in future growth resulting from mortality or weakening of trees, but it may often be more serious. This is harder to measure, and there may well be greater differences of opinion as to the future effect of a present or recent damage to the forest. The Forest Service has sought to measure "growth impact," but this idea has not yet had universal acceptance among foresters.[8]

In 1952 insects caused the greatest mortality both in growing stock and in sawtimber (Table 44). The next most common cause was weather, with tree diseases in third place. Fire was not a major cause of tree death, accounting for only 7 per cent of the deaths of growing stock and 6 per cent for sawtimber. More efficient fire control in recent years has materially decreased acreage burned. The relative importance of different factors in growth impact was different: forest tree diseases accounted for nearly half of the total, followed closely by insects and fire. This comparatively minor role of fire may surprise many people accustomed to the widespread publicity campaigns waged against fire in the forests. However, because fire can expose trees to the depredations of insects, disease, and weather, its indirect importance is somewhat greater than would appear. In addition, the well-organized campaigns to prevent and control fires that have been carried on for many years by now have become very effective. In their absence, or should they slacken, fire might again become a greater destroyer. Because efforts to combat insects and diseases are of comparatively recent origin their scope is not yet as great.

Compared with the total volume of standing timber in 1952, the mortality of growing stock was 0.53 per cent, and of sawtimber, 0.47 per cent; growth impact was 2.02 per cent for growing stock and 1.98 per cent for sawtimber. It is worth noting that growth impact is reckoned to be roughly four times mortality loss or, in other words, that mortality at the time is only a quarter of ultimate growth loss from any unfavorable event, such as a fire or a disease attack.

The growth of timber in 1952 amounted to 14 billion cubic feet of growing stock and to 47 billion board feet of sawtimber (Table 45).

[8] John A. Zivnuska, *Timber Today—and Tomorrow*, prepared for the Forest Industries Council (Berkeley: University of California Press, 1956).

Table 44. Timber mortality and growth impact resulting from 1952 damage, by cause, and in relation to standing volume and cut

Item	Growing stock				Live sawtimber			
	Mortality		Growth impact		Mortality		Growth impact	
	Million cu. ft.	Per cent	Million cu. ft.	Per cent	Million bd. ft.	Per cent	Million bd. ft.	Per cent
1. Cause of timber loss								
Fire	240	7	1,690	15	780	6	7,370	17
Disease	770	22	5,050	45	2,240	18	19,890	45
Insects	1,000	28	1,780	16	5,040	40	8,620	20
Weather	840	24	960	9	3,390	27	3,870	9
Animals	70	2	1,000	9	190	1	2,720	6
Other	590	17	730	6	1,030	8	1,360	3
Total	3,510	100	11,210	100	12,670	100	43,830	100
Salvage	770	—	770	—	3,090	—	3,090	—
Net loss	2,740	—	10,440	—	9,580	—	40,740	—
2. Standing volume (million bd. ft.)		517,000				2,057,000		
3. Net loss as percentage of standing volume	.53		2.02		.47		1.98	
4. Timber cut in 1952 (million cu. ft.)		10,760				48,840		

SOURCE: U. S. Forest Service, *Timber Resources for America's Future* (Washington: U. S. Government Printing Office, 1958), Tables 36 and 39, and p. 45.

Table 45. Timber growth, 1944 and 1952, and in comparison to standing volume, 1952

Item	Growing stock			Live sawtimber		
	1944[1] (billion cu. ft.)	1952 (billion cu. ft.)	Percentage change 1944-1952	1944[1] (billion bd. ft.)	1952 (billion bd. ft.)	Percentage change 1944-1952
1. Annual growth in						
Eastern[2] softwoods	3.8	4.4	+16	15.2	17.0	+12
Eastern[2] hardwoods	5.9	7.1	+20	16.6	19.1	+15
Western species	2.8	2.7	− 4	11.6	11.2	− 3
All species	12.5	14.2	+14	43.4	47.3	+ 9
2. Standing volume in 1952						
Eastern[2] softwoods	—	74	—	—	242	—
Eastern[2] hardwoods	—	151	—	—	381	—
Western species	—	292	—	—	1,434	—
All species	—	517	—	—	2,057	—
3. Growth as percentage of standing vol.						
Eastern[2] softwoods	—	5.9	—	—	7.0	—
Eastern[2] hardwoods	—	4.7	—	—	5.0	—
Western species	—	.92	—	—	.78	—
All species	—	2.75	—	—	2.30	—

[1] Adjusted to 1952 basis.

[2] As used in this table, "eastern" includes North and South.

SOURCE: U. S. Forest Service, *Timber Resources in America's Future* (Washington: U. S. Government Printing Office, 1958), Tables 21 and 28.

Exactly half of the growth of growing stock was in eastern hardwoods, and 40 per cent of the growth in sawtimber was in the same types in the same regions. Unfortunately, a great deal of this growth is of species that are in poor demand, or in trees of poor form and quality, or in areas where local markets are limited for the particular species. Thus, while this growth is desirable, it has severe limitations. On the other hand, only 20 per cent of the growth in growing stock and not much more than this of sawtimber was in the West, where much of the commercial forest harvest now is, and where the potentials for high-quality timber are great.

Fortunately, the annual rate of growth in 1952 seems to be higher than in 1944, by 14 per cent for growing stock and 9 per cent for sawtimber. For both softwoods and hardwoods, the relative increase from 1944 to 1952 was higher than this in the North and South, but in the West an actual loss seems to have occurred. The latter is the more surprising because the comparatively heavy timber cutting in this region was opening up old mature timber stands which were making little or no net growth, and thus permitting new growth to occur. However, this region suffered unusually high mortality due to bark beetles.

In 1952 the Eastern softwoods added 5.9 per cent in growth to the volume of growing stock then standing, and for sawtimber added 7.0 per cent. The latter comparison in one sense is not valid, because some small part of the growth was due to smaller trees increasing in size to sawtimber size. The eastern hardwoods grew slightly more slowly, 4.7 per cent and 5.0 per cent, respectively. In the West average growth is well under 1 per cent of volume of standing timber. The basic reason for this is that much of the western forests are still mature and so produce little or no net growth.

More than half the sawtimber growth and nearly as much of the growing stock growth is in the South; of this, roughly 80 per cent is southern yellow pine. Oak is responsible for a larger volume of growth than any other species except the southern yellow pine. In contrast, Douglas fir constitutes less than 10 per cent of the growth.

As Table 44 shows, timber cut in 1952 was four to five times mortality from all causes, and roughly equal to the estimated growth im pact. Almost half of the sawtimber cut was softwood in the West; softwood in the South was the next largest volume, and hardwood in the South, third (Table 46). When all growing stock is considered, the extensive harvesting of pulpwood in the South brings the softwood harvest there almost up to the softwood cut in the West. For all sawtimber, the cut was only 3 per cent above growth; for all growing stock, growth was 32 per cent above cut. This is the first time since data on growth and cut have been collected—and probably for well over a hundred years, since cutting began on a large commercial scale—that such favorable relationships have existed. There is considerable cause for gratification that sawtimber growth has at last about caught up with cutting.

Table 46. Comparison of net annual timber growth and timber cut in the United States including coastal Alaska, 1952

Species group and section	Growing stock			Live sawtimber		
	Growth	Cut	Ratio of growth to cut[1]	Growth	Cut	Ratio of growth to cut[1]
	bil. cu. ft.	bil. cu. ft.		bil. bd. ft.	bil. bd. ft.	
All species:						
North	4.66	1.94	2.40	12.07	6.70	1.80
South	6.80	5.06	1.35	24.02	19.60	1.22
West and coastal Alaska	2.78	3.76	.74	11.31	22.54	.50
Total	14.24	10.76	1.32	47.40	48.84	.97
Softwood:						
North	.82	.70	1.17	2.47	2.37	1.04
South	3.56	3.05	1.17	14.50	11.72	1.24
West and coastal Alaska	2.63	3.74	.70	11.04	22.46	.49
Total	7.01	7.49	.93	28.01	36.55	.77
Hardwood:						
North	3.84	1.24	3.10	9.60	4.33	2.21
South	3.24	2.01	1.62	9.52	7.88	1.21
West and coastal Alaska	.15	.02	6.48	.27	.08	3.31
Total	7.23	3.27	2.21	19.39	12.29	1.58

[1] Ratios computed before rounding.

SOURCE: U. S. Forest Service, *Timber Resources for America's Future* (Washington: U. S. Government Printing Office, 1958), Table 36.

However, these national totals conceal some rather unsatisfactory situations. The over-all national balance is not duplicated in any region, or for any important species, or by size and quality groupings of trees. Softwoods are being cut for sawtimber much faster than they are growing; there is a relatively ample supply of hardwood sawtimber, but much of this is of species or grades not in demand. More timber is being grown in the North, and to a lesser extent in the South, than is being harvested; but this again is largely a quality and species problem. On the other hand, the timber cut in the West is still greatly in excess of growth; given good management after cutting the mature trees, however, the growth rates on these lands will increase materially. In general it may be said that, as far as sheer volume is concerned, most of the eastern species show growth rates for both growing stock and sawtimber in excess of cutting—quality and species distribution is another story—and that most of the major commercial softwoods of the West show unfavorable growth-relationships, both for growing stock and sawtimber.

The total cut of live wood in 1952 was divided into three groups of roughly equal sizes: eastern hardwood, eastern softwood, and all western timber (Figure 44). To this was added dead and cull trees, and some imports. Amazingly enough in this modern age, the largest single use of wood was for fuel—mostly residues of various kinds burned locally and as the plants were available—to generate steam and for other relatively low-value uses. The largest commercial use was for lumber, with pulpwood a close second. Total unused residues, at plants and in the woods, are estimated at just 20 per cent of the total volume of wood fiber. These volume relationships would, of course, look very different if placed on a value basis.

Some Economic Relationships in Present Forestry

Thus far in American history forestry has consisted chiefly of harvesting mature virgin timber or second or third growth which, to use an agricultural analogy, might be called "volunteer"—that is, one which seeded itself and grew up without specific cultural practices by the farmer or forester. Most of the immense volume of physically superb timber that once grew in this country has been harvested, but some of what remains will await cutting for two or three decades. Forestry has been dominated by this original virgin forest situation.

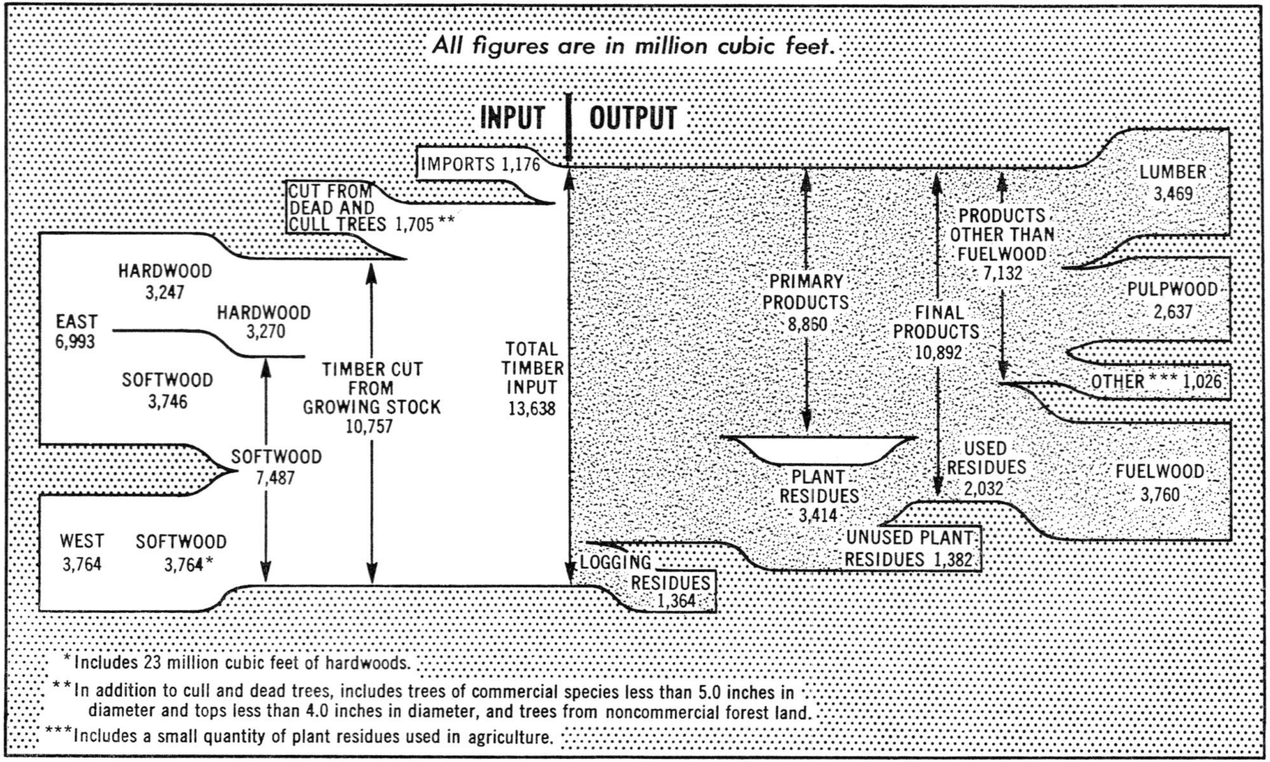

Figure 44. Input and output in the timber economy of the United States, 1952. (Chart from *Timber Resources for America's Future,* U. S. Forest Service.)

To some extent it still is. The nation and forestry as an industry have not yet fully faced up to the technical and economic problems of growing trees purposefully, beyond merely harvesting them. Some important beginnings have been made, it is true, and much further progress will surely be made in the next decade or two.

Logs and other products from the woods are raw materials roughly comparable to petroleum from the field or ore from the mine or unginned cotton from the field. They must be processed, sometimes through several steps, before the wood fibers they contain are usable to any ultimate consumer. Many of the primary products of forest processing are themselves raw materials for other processes. Lumber, for instance, the major product from forests, is in turn a raw material for building and other purposes. The log as cut in the forest is heavy and bulky in proportion to its value; both it and the tree from which it was cut contain a considerable amount of waste material, although the proportion of waste has declined as more complete utilization has developed in the forest industry. For these reasons, it is usually most economic to process the logs or other forest products, at least through the first stages and increasingly through later stages as well, reasonably close to the site where they grew. The average haul for lumber, it will be remembered, is over 1,000 miles, and for paper, pulp, and paperboard, over 700 miles. Comparable data for the logs from which these products were made are not available, but it is probable that they move an average distance of between 50 and 100 miles. Occasionally raw woods materials are shipped quite long distances, but the larger part of the sawlogs goes through a sawmill at the edge of the forest.

Any organized market for logs and other raw woods products hardly exists, in the economists' sense of a market where either the goods are physically traded or where title to them changes hands. There have at times been log markets in areas where logs were floated down streams; this was once true along the Columbia River and in the Puget Sound region. But these have been exceptions in forest history. In some areas today pulpwood buyers from competing mills buy pulpwood from growers in the same area, thus providing alternative outlets. Aside from the fact that the more usual situation is one where potential buyers come singly or not at all, this does not really constitute a market in the classical sense. The goods are neither standardized nor physically present at a central point for inspection—the latter would usually involve excessive transportation costs—and the buyers and sellers or their representatives do not come together at one point in space and time. Logs and other woods raw materials either move from forest to primary processing point while under one ownership, or sale takes place under conditions which do not, and perhaps could not, equal the ideal economic market.

Nonforesters often do not realize that the products of the forest are varied as are the products of agriculture or of industry. Not only are there different kinds of trees, each with its particular physical properties of hardness, brittleness, resistance to rot, workability, appearance, and so on, but there are great variations in grade or character of individual trees of the same species. Some will produce logs free from rot, defects, and knots, with wide growth rings and attractive colors; others will have very different characteristics. Moreover, different parts of the same tree will have different physical usefulness and economic value. The butt log may be more valuable than the uppermost log; and the limbs may be worthless. Some logs will have high value as peelers for the best grade plywood. The sizes and quality of the lumber that can be sawed from logs differs according to the quality of the log itself. Sound logs, even if crooked, may make pulp without penalty, but crookedness is a serious deficiency for lumber. The trimmings, edgings, and sawdust from lumber manufacture are likely to have limited value unless they can be used for pulp manufacture. While the best grade and highest value logs can be used for low-value purposes, this would ordinarily be highly wasteful—rather like using a Cadillac limousine as a pickup truck. The most profitable forestry involves skillful sorting of the logs and parts of logs coming from the forest, so as to put each to its highest economic use.

Because of these facts, there is demonstrable economic advantage in the integration of woods and mill operations, in diversification of the latter to take advantage of all the major qualities of the wood fiber, and in horizontal as well as vertical integration of the wood processing and selling operations. There has been a major trend toward integration of forest operations in recent years, most notably in the Pacific Northwest and in California. Integration of woods and mills may in time significantly influence the species, sizes, ages, and quality of the trees grown and harvested; but for the present the line of economic influence runs predominantly the other way—given what the woods produce, integration, combined with diversification of mill operations, is necessary to get the full value out of the logs. In some areas it is possible to have industry diversification without diversification, necessarily, of each individual firm; plywood manufacturers, sawmill operators, and pulp mill operators may trade or buy and sell logs or the pulp mill may buy residues from other mills, so as to put the logs and their various parts to the greatest economic use. However, this type of trading has its problems, and integration within the same firm seems to have more advantages.

From the viewpoint of many larger mills, some control over a significant part of the raw materials is needed. In the recent past there has been more profit from processing raw wood than in growing it. Major capital gains have occurred from buying timberlands at low prices and holding them for eventual price rises, but growing wood fiber is different. By owning or controlling a part of its sources of supply, a forest industry is better able to be assured of the needed volume and quality of raw material at prices less subject to market fluctuations; this has been one major force behind many of the consolidations and acquisitions in all forest regions since the war. Ownership of at least part of its supply makes the mill less dependent upon seasonal or erratic changes in supply from woods owners, and gives greater control over short-run price fluctuations. It is possible, though direct evidence is lacking, that ownership of a considerable part of raw material supply gives the processing firm a major bargaining advantage over small forest owners.

As forestry grows more intensive, there will be much more integration of woods and mills, more diversification of the latter as far as processes and products are concerned, and larger units and greater vertical integration as far as marketing is concerned. But it still remains true that by far the greater part of the *area* of forest land is not within integrated operations. Only 13 per cent of all forest land is owned by forest industries, and some of this is not closely integrated with processing plants. Various arrangements are being developed whereby owners of small forests tie their operations more closely to those of particular processing mills, but these are not yet extensive as to total area. Most of the farm, "other," and public forests are still not integrated with particular mills.

Some hypothetical or generalized relationships of timber volume, tree values if cut, and costs are shown in Figure 45. These might apply to a single tree in a stand of mixed ages, or to a whole stand of even ages. The volume of wood in the tree or stand would grow slowly at first, then faster, then growth would slow down, then reach a virtual plateau, and finally drop to zero when the tree or stand

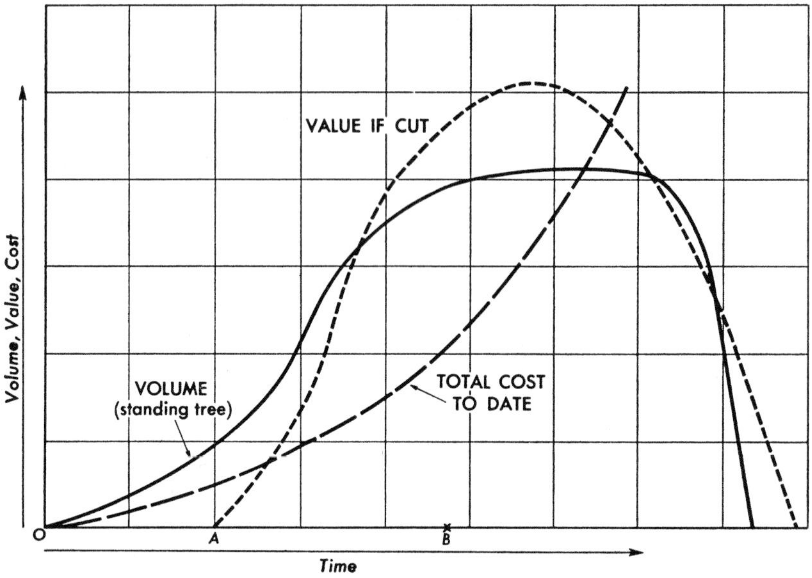

Figure 45. Hypothetical volume, value, and cost curves for forestry in the United States.

blew down or fell from decay. The length of the time periods in each stage, the absolute peak of total volume, the rates of change at different periods, all might vary greatly from species to species and from site to site. Southern pines cut for pulpwood might reach a point well up on the growth curve in twenty years and a peak in fifty to seventy-five years; Douglas fir might not reach comparable growth stages for a hundred and six hundred years. If we were considering a stand of even-aged trees, such as would normally result from planting or from natural seeding of those species that come in after fires, the numbers of trees in the stand would change greatly as the total volume of growth increased.

The value of the trees, if cut or for cutting, would follow a somewhat different course. Until the trees reached some size they would have no value for cutting; in the Figure, we have shown this by point *A*. For Christmas trees, this point might be reached in six years; for southern pine for pulp wood, in fifteen years; and for Douglas fir, in fifty years. These are not, it should be noted, either the point of greatest value or the most economic point of harvest; they are rather the points at which *any* net value to the stumpage arises from cutting. It is true that the younger stand may have value, because it is growing and may in time reach a merchantable age and size. But the value of growing stands before merchantable size is notoriously low and variable; the owner could not realize on this value without sale or mortgage of the property, and either may be difficult. And however thrifty the young growing stand may look, there are always unpredictable risks from fire, disease, and insects to bear in mind.

Once the tree or stand acquires value if cut, its value rises much faster than the physical volume of the tree or stand rises. The curve of value may rise farther and longer, relatively, than the curve of volume; even after the tree or stand reaches virtual maturity, the quality of its timber may increase for some time. However, the value curve will almost surely turn downward before the volume curve does; although the tree still stands, over-ripeness and finally rot begin. The precise relations between the volume and value curves in different situations and for different species may be very different from what we have shown. If Christmas trees are grown, for instance, value may reach a peak long before volume; there is a rather sharp limit to the size of Christmas tree that most people want. Value for pulpwood might also reach a peak before volume did; the largest trees would be too difficult to handle in ordinary pulpwood operations. For lumber or plywood, however, value and volume relations would generally be as is shown on the figure. It is assumed there that the tree or stand would have some salvage value after it blew down or fell, and in fact for some time thereafter; but it is possible that rot would completely destroy the value of the tree while it still stood.

The total costs of the tree or stand follow a still different course. For this purpose we have assumed that costs include taxes on the land and timber, fire prevention and control costs, current management costs before harvest, interest on the bare land value, and interest on the value the tree or stand would have if cut. The proportions among these various costs might vary greatly from one area to another, and the proportion borne by the general public might also vary—fire costs in particular might be almost wholly public, and taxes might include a measure of subsidy. Under many situations, however, a major, if not the dominating, cost would be interest on the value of the tree or stand that could be realized by cutting or by selling the land and stand of trees. We have assumed that the latter was moderately large; hence total costs rise at an accelerating rate, as total value rises. Total costs of the tree or stand do not reach a peak, but continue to rise until liquidated by cutting, after which they fall sharply.

The relationship between the cost and value curves may vary greatly from situation to situation. At the extreme, the costs may be so high that they at all times lie above the value curve. In Figure 45 the maximum net value of the tree or stand is reached at point *B* in time. It will be noted that this is before either volume or value has reached a peak. The time period up to point *A* is obviously the minimum production and planning period, unless one hopes to sell the property while the trees are smaller. The costs and returns for that whole period must be considered. After that, theoretically every year but in practice at longer intervals such as five years, a cutting program might be undertaken and the tree or stand harvested.

Each such period must therefore be considered a production or planning period; the decision facing the owner is to cut or to leave for further growth. On the income side will be the gains in value, due to physical growth and to changes in price per unit of volume; on the cost side accumulated costs will surely rise, due to all the factors previously mentioned. For each planning period, it is the comparison between added income and added costs that is controlling. As the tree or stand acquires relatively large value, the interest on the money that could be realized from harvest will become larger and larger, and may be the dominant part of total added costs. If maximum net income is the objective, then harvest should take place when marginal income is just sufficient to meet marginal or added costs from holding; and large among the latter will be the relevant interest rate to the owner of the land. The maximum profit point will lie to the left of point *B*. If he is a farmer, with a serious shortage of capital, he may choose to liquidate the tree or stand in order to use the capital elsewhere; if he is an "other" forest landowner, not dependent upon the forest for his main income and perhaps lacking a good alternative investment outlet for his money, he may choose to leave his capital in the growing trees for a much longer period. If the tree or stand is part of an integrated and well-run total forest operation, the owner may choose to liquidate at the earliest date in that wide range of tree size and volume where profitability is often approximately constant, in order to keep his capital as liquid as possible and to have a turnover as short as possible.

The above discussion assumes that the landowner continues to own the land and to use it for forestry. Perhaps a prior question should be: Is it profitable to own forest land? If someone were considering a piece of forest land as a business or as an investment, he would surely want to consider the growth, value, and cost relations that probably would exist on the land; and then to decide if any age of harvest would promise returns on his capital equal to his next best alternative investment, and, if so, what age of harvest seemed best. In fact, however, very little forest land is owned under these simple conditions. Certainly, that 27 per cent which is publicly owned is guided by different considerations. Even for the 13 per cent owned by the forest industries the chief consideration is not the woods alone, but the woods and mill in combination; and the need to have a supply of wood fiber in order to operate a mill profitably may be dominant. For much of the 61 per cent that is owned by farmers and "others" there seems good reason to believe that relative profitability of the woods operation is only a minor consideration. Farmers often have woodlots because they are part of the farm as they bought or inherited it; the land is not suitable for farming or will not justify the costs of clearing, and there is no easy way to sell the woodlot away from the rest of the farm. It is often difficult to devise a farm plan which offers net income from woodlot operations as high as from use of the labor and capital elsewhere on the farm.

If we assume that the forest landowner will continue to own the land almost irrespective of the costs and returns from doing so, then the value-cost relationships change considerably. Some of his costs, such as taxes on the land, interest on the bare land, and others, are no longer relevant to decisions about how to operate the forest—they are incurred, and will be met, anyway. He need only consider what extra costs and extra returns he will receive from a particular forest management program. Moreover, he must consider what proportion of the costs are or might be publicly borne; if he can succeed in shifting some of the costs, or getting subsidies to meet some of them, this also affects his level of profitable operations.

As a matter of fact, very little is known about the volume-value-cost relationships for forests of different species, on different sites, and operated in different size units. Agricultural economists are far from satisfied with the data on output, costs, and income for farms of different sizes and types in different parts of the country; but for the forest economy infinitely less is known. We lack even the physical data on growth rates and value curves under different conditions and methods of management; and data on forest operating costs are conspicuous by their absence.

As Table 45 shows, for all forests in the country the average rate of physical growth in 1952 was only 2.75 per cent for growing stock and 2.30 per cent for sawtimber. These averages are distorted by the large volume of old-growth timber in the West, where little or no net growth is taking place. If the West is excluded, the rates range from 5 to 7 per cent on the average, depending on whether it is softwood or hardwood and whether it is growing stock or sawtimber. Since the average rate of stocking for all forests is only about half of capacity, it can well be argued that growth rates under full stocking would be higher. However, full stocking would also mean a larger volume of standing timber, so that in spite of more intensive forestry practices the rate of growth in relation to standing timber might not be much greater.

While these data on physical volume growth rates do not directly indicate the growth in value, for large areas of forest of all ages and conditions it may well be that, on average, growth in value is not far different from growth in volume; some trees or tracts would be growing faster, some slower, in value than in volume. If the annual growth in value is no more than 5 to 7 per cent, it should be obvious that forestry of this type is a poor investment. Such a rate of *gross* return would scarcely pay interest on the realizable value from liquidation, and in addition there are some other costs. There are situations where the liquidation value of nearly peak stands of old timber is greater than the value of all future income from forestry on those lands, discounted back to a present worth basis. This is true for areas in central Arizona[9] and southern Oregon,[10] for instance. Such situations could

[9] George W. Barr, Carl A. Anderson, and Charles F. Cooper, "Recovering Rainfall—Arizona Watershed Program," University of Arizona, State Land Department, Water Division, and Salt Valley Water Users' Association, Tucson, 1956.

[10] *Hearings before the Subcommittee on Indian Affairs of the Committee on Interior and Insular Affairs,* United States Senate, 85th Congress, First Session, on S. 2047, October 2 and 4, 1957. See statement by George Weyerhauser, p. 156 ff.

obviously exist only on publicly owned land, because no private owner could afford to forego the greater income resulting from liquidation.

There are many sites in the United States where the growth rate typically is low. It is difficult to see how intensive forestry in such areas can produce wood fiber economically; at the best, such areas may be allowed to grow such timber as they will "naturally," with a minimum of outlay, and then be harvested when enough volume is attained for an economical harvesting operation. Many forest landowners have high costs of management and operations, or are unable to get reasonably full value from their products, primarily because their operations are so small.[11] In areas where output is low and costs are high under present sizes of ownership it seems unrealistic to expect high wood output as long as the lands are in private ownership. While more intensive and perhaps better organized public programs to aid these owners would produce some results, the basic difficulty may be that the areas are simply too low in productivity to warrant better operation.[12]

One basic problem of forestry for the future—perhaps *the* most important one—is a greatly widened and deepened understanding of these economic relationships for all important species, sites, and types of forestry operations. In discussing the unfavorable situations, as we have, one should not lose sight of the fact that some situations are well above average; growth rates are high, value rates are high, and costs are low, all comparatively. Under these circumstances, private forestry can produce—and to a large extent is producing—wood fiber efficiently and profitably.

Taxes on Forest Land and on Forestry Operations

Taxes are often a major cost in forestry. This is especially true for the growing of trees where, aside from interest on the value of the growing stock, taxes are often the largest cost. But taxes may also be important to forestry operations, and operation in a manner to mini-

[11] W. M. Carroll, C. E. Trotter, and N. A. Norton, *Marketing Forest Products in Pennsylvania,* Progress Report No. 131 (University Park: Pennsylvania State University, 1955).

[12] Marion Clawson, "Economic Size of Forestry Operations," *Journal of Forestry,* Vol. 55, No. 7, July 1957.

mize the tax burden may be as important for forestry as for any other business operation. And, with today's taxes, this is often very important indeed.

As far as forestry is concerned, taxes take three major forms: (1) taxes on real estate, including the value of the standing timber, levied in proportion to assessed values per acre; (2) yield or severance taxes, based on the value of the timber, and payable when the timber is cut; and (3) income taxes, including taxes on capital gains, paid out of income from forestry operations. Various combinations of these taxes exist; typically, forest landowners and operators pay more than one of them. In an earlier period it was often contended that annual taxes based on the assessed value of the land and growing timber were not only a serious burden on the forest landowner but also an incentive to early liquidation of the timber. In partial response to this argument, yield and severance taxes have been enacted in several states. In general, taxes under these plans are deferred, in whole or in part, until the timber is cut, when a substantial proportion of the revenue from timber sale (one-fourth, in some instances) is paid in lieu of the postponed taxes. A real estate tax on the value of the forest land and a severance tax on the timber when cut are sometimes combined. In most states use of the yield or severance tax instead of the real estate tax is optional with the forest landowner. While the area of forest land under deferred tax plans is relatively low, it may be true that the existence of such an arrangement exerts a moderating influence on the level of real estate taxes.

As to income taxes, where individuals pay according to the level of their income, corporations pay 48 per cent of net income. Particularly important for forestry is the fact that income from gains in value of timber, including from growth of timber, is considered a capital gain, and hence pays a tax of 25 per cent instead of the higher corporation income tax. This provides a substantial tax incentive to the growing of trees, especially for corporations, and especially for those with high incomes. Were gains in value of timber from growth not considered as a capital gain, timber companies would be tempted to sell their stands of timber, take the capital gain, and buy other timber at a higher price. This would be disruptive to the continued management of timber stands, essential for the best management.

In the forest taxation field there are at least three major policy issues: (1) What is forestry's appropriate share of the total cost of government? (2) What effect do the level and method of establishing taxes have upon the management of the forest resource? and (3) What effect do different tax systems have upon the sizes and kinds of forest enterprises? The first is a general governmental problem, but the way in which it is solved may have important effects upon forestry. In practice, real estate taxes on forest property have often been so high as to absorb the entire net income of timber growing. This obviously discourages forestry, and it would seem to argue that such taxes are much too high.

Taxes may affect the intensity of forest operations, or the length of the cutting cycle, or both. Their effect may be to encourage more intensive forestry and longer cutting cycles, or the opposite. It can be argued, for instance, that severance taxes tend to penalize intensive forestry, because the greater output must repay not only the costs of producing it but also a substantial margin for deferred taxes. The actual practices in levying and collecting taxes are often more important than the theoretical advantages or disadvantages.

It is often contended that treatment of income from timber growth as a capital gain, plus the possibilities of shifting much of the total income to the woods operation as against the processing operations, provide substantial net advantages to large integrated forestry operations. This may well be true and, as has been noted, there are other advantages to large-scale integrated forestry operations. It could also be argued that the small-scale private forest owner, whose income taxes are probably low, has a substantial tax advantage in producing timber. Yet, on the average, this group seldom practices any but a very extensive type of forestry.

No thorough study of the entire forest taxation field has been made in recent years, although there have been several studies of forest real estate and severance taxes. Since taxes necessarily involve public action, this would seem one place where the kind of forestry desired on grounds of general public welfare might be influenced by public action.

Summary of Forestry Today

Before we consider the problems and opportunities of forestry as a land use in the future, it seems worth while to summarize some of the main points brought out in the preceding pages:

1. Forestry is a major land use in the United States. Commercial forests occupy somewhat less land than agriculture, as defined in this

study, and much less than grazing, but far more than any other major land use.

2. Forestry in the United States is in transition from an industry primarily concerned with the harvesting of virgin old-growth timber, to one primarily concerned with the growing of new timber under relatively intensive management. The former situation prevails in some regions, and still shows its mark everywhere; but the second situation is coming increasingly into importance.

3. There are many and diverse situations within forestry; it is not physically or economically homogeneous. As to ownership, there are public forests, forest industry lands, farm woodlots, and "other" categories, each with its own characteristics and problems. The situation differs considerably by regions, from the West with its large volume of old-growth timber, to the South with its rapidly growing southern pine, to the North. The situation also differs as to major species and forest types. Caution therefore is required when making generalizations.

4. By and large, the output per average acre is low—low when compared not only to biological potential, but also to moderately good commercial practice. Much of the forest land is in far below optimum condition.

5. At present, growth and drain are about equal for sawtimber, and growth exceeds drain for growing stock. No national "famine" is present or imminent. On the other hand, the situation is not so nicely balanced by regions, by species, for hardwoods and softwoods separately, or by timber quality.

6. The productive potential of the forests is much greater than is now realized.

7. The most serious problems are on the more than 4 million small forest land holdings.

8. Perhaps the basic question is, how much additional output is economic—economic for the nation, economic for the individual producer?

9. Lastly, the lack of really appropriate information about output, income, and costs of forestry for different species, under different regional and site conditions, and for ownership or operating units of different sizes, is critical. We know very little about the economics of wood fiber production, yet this question is likely to be all-important in the future.

FORESTRY IN THE FUTURE

What seems the most probable future use of forest land in the United States, in the light of the past and present situation as we have analyzed it? Perhaps the best place to start is with the demand for wood fiber for its many uses. Demand will be a major influence, perhaps *the* major influence, affecting forest land use in the future. The physical potential exists for much greater output of forest products; the extent to which it will be economic to increase output will depend in large part upon the demand. This is not to deny the importance of supply factors, which will be important indeed; but their effect will be operative within the limits set by the potential demand.

Demand for Wood

When it comes to estimating the future demand for forest products, many factors may enter. There is, first of all, the matter of the total population and its various components. Perhaps the most important population factor is the rate at which new households are being formed. One major use for wood is for dwelling construction; this, and perhaps other uses as well, are more directly correlated to the number of new households formed each year than to the total number of people. Some building is replacement of structures no longer usable. But it has been new or additional dwelling construction, rather than replacement building that has provided most of the demand for lumber.

The size of the gross national product and of personal disposable income also affects the demand for forest products. Except for firewood, there seems to be no forest product whose consumption declines as incomes rise. With higher incomes and larger industrial output, more forest products will be consumed if supplies are adequate and relative prices are unchanged. Then there are questions about the level of various economic and social activities which use forest products heavily, or could do so. Even with the same level of population and total economic activity, there might be differences in the rate of specific activities.

Lastly, and perhaps most important of all, what will be the role of wood and its products, compared with the role of its substitutes? We have seen that coal, oil, and gas have largely but not wholly displaced wood as a fuel; concrete, brick, and other building materials have displaced lumber to some extent as a building material; and synthetic plastics compete with paper as a wrapping material. The future competitive relations among these and perhaps newer products may be critical to the demand for wood fiber. There are almost innumerable combinations of assumptions as to the future course of each of these, and perhaps of other, factors.

Fortunately for our purposes, two careful studies of the future demand for wood fiber have been made. We simply summarize them without attempting original analysis.

The Forest Service has made careful projections of the future demand for timber products under various assumptions.[13] They are projections not only according to the terminology used in the Forest Service study, but also according to the definitions we have been using in this book. That is, the authors started with certain basic assumptions which seemed reasonable to them; from these they proceeded with a careful and detailed analysis of each sector of demand for timber products. They considered such matters as probable residential construction, nonresidential construction, building maintenance and repair, railroad, farm, mine, furniture, packaging, and other uses of timber products, each separately and in some detail. While judgment necessarily entered at many points in the analysis, yet it is basically a set of projections rather than of forecasts as such. Other analysts starting with the same premises might reach somewhat different conclusions, but large differences of opinion would be much more likely to arise over the basic assumptions.

In order to keep the analysis within reasonable limits of work and space, the Forest Service chose three major sets of assumptions for its projections, as follows:

1. *Lower.* This uses the Bureau of the Census Series B projections[14] for population for 1975, and estimated population for 2000 by extension of those projections. This results in a projected figure of 215 million for 1975 and 275 million for 2000. Gross national product is estimated on the basis of certain assumptions as to size of working force, hours of work per week, and real output per man-hour; in general, this means a slightly slower rate of increase in gross national product

[13] *Timber Resources for America's Future,* cited above.

[14] In Current Population Report Series P-25, No. 123, October 1955.

per capita in the future than in the past. Total gross national product on this basis would reach $630 billion in 1975 and $1,200 billion in 2000. The third and crucial assumption for this projection is that "future prices of timber products will rise substantially faster than prices of competing materials; with resulting extensive price-induced substitution of non-wood materials for timber products."

2. *Medium.* This is based on exactly the same estimates of population and gross national product as the lower projection. The crucial assumption in this one is "no change in relative prices; trends in future price of timber products will, in general, parallel price trends of competing materials."

3. *Upper.* This projection uses exactly the same price assumption as the medium projection, but it employs instead the Census AA projection (extended to 2000), which yields a total population of 228 million in 1975 and of 360 million in 2000. Gross national product rises because population is greater; but not proportionately, because the age distribution, and hence the number in the working force of the larger population, would be different. Because the upper projection yields results only insignificantly different from the medium for 1975, this analysis is not presented for that year, but only for 2000.

It should be pointed out that the medium and upper projections, while assuming no adverse price movements for timber products, do *not* assume constant per capita consumption of them. Even with constant price relationships, some differences in consumption seem probable. If present per capita uses of timber products were continued until 2000, total wood demand in that year would be considerably above those projected.

It will be remembered that our own projections for population (pages 6–13) are (*a*) 240 million people for 1980, compared with 215 million for the lower and medium Forest Service projections and 228 million for the upper for 1975; and (*b*) 310 million for the year 2000, compared with 275 million and 360 million, respectively. For gross national product, our projections are $890 billion for 1980 and $1,670 for 2000, compared with the Forest Service's lower and medium projections of $630 for 1975 and $1,200 for 2000. Thus, our projections for population are about the same as the Forest Service's upper projection for the middle period but lower for the later years, and definitely above the Forest Service lower and medium projections. Our projections for gross national product are higher at all periods than any Forest Service projection. Comparatively, then, the Forest Service projections are somewhat more "conservative" than ours; but the differences are not so great as to make them wholly incomparable.

Another excellent study has been made by the Stanford Research Institute.[15] From the introductory discussion in its report, it is not clear the extent to which the study is a forecast or a projection, in terms of our earlier distinction. Apparently it is more nearly a forecast, but the methods used are closely similar to those of the Forest Service for its projections. The analysis is made on only one set of assumptions, and only as far as 1975. Population is projected on the assumption of a gradual decline in birth rate to the 1946 level; as a result, the projected total population for 1975 is 212 million. Gross national product, projected through much the same method as that used by the Forest Service, is estimated at $586 billion for 1975. The Stanford report thus uses lower figures for total population and for gross national product than does the Forest Service study, and still lower ones than we use for our projections.

Starting with these basic assumptions, the Institute examines each major use of wood in some detail. Trends in past use and technology are examined carefully and their effect upon use of wood are evaluated. The resulting projections of probable future demand for wood are thus built up, item by item, into totals. While this study, which was completed first, differs in many details from that of the Forest Service, the general approach of the two is closely similar.

The results of these two sets of projections are shown in Table 47. The basic assumptions as to total population and gross national product are shown for each projection, and in addition one other basic assumption, which unfortunately was in different terms for the two studies, as to the level of economic activity in the wood-using field.

The two studies are in agreement in projecting a higher level of lumber consumption in 1975 than in 1952; the Institute study projects a 7 per cent higher level, the Forest Service a 15 per cent and a 34 per cent higher level for the lower and medium projections, respectively. For the year 2000, the Forest Service projects still higher consumption of lumber—under its upper projection, a more than doubling over the 1952 level. Lumber consumption in the United States seems to show no clear trend since 1900 (Figure 46). Total consumption trended steeply downward until 1932, and of course per capita consumption

15 Stanford Research Institute, *America's Demand for Wood, 1929-1975,* a report to the Weyerhauser Timber Company, Tacoma, Washington, 1954.

Table 47. Projected demand for forest products, 1975 and 2000, under alternative assumptions

Item	Unit	1952	Projection 1975: Forest Service Lower	Projection 1975: Forest Service Medium[1]	Projection 1975: Stanford Research Institute	Projection 2000: Forest Service Lower	Projection 2000: Forest Service Medium	Projection 2000: Forest Service Upper[1]
Total population	Million	157	215	215	212	275	275	360
Gross national product	Billion $ (1953 prices)	350	630	630	586	1,200	1,200	1,450
New construction expenditures	Billion $ (1952 prices)	32.3	[3]	[3]	44.8	[3]	[3]	[3]
Input of physical structure	Billion units[2]	5.9	8.3	8.3	[3]	12.2	12.2	14.7
Domestic consumption of:								
Sawlogs for lumber	Billion bd. ft.	41.5	47.6	55.5	44.6	54.8	79.0	90.0
Pulpwood[4]	Million cords	35.4	65	72	60.7	90	100	125
All timber products, including fuelwood	Billion cu. ft.	12.3	14.2	16.2	13.7[5]	17.9	22.4	26.2

[1] The Forest Service had an upper estimate for 1975 based on 228 million people and $645 billion gross national product, but the difference in timber consumption, compared to the medium estimate, was too small to be considered significant.

[2] Quantities of various physical structure materials purchasable at 1935-39 national average prices.

[3] Data not available.

[4] For Forest Service estimates, includes pulpwood net imports and pulpwood equivalent of woodpulp and paper. Since Stanford Research Institute data were on a somewhat different base, the figure given here was estimated by the authors, measured from Forest Service base and using absolute increase projected by Stanford Research Institute.

[5] Stanford Research Institute projections are on a different basis than Forest Service projections; the authors have taken the difference reported by Stanford Research Institute between 1952 and 1975, and added it to Forest Service base figure for 1952, to get a roughly comparable figure.

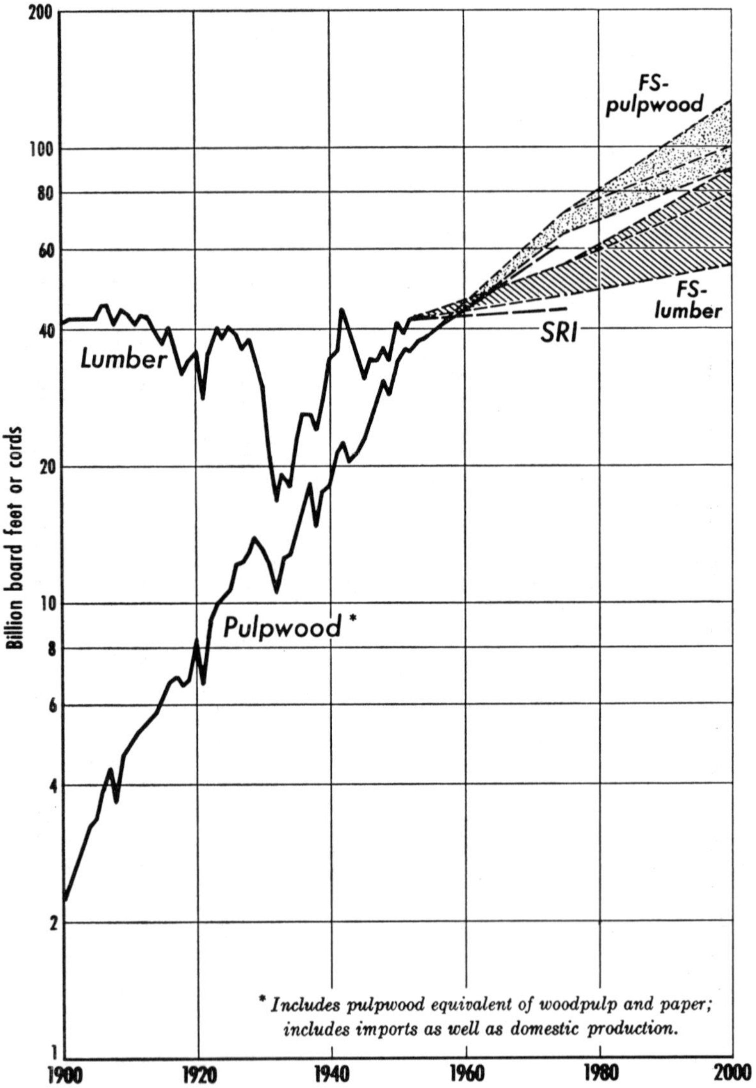

Figure 46. Lumber and pulpwood consumption, 1900–1955, and alternative projections, 1975 and 2000.

was even more rapidly downward; since that year, total consumption has been steeply upward, and per capita consumption has about doubled from the extreme low point. Anyone would agree that many of the year-to-year changes shown in Figure 46 were due to unusual conditions—the Great Depression, World War II, and the like. What would have been the trend in lumber use had peace and full employment been constant? No one knows, or can know. It might be argued that the situation in the 1920's was not too abnormal—general economic conditions were fairly good, building activity was high, and there was peace; and similarly it might be argued that the late 1940's and early 1950's were not too abnormal. But there was almost no trend in total lumber consumption between these two periods. Nevertheless, these two groups of specialists found good reason to believe that total lumber consumption will rise; and the Forest Service medium projection indicates that if lumber prices do not rise proportionately to other prices, there will be a large increase in total lumber consumption.

The two studies are in agreement in projecting a much larger rise in pulpwood consumption, although they disagree as to the extent of this increase. The Forest Service projects a rough doubling by 1975 compared with 1952 even on its lower estimate, and somewhat more on its medium one; the Institute projects over a 70 per cent increase. The Forest Service projects further large increases by the year 2000. The total consumption of pulpwood has been rising rapidly for many years, yet both studies agree in projecting a future growth in pulpwood consumption at a slower rate than in the past. Figure 46 should make it clear that neither study was a mere "trend extender," but that other factors entered the analysis.

Total wood consumption for all purposes is projected to rise in each study. The Institute study projects an 11 per cent rise by 1975; the Forest Service, 15 per cent and 32 per cent for its low and medium projections, respectively. For 2000, the Forest Service projects a 45 per cent, 82 per cent, and 113 per cent rise for its lower, medium, and upper projections, respectively. Total industrial wood consumption shows no major trend over the past if one ignores the experience of the depression years (Figure 47). Full data are lacking, but it appears that the amount of wood used for fuel has declined considerably. For the future, further large decreases of wood used for fuel seem probable. Much of the wood that has been going into this use—trimmings,

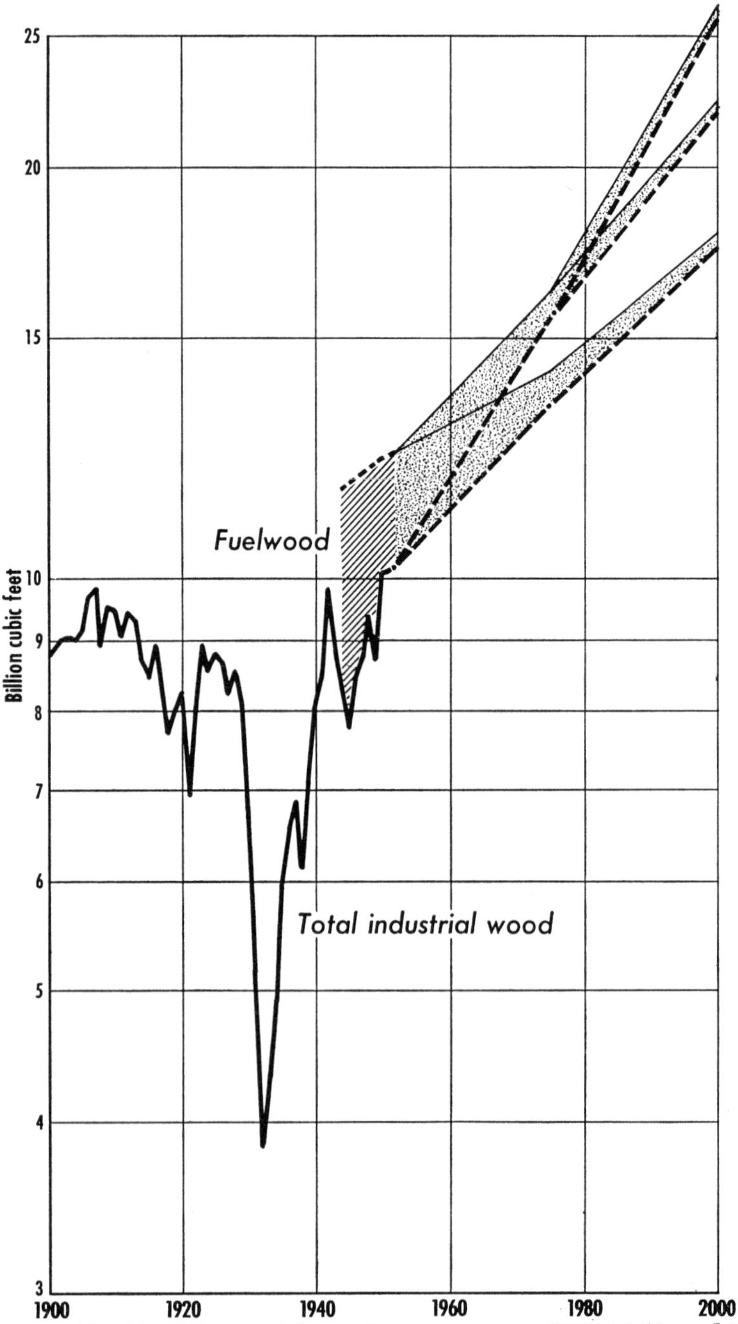

Figure 47. Total industrial wood consumption, 1900–1952, and U. S. Forest Service projections for 1975 and 2000.

edgings, and other sawmill or woods waste—will in the future be used for pulp. If, for all industrial wood, one accepts our previous hypothesis that the 1920's and the later 1940's and early 1950's were not too far from normal, then the Forest Service lower projection is not much out of line with an extension of the trend between these two periods.

The Institute apparently assumed that wood supplies would be adequate to meet the demands it projected; this is not explicitly stated, but is implied if the demands are in fact forecasts. Certainly the prices at which products would sell were considered in arriving at demand estimates. The Forest Service, however, definitely did *not* take for granted that timber supplies would be forthcoming to meet potential demands, but rather that special steps were necessary if potential demand was to be met, and that even then it might not be met.

Supply Possibilities

Since potential demand is well above present output, how can output of forest products be increased to meet it? Increases in supply might come from increases in area of forest land or from increases in output per average acre, or from imports or from some combination of these.

How much is the area of forest land likely to change in the future? Where and how might additional acreage be added, and where and how might some acreage be lost? It must be recognized that most forestry is a comparatively low economic use of land; if the land is physically suited for other uses, and comes into active demand for them, then forestry usually is unable to offer stiff competition in bidding for the land.

The chief process, in fact almost the only process, by which additions will be made to commercial forest land is by conversion of poor, unneeded farm land to this use. Our analysis of the use of land for agriculture, in Chapter IV, shows an actual or a potential surplus of agricultural commodities in prospect for some years into the future and that some land now used for farming might be shifted to other uses. The difficulties of making such shifts are often great, but it is both possible and probable that some shifts of this character will be made. The maximum area that might be shifted out of agriculture and into forestry is in the range of 30 to 50 million acres. Most of this is in the South, primarily in the hilly areas and areas of poor soils, but there is some also in New England and elsewhere around the fringes of the better commercial farming areas. It is unlikely, however, that any shift as great as this will actually be made.

Some of the land now classed as noncommercial forest has excellent timber, but is reserved for parks or other purposes. It seems highly unlikely that any significant area of this type will become available for commercial forestry in the future. Some of the better noncommercial forest land, such as that with stands of lodgepole pine, has been or might be shifted into a relatively low-grade commercial forest classification; but the area and volume of timber so involved will be small.

Some land will be lost to commercial forest use in the next several decades. To the extent that cities grow into presently forested areas, this land will shift in classification. Such changes will be primarily in the eastern half of the country and not exceed a few million acres. Some presently commercial forest land may be taken into parks in the future. It is clear from the discussion in Chapter III that a large expansion of parks or recreation areas is needed, especially of those areas we have called "intermediate." Some of this expansion may be on lands now classed as commercial forest land. However, in such cases the best forests would rarely be required, and it might be possible to work out systems of management which would permit at least some of the timber growth to be harvested. Even if state park and other intermediate-type recreation areas expand in line with the potential demand, it seems unlikely that this would take as much as 30 million acres out of the commercial forest classification by the year 2000.

Some land now in farm woodlots and classified as commercial forest land is almost certain to be cleared before 2000. The farm woodlot, which once performed an economic function for the farm family as a source of building material and fuel, no longer does so in many areas. In the midwestern states there are 35 million acres of commercial forest in farm woodlots; some 13 million acres of this are on land which the Soil Conservation Service classifies as Classes I, II, or III—suitable for continued cultivation without excessive difficulty.[16] It seems probable that most of this land will in time be taken from forestry and put into agricultural use. By 2000, the total loss of Middle

[16] H. H. Wooten and James R. Anderson, *Agricultural Land Resources in the United States,* Agriculture Information Bulletin No. 140, U. S. Department of Agriculture, June 1955, p. 37.

West farm forests to agriculture might run as much as 15 million acres. There are even larger acreages of good land in woods in the South. These could be cleared and put into farming, if the land is needed there. There will probably be some shifts of land out of agriculture into urban, recreational, transportation, reservoir, and other uses; to the extent the area in agriculture remains more or less constant, this is likely to mean some losses and some additions, much of the latter coming from forestry.

Where roads and other rights of way go through forested lands, the land will be lost to forestry. High voltage power lines sometimes take considerable areas out of forestry, because their rights of way must be kept clear of the larger trees. In Chapter VII we estimate that the area used for reservoir and other water management programs will increase somewhat during the next few decades. Much of this land is now used for forestry.

On balance, it appears unlikely that the shifts into or out of forestry will be large on a net basis, compared with the present area of commercial forest of 484 million acres. The losses will be due to the demand for other uses. While none of them individually is large, it is probable that in total they will overbalance any gains in forested area. The best outlook is for some decrease in commercial forest area; we provide a more specific estimate in Chapter VIII. If the potential demand for forest products is to be met, it must come primarily from increased output per acre.

The Forest Service has calculated the output of forest products in 1975 and 2000 under various assumptions.[17] These, like the Service's studies of demand, are essentially projections; that is, starting with certain assumptions, they are calculations of what will happen, and are not necessarily forecasts of what is most likely to happen. Two basic relationships should be kept in mind when considering future supply:

1. The volume of timber cut and the volume of growth may diverge greatly in any single year or even over a considerable span of years; timber cut may run ahead of growth by drawing upon present stands, or cut may lag if timber volumes are being built up. In the truly long run, however, cut cannot exceed growth, but, in view of the heavy demand for forest products in this country, neither is it likely to lag far behind growth, at least for those species most in demand.

2. The annual growth and the volume of growing timber at any

[17] *Timber Resources for America's Future,* cited above.

time depend on decisions taken many years before; the situation in 2000 will depend largely upon decisions taken in the next decade or two.

Actual annual growth of forest products is far below potential for many reasons. The Forest Service estimates of three important levels are as follows:

	ANNUAL GROWTH	
	Total volume (billion cu. ft.)	Sawtimber only (billion bd. ft.)
1. Presently biologically possible "growth that could be obtained if there were proper distribution of age classes and if every acre of forest land in each type and site were producing as much as the most productive timber stands are today for the respective types and sites."	50	200
2. Realizable growth—"growth that ultimately would be attained if the commercial forest land in each region were placed under the better forest management currently in effect in that region."	27.5	100
3. 1952 actual growth	14.2	47

These estimates indicate the great physical possibilities for increased output that can be achieved from the present area. The higher estimate, while of significance as indicating a potential, is most unlikely to be achieved. The Forest Service estimates of realizable growth are based upon the second estimate above; and this includes the further assumption that recent trends toward better forestry will continue in the future, so that in 2000 forestry would be considerably more intensive than in 1952.

If the timber demands of the medium-level projections (which, it will be recalled, are on a basis of greatly increased but still far from maximum possible population and income for the future, but assume no further rise in prices of timber relative to other prices) are met by cutting the full volume demanded, regardless of the effect upon growing stock, the result would be a modest degree of overcutting up to 1975 and an increasingly severe one from then to 2000, so that by the latter year annual growth would have declined drastically (Figure 48). Growth would fall short of removal by 14 per cent in 1975, but, by 2000, due to continued excessive cutting which would deplete growing stock severely and hence reduce annual growth, removal would be about four times growth. These projections are highly unrealistic, as the Forest Service recognizes; what they do indicate, if the estimates of growth rates are reasonable, is that the projected medium-level demand cannot be met. Long before depletion would have proceeded to these levels, the price of forest products would have risen materially, thus checking consumption and stimulating more intensive forestry. This, as is indicated by the intermediate lines on Figure 48, is regarded as being a more reasonable situation.

On the other hand, if the low-level demand projections (which are based on the same population and income assumptions but assume further a substantial rise in timber prices relative to all other prices) are taken as the standard, timber removal and growth would be about in balance until 1975 but thereafter removal would exceed growth by 16 per cent (Figure 49). This overcutting, while not as drastic as would be necessary if the high level demand were to be met, would nevertheless mean some liquidation of growing stock and hence some

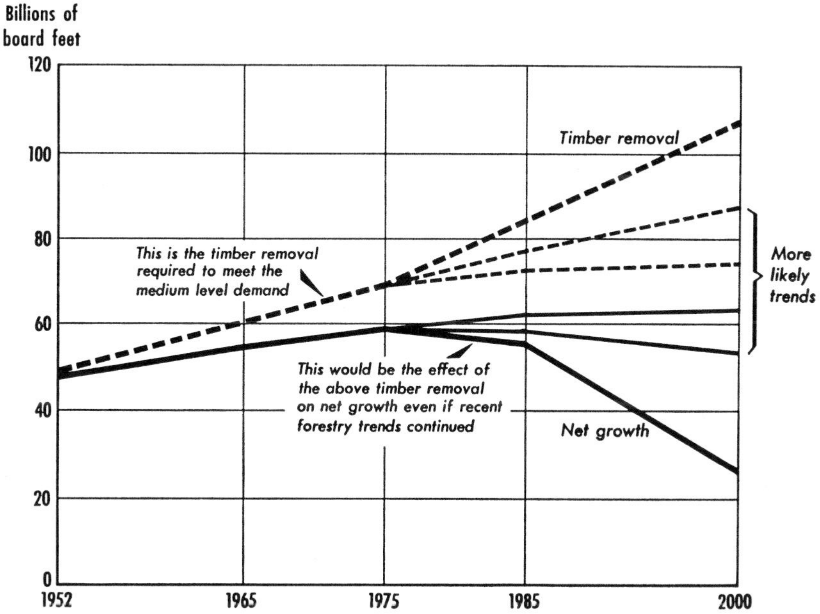

Figure 48. Trends of timber removal and growth of sawtimber under medium level assumptions. (Chart from *Timber Resources for America's Future,* U. S. Forest Service.)

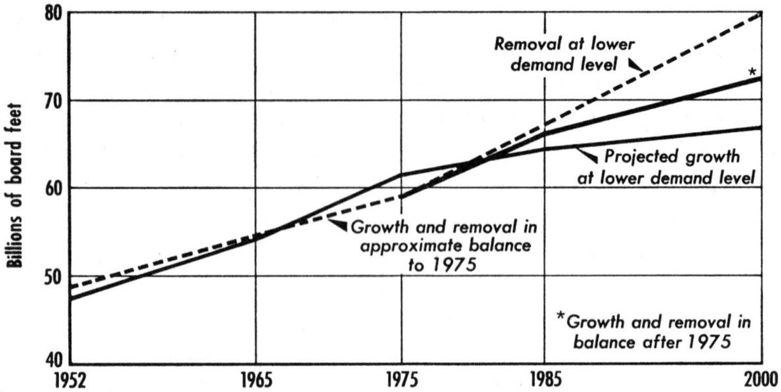

Figure 49. Removal of sawtimber under approximate balance of removal and projected growth. (Chart from *Timber Resources for America's Future,* U. S. Forest Service.)

reduction in annual growth. Comparatively modest changes in total cut would bring removal and growth in balance, as the middle lines on the chart suggest. Thus, given a continuation of past trends toward better forestry, the low-level demands can be almost but not quite met up to 2000. If the low-level demands are to be met fully, growth must somehow be stepped up more than recent trends lead one to expect; and a vastly accelerated forestry program beyond anything now probable must be quickly undertaken if the high-level demands, or anything closely approaching them, are to be met.

These total or over-all relationships conceal a great deal of variation between species groups and regions. The probable growth of eastern hardwoods, in relation to demand for them, is high; under some projections, a surplus of this type of wood would be available. On the other hand, the situation is least favorable for softwoods, East and West. In the East the problem is to sharply increase the growth rate and volume of softwood growing timber, so that cut may ultimately increase; in the West, the problem is to cut the present relatively large stocks of softwood as to promote maximum realizable annual growth rates. Under the most favorable conditions, it will be difficult to attain a balance of removal and growth for the softwoods. In the line on Figure 49 showing a balance of growth and removal after 1975, it has been assumed that a substantial shift from softwoods to hardwoods would occur—a shift relatively greater than any that has occurred in the past.

These general comparisons also conceal important differences in the outlook for materials of different qualities. While technology has done much to make lower quality wood suffice where the better quality would once have been used, there is reason to believe that the total demand for high-quality wood will increase, even if relatively it is less important. On the other hand, the quality of wood in the growing stock seems to have been declining. A relative shortage of the better grades of wood is therefore likely.

Major Problems in Increasing Timber Output per Acre

The foregoing analysis of the possibilities of increasing output per acre of forest land is a brief summary of the Forest Service analysis; at the best, it is in general or over-all terms. We need a closer look at the problem of increasing output for forests of different sizes and ownerships, but unfortunately some of the critical data for doing so are lacking. The Forest Service study includes data on the number and acreage of forests of different sizes and ownerships; and it also provides some limited data on volume of growing timber by broader ownership classes, but this is not detailed for each kind of private forest ownership, nor is it given according to size of ownership. The study gives no data at all on growth according to ownership classes, either by kind of ownership or by size of ownership unit. In the absence of these important items, we must necessarily speculate to some extent on the situation.

It is generally agreed, however, that the farm and "other" forest land ownerships are the key to the situation; unless their output can be materially stepped up, total or average output can be raised only within rather narrow limits. This conclusion seems valid for three reasons:

1. Farm and "other" ownerships include a large area of commercial forest land—165 and 131 million acres or 34 per cent and 27 per cent, respectively, of the total. With over 60 per cent of the total in these two classes, anything that happens to them inevitably has a major influence on the total for all forests.

2. There are very large numbers of each class of owner, hence the average unit per owner is small. There are 3,382,000 and 1,105,000

owners, respectively; and they average 49 and 118 acres of forest land. Some ownerships in each group are much larger than this, but far more than half of each class are below the arithmetic average. With such very small tracts of forest land, how can they be expected to do a good job of forestry?

3. The productivity of the recently cut lands of these two classes is low. This is primarily a matter of size of holding rather than type of ownership—that is, the few large units of each of these two types do about as well as the more typical large units of other types.

How can the problems of these small forest land owners be overcome, so that their lands can be more productive? Comparisons must in all cases be made with forests in larger ownerships and in other types of ownership. For instance, productive forestry is almost impossible where there are no markets for the products grown; few owners of any type or size will practice good forestry in the hopes that the output of their land will some day be in demand. In other areas, natural conditions limit the annual growth per acre. But we are concerned here with the factors that result in a lower output per acre on farm and other forests than is achieved on forest industry and public forests, where conditions are comparable.

Several factors make for low output per acre of the typical farm woodlot. In the first place, even if the woodlot is in highly productive condition and carries a good stand of growing stock, its total annual output, measured in monetary terms, would be small. With an average woodlot of 49 acres, one of 20 acres or less must be typical of at least half of the farm woodlot owners. A gross stumpage return of even $6.00 per acre (based on 200 board feet growth at $30.00 per 1,000) does not produce a large sum per farm under these conditions. The typical farm has a much lower output per acre of forest land than this, and a poor chance to raise its output, at least immediately. A more common situation is that the woodlot is poorly stocked with relatively low-grade trees; both volume and quality of growth are low.

In the second place, the typical farmer is not greatly interested in good management of his woodlot, nor does he regard it as a likely place to increase his net income. The larger commercial farmers are more interested in other farm enterprises, believing—often correctly—that time and attention devoted to them will yield more income than if used on the woodlot. Many of the smaller farmers are more likely to look for outside off-farm work, which, as we have noted in Chapter IV, has become increasingly common. Many farmers are uninterested, or uninformed, or too uncertain as to continued tenure on this land, to be willing to devote much time and energy to improvement of the woodlot, which is likely to be depleted anyway. The farmers who practice good forestry, because they like woods work, or because for them its income possibilities are fairly good, or for other reasons, are in the minority. And we fear that the forces impinging upon modern agriculture are likely to result in less, not more, of the farmer's attention given to the woodlot.

Studies in recent years have given some insight into the nature and conditions under which nonfarm "other" forest lands are owned and operated.[18] The central theme running through much of the information is along the following lines:

1. Nonresident owners are generally middle-income people with a wide variety of occupations who have held their land for more than a decade.
2. Objectives of ownership vary widely; in the South investment for income is an important factor; in the northern Lake states and New England recreational use leads by a margin; in all regions considerable areas are held for no clear reason, although speculation and sentiment are found to be influential.
3. The average length of time owners hold their lands varies widely; in most forest regions the majority of nonresidents had owned their lands ten to fifteen years.
4. Forest cutting practices are considered to be less than satisfactory to produce an early regrowth after logging on a large majority of holdings studied.
5. Only recently has the fact that most nonresident individual owners lack the time and ability to supervise the management of their

[18] Lee M. James, Wm. P. Hoffman, and Monty A. Payne, *Private Forest Land Management in Central Mississippi,* Mississippi Agricultural Experiment Station Bulletin 33 (State College: Mississippi State College, 1951); Charles H. Stoddard, "The Future of Private Forest Land Ownership in the Lake States," *Journal of Land Economics,* Vol. 18, No. 3 (1942), pp. 267-83; Solon Baraclough, "Forest Land Ownership in New England" (Ph.D. thesis, xi, Harvard University, 1949), mimeo.; A. D. Folweiler and H. J. Vaux, "Private Forest Land Ownership and Management in the Loblolly-Shortleaf Type of Louisiana," *Journal of Forestry,* Vol. 42, No. 11 (1944), pp. 783-90; H. H. Chamberlin, L. A. Sample, and Ralph W. Hayes, *Private Forest Land Ownership and Management in the Loblolly-Shortleaf Type in Southern Arkansas, Northern Louisiana and Central Mississippi,* Bulletin 393 (Baton Rouge: Louisiana Agricultural Experiment Station, 1945); Lee M. James, "Determining Forest Landownership and its Relation to Timber Management," *Journal of Forestry,* Vol. 48, No. 4 (1950), pp. 257-60.

forest property come to light as a factor militating against the application of intensive forestry measures. It is becoming increasingly clear that small owners who live at a distance from their lands are unable to give the close attention required for the practice of reasonably good forestry. The cost of frequent travel is often prohibitive when weighed against the possible returns. Furthermore, the volume of timber available for removal from small holdings under good forestry is frequently inadequate for purchasers to pay the best prices.

6. Infrequent sales and lack of market knowledge on the part of sellers frequently result in less than current stumpage or product prices and hence low returns to owners.

7. The better managed lands appear to be those on generally good sites in larger blocks and under the supervision of either a resident part-time manager or a consulting forester who tends to the operating details as they arise. These are a very small minority.

A considerable change in the present organizational structure seems to be needed before smaller holdings of "other" forest land will be efficiently and effectively managed. During the last several decades much attention has been given to the problem but clear-cut solutions have not emerged. Proposals for federal leasing and management of private lands were discussed frequently just prior to World War II, but little has been heard of this idea in recent years. During the 1930-40 decade a number of experimental co-operative management and marketing associations with various kinds of services performed for owners were tried out, and much was learned from them. The problem of adequate managerial direction, the difficulty of getting owners to operate their woodlands, and the problem of getting adequate quantities of high-grade material needed for profitable operation often proved to be major obstacles to co-operatives. Financing, both fixed capital and production credit, frequently caused difficulties because of the need for waiting the several years required for timber harvest. The infrequency of cutting on any one small ownership often resulted in a loss of owner interest in participation in the co-operatives.

More recently, consulting foresters in the South and elsewhere have undertaken to provide group managerial service for numbers of owners on a retainer basis. They have tended toward the larger ownerships where owner interest is greater and volumes of timber removed are sufficient for reasonable profit. The smaller ownerships have only been included in this sort of program when they lay close to larger tracts under agreement.

Other proposals are developing as the nature and characteristics of the problems and status of small owners are more clearly understood. Recent studies have shown that large numbers of owners are not equipped or are unable to find the time to conduct forest management operations.[19] In this position they are most likely to take the simplest way out, i.e. sell off their merchantable timber in a "lump" to local timbermen whose incentive is to remove everything possible. Since most of these smaller tracts held by nonresidents are too small to be economic units they must necessarily be cut at infrequent intervals—the minimum being a five-year cutting cycle—under recommended forest management procedures. Lacking the time, knowledge, and equipment needed for continuous forest management of their tracts, many of these owners have indicated a willingness to turn over the full management of their forests to specialists.

Undoubtedly future developments in this direction will include a variety of operational combinations which will bring enough of these small holdings into a central management group to permit full-time economic operation. Those that have been suggested include private, nonprofit associations, and public arrangements. The private arrangements proposed are joint stock companies in which the owners pool their holdings under one company with some rights retained, simple management agreements with timber processing companies or consulting foresters to manage their lands, and outright leasing to private concerns who manage the properties over the time of a contract. The nonprofit association management forms have included co-operative associations, the type of management assistance association that is found in the New England Forestry Foundation and the Trees for Tomorrow group in Wisconsin. Public agency management of private holdings has never been seriously developed except for the work of the Soil Conservation Districts. Considerable success was developed in this program while the Districts were able to perform direct management operations for their owners. However, in recent years most of the assistance has been limited to advisory work, the owners being left with the problem of carrying out the technicians' advice.

Whatever forms of group management eventually evolve will require combining under single operational direction areas of 10,000-20,000 acres as a minimum needed for economic management in an operating

[19] J. G. Yoho, Lee M. James, and Dean N. Quinney, *Private Forest Landownership in the Northern Half of Michigan's Lower Peninsula*, Technical Bulletin 261 (East Lansing: Michigan State University, 1957).

radius.[20] Such an acreage of adequately stocked land would supply sufficient timber cutting and other work to keep a woods crew and technical overhead fully employed the year around. Until recent high stumpage prices, there has been little incentive for smaller owners to put their forests under continuous management programs. The development of group management efforts will be accelerated in the next few years as economic incentives improve, and as assistance facilities become available. Until very recently when limited amounts have become available to low-risk borrowers, credit for forestry has not been available on terms and conditions adapted to the peculiar nature of tree growing.[21] Financing of the many kinds of group arrangements will require credit in moderate amounts and for periods sufficient to enable borrowers to establish going businesses.

An element of inertia seems to be present in attempting to overcome the obstacles facing efforts of group management. The relatively low profits in early stages of management of understocked lands have held back some who would venture forth in this type of undertaking. Two courses seem to be called for if this inertia is to be overcome: an initial underwriting of developmental costs until such time as the ventures are self-sustaining; and prior to this, some experimental efforts in development of the various proposed types of organizations to find the ones best adapted to the conditions presented by small management operations. While favorable credit may be helpful in more rapid development of such group efforts once they are under way, it is not likely to provide the initial implementation of this new form of forest management enterprise.

Economic Feasibility of Increased Investment in Forestry[22]

If a greater total output from forests is to be achieved by increasing output per acre, it will require more intensive forestry; and this in turn will require greater investment and larger current outlays. Will the investment and outlays produce results that more than compensate?

20 *Forest Credit in the United States,* report by Committee (Washington: Resources for the Future, Inc., 1958).

21 *Ibid.*

22 This subsection is based upon and summarizes: Charles H. Stoddard, "An Estimate of Capital Needed in Forestry to meet Projected Timber Requirements for the Year 2000," *Journal of Forestry,* Vol. 56, No. 7 (July 1958).

Many of the costs of forestry are now being met in one way or another; intensified forestry would not add to their amount. The additional costs of more intensive operations can thus be balanced against the additional returns from them, without considering whether all the costs can be met by all the income.

Major ways to increase output of forest products from the present area include: (1) reforestation of all or part of the 50-odd million acres of idle commercial forest land; (2) acceleration of growth through thinnings and other stand improvement measures; (3) intensified insect, disease, and fire protection; (4) elimination of premature clearcutting of better species of young sawtimber; (5) harvesting of cuttings in mature stands so as to assure maximum wood production of desirable quantity, quality, and frequency interval; and (6) shifting of a portion of the present cut to poorer species and stands, with resultant lowering of the quality of the annual cut for a time, in order to build up growing stock inventory.

Achievement of the Forest Service estimated "medium" demand (21 billion cubic feet) would require an investment of about $2.1 billion and an increased annual outlay of about $40 million for intensified protection measures. The estimated increases in growth rates would repay the added annual outlays and offer a return of about 5 to 6 per cent on this added investment. While rough and perhaps arbitrary estimates enter both cost and income sides of this equation, the over-all results seem reasonable and hopeful. There would be many difficult problems in achieving the higher level of management called for, especially on the smaller land holdings. But the economic feasibility of intensified forestry seems promising, if yet unproven.

Output under intensive sustained yield management. If these measures to increase forestry output per acre were adopted to the extent necessary to build up forest productive capacity so that it could sustain an annual cut of 21 billion cubic feet by the year 2000, what would the annual cut and demand picture look like in the intervening years? The Forest Service report *Timber Resources in America's Future* does not attempt to estimate a maximum allowable cut under intensive sustained yield management, but some unpublished studies by Resources for the Future do make such estimates. Under the management assumptions we have outlined, calculation of an allowable cut involves primarily an effort to achieve "normal" distribution of trees by size and age classes. In the generally unmanaged condition of most American forests, this means building up growing stock in the East, particularly in the softwoods, while replacing the overmature timber of the West with younger growing stands in various stages of development. The present large volume of overmature timber could be liquidated over a variable period of time; we have used a twenty-year period for liquidation of some 90 billion cubic feet of excess old-age stands.

The results of these calculations are shown in Figure 50, in comparison with two of the projected demand levels estimated by the Forest Service. Demands during the next twenty years can be more than met, on our assumptions as to rate of liquidation of the present overmature stands. For the longer period, the lower level of demand can be more than met. A portion of the surplus in allowable cut during the first twenty years could be carried over into the next two decades, and thus meet the medium level of demand defined by the Forest Service. While this is a generally optimistic situation, assuming that the necessary intensive management steps are taken, some imbalances are submerged in it. In particular, problems of timber quality

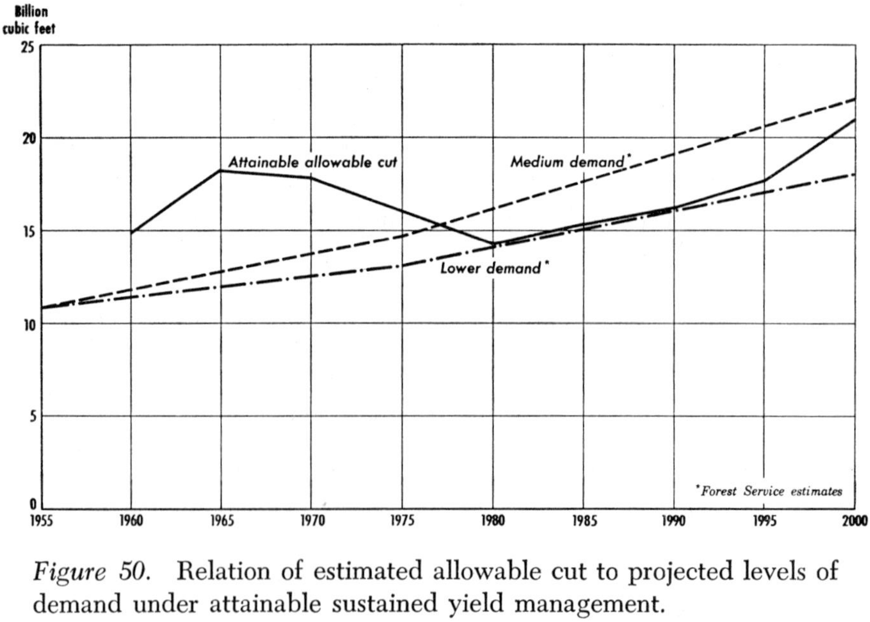

Figure 50. Relation of estimated allowable cut to projected levels of demand under attainable sustained yield management.

would still remain, as would the need for major shifts in utilization between softwood and hardwood species. Moreover, the regional balance would still be imperfect.

However, in summary, these projections show that it is possible to manage the forest resources so as to increase annual output from the present 14.8 billion cubic feet to 21 billion cubic feet by 2000, while at the same time meeting the medium demand levels during the next four decades. Intensive management will be needed to achieve these results. If less intensive practices are followed, as assumed in the Forest Service projections, then deficits in growing stock will arise.

Forest Purposes Other than Wood Fiber Production

The discussion in this chapter has focused on the commercial aspects of forestry, or upon the forest products output of forests; earlier chapters have focused similarly on the recreational use of park lands, and upon the agricultural use of farmland. Yet in each case uses other than the major one occur and may be important. This is nowhere more evident than with forests. Forests have high values for recreation, watershed management, wildlife, sometimes for grazing and minerals, and these to a degree not often found for land used primarily for other purposes. In fact, the primary objective of much ownership of small forest tracts classified as commercial forests, is not to produce wood fiber, but to provide recreation or in other ways satisfy the human values of land ownership.

In bringing to a close this consideration of the future of forestry we wish to re-emphasize the other roles of forests. There are many reasons for thinking that these roles will become more important during the next several decades. As the demand for recreation rises, for instance, the area of land devoted primarily or exclusively to recreation may not increase proportionately, and there will be a greater demand to use forests for recreation. As has been suggested in Chapter III, if the acreage of outdoor recreation areas increases as much as we think necessary, some of this land may continue to be used for production of wood fiber in one form or another. The distinction between a commercial forest also used for recreation, and some types of recreation areas upon which some forestry is practiced, may become much less sharp than today. The same general situation may exist with respect to wildlife. Specialized areas for wildlife will perhaps increase in acreage, but the importance of wildlife on forests of all ownerships may also rise.

This matter of multiple use of land is one to which we return in Chapter VIII. Its importance for forestry is great.

Forest Policy Directions

The decisions affecting forest unit reorganization as well as many other phases of private forestry will largely be made by private owners; but federal, state, and local governments will have a strong influence in determining the economic climate for private forestry practice. Property tax rates, inheritance and income taxes, fire, insect, and disease protection programs, availability of market and price information, forest cutting regulations, and many other policies are largely determined by public bodies. Subsidies have been extended to private forest owners in the past and are now, and might be increased in the future. On the other hand, decisions affecting the public forest lands will be made by the agencies and legislative bodies to which they are responsible.

Forestry programs in this country to date have flowed directly from democratic legislative processes and presumably have fairly well met the demands of the forest-oriented public. A continuance of similar pragmatic policy determination in the future will probably develop to meet needs and anticipate, but only in part, future problems before they arise. Because forestry is unique in its long production cycle this is likely to result in a definite lag between the time a policy is adopted and the results that are forthcoming. But forests have a way of growing if only fire protection is provided and thus to some extent make up for man's lack of foresightedness.

Among the more important policy considerations likely to develop during the next quarter century and beyond are those centering around taxation, insect and disease control, size of national private and public investment in forestry, and more intensive management of small private holdings. A continuation of the struggle over control of uses of public forest lands between large and small timber manufacturers, sportsmen and other outdoor users, and those interested in grazing and watershed values, seems probable. It is becoming increasingly clear that the noncommodity benefits of the forest are assuming a greater place in the public mind than at any time heretofore. The pressures on the forest exerted by these groups are not in the marketplace but are expressed in terms of political processes. One example of this may be found in the efforts of those who feel that undeveloped natural and wild areas should be maintained in their pristine state without roads, logging, grazing, mining, resort buildings or other commercial development of any kind.

On public lands the co-ordinating of these separate and specialized uses is a difficult task but one in which some experience has already been gained. The balancing of the various use-interest groups while at the same time carrying out sound long-term management programs calls for a high degree of managerial skill. Yet public ownership in these situations has proven itself to be a workable and flexible form of tenure.

Private lands, on the other hand, are ordinarily used primarily for a single commodity use. In the case of forest lands this is principally timber. Such other uses as may be made are usually incidental and are limited by the owner's control. Consequently, in areas where little public land exists for recreational use by the public, conflicts over forest land use may develop between farmers and hunters, picnickers and others. Although tenure and other arrangements which would facilitate other types of uses of private lands are yet uncommon or nonexistent, increasing pressure and the possibilities of profit from recreational and other uses may gradually induce many owners to permit access and provide facilities. Here a new type of experimental research which blends the social and physical sciences and relates them to land use can contribute new answers to many of the problems.

Unlike many modern industries, the lumber and logging enterprise is scattered among thousands of owners rather than controlled by a few large concerns. The relatively low capital investment required in logging and sawmilling has made it possible for many people to enter this business during periods when demand for lumber is high. But minimum necessary investment will probably increase in the future as mechanization proceeds. This has been true since World War II. Competition for diminishing supplies of good sawtimber stumpage will be great; and conflicts will probably continue to arise over publicly owned timber, between smaller independent loggers and sawmill men and the larger better established concerns. The same general situation has been true of the paper mills with their own forestry programs and the numerous independent suppliers of pulpwood. Larger blocks of stumpage have tended to go to the larger concerns, but the smaller producers with greater political power will continue to force public agencies to sell publicly owned timber in smaller lots.

Any forecasting of the trends in policy decisions must of course be highly speculative. The increasing intensity of American forestry will force the facing and making of decisions on policies and programs which will encourage wider applications of applied forestry. To the extent that these decisions are postponed, the production of timber will be seriously delayed because of the time lags due to long growth cycle. Government may have to stimulate needed capital investments on private lands and must make direct investments on public lands. The shorter term credit needed in private forestry will be supplied in part directly by the owners. Additional longer term credit may not be forthcoming without some form of public underwriting of risks and even interest rate.

In many ways these decisions will be a test of our national foresight and ability to discipline ourselves as a social group. The Scandinavian and other democratic nations have undertaken similar measures for forestry without resort to autocratic controls of central authority, but they have accepted locally developed and enforced land use regulations as a means of assuring owner responsibility. Everything in our tradition would indicate that we can do as well.

VI

Land for grazing[1]

Grazing as a major land use is difficult to define. Partly this is because land used for grazing is covered with a variety of vegetation, not all of it suitable for grazing, and because grazing frequently involves special land ownership and management situations. And partly it is because on considerable areas grazing is one but not the dominant use, and our scheme of analysis calls for consideration of land according to its dominant use. In spite of difficulties of definition, we have decided to consider grazing as a separate land use here, because (1) the area concerned is large—larger than for any other major land use in our terminology, and (2) because grazing has special characteristics as a land use. We recognize, however, that this is largely an expositional matter; grazing could as readily have been considered a subdivision of agriculture.

Land uses, like many other phenomena growing out of human activity, do not fall into neatly divided categories. Instead, there is something of a continuum; certain ways of land use are clearly within

1 Some of the general sources for this chapter are: Laurence A. Stoddart and Arthur D. Smith, *Range Management* (New York: McGraw-Hill Book Co., 1943); A. W. Sampson, *Range Management, Principles and Practices* (New York: John Wiley and Sons, 1952); Marion Clawson, *The Western Range Livestock Industry* (New York: McGraw-Hill Book Co., 1950); U. S. Department of Agriculture, *Grass, The Yearbook of Agriculture, 1948* (Washington: U. S. Government Printing Office, 1948); *The Western Range*, Senate Doc. 199, 74th Congress, 2nd Session, 1936; Edward Everett Dale, *The Range Cattle Industry* (Norman: University of Oklahoma Press, 1930); *Journal of Range Management*, all issues from Vol. 1, No. 1 (October 1948) to 1957; *Agricultural History*, all issues 1939-57.

a particular category, others are on the fringe. Some land is in cultivated crops every year; in other cases, crops and seeded pastures are rotated, more or less regularly and on relatively short rotations. Some of the seeded pastures may be cut for hay, or grazed by livestock, according to relative need. Even the crop fields may provide substantial aftermath grazing. These clearly cropping uses of land gradually shade off into seeded and improved pastures, where the land is plowed at longer and longer intervals, and cropping to cultivated crops becomes less and less common. Up to the point where the land is plowed, perhaps alternated with other crops, and reseeded to grass, at more or less regular intervals, we would clearly classify this use of land as agricultural. One of the crops is grass, legumes, or other forage; it may be harvested mechanically or by animals.

In contrast, some land is clearly in forest as a dominant use. The forest may be commercial, noncommercial, farm, nonfarm, privately or publicly owned; or may differ in other ways. But if forestry is the dominant use, we have considered it here as forest land. This is true whether there is no grazing or, comparatively, a good deal of grazing, so long as grazing is not dominant. Clearly, there are situations in which it is hard to say whether forestry or grazing is dominant.

When agriculture and forestry are thus excluded, we have left several rather distinguishable situations in which grazing is the dominant land use. First, there are the extensive range lands of the West; excluding forest and woodland, they amount to over 500 million acres, one-fourth of the total land area of the nation. These are predominantly native forage areas, with comparatively little improvement through plowing, seeding, or otherwise replenishing the vegetation. A major part of these lands is federally owned, and an additional acreage is owned by other governments; and these publicly owned areas are used by private ranchers under lease, license, or permit. Second, there are relatively large unforested tracts, chiefly in the South, where the forage is mostly of native plants, and where grazing is the primary use.[2] The area of such lands is less than many estimate

[2] In recent years a number of excellent articles on southern grazing areas have appeared in the *Journal of Range Management*. Some of special interest are: W. O. Shepherd, "The Forest Range in Southern Agriculture," Vol. 3, No. 1 (January 1950); H. H. Biswell, "Studies of Rotation Grazing in the Southeast," Vol. 4, No. 1 (January 1951); John T. Cassady, "Bluestem Range in the Piney Woods of Louisiana and East Texas," Vol. 4, No. 3 (May 1951); B. L. Southwell and L. K. Halls, "Supplemental Feeding of Range Cattle in Longleaf-Slash

it to be, since most of the southern grazing is on forest land. Third, throughout the commercial farming areas of the country, but especially in the more rugged areas, there are many relatively small tracts of land that are usually covered only with native plants, are not forested to any great extent, and are grazed seasonally by farm livestock. In some parts of the country such land is called permanent pasture.

In this chapter we deal mainly with the first of these categories. It is the largest in area and most markedly shows the special problems and characteristics of grazing land.

Wooten and Anderson estimate that there are 633 million acres of grasslands used for grazing.[3] This presumably includes shrub areas as well. It does exclude cropland used only for pasture, which we have included in our consideration of agricultural use of land. Of this grassland area, 460 million acres are in farms. In addition, 301 million acres of forest and woodland are grazed to some extent, 121 million acres of which are in farms. As nearly as we can judge, about 67 million acres of noncommercial woodland probably is more valuable for grazing than for forest purposes, and hence we class it as grazing. The rest of the forest and woodland we consider to be either commercial forest land or miscellaneous land. Thus, in round terms, there are about 700 million acres used primarily for grazing. The 200 million acres that are not in the West are found chiefly in the South. This is by no means the total area grazed; in addition to grazing on forest and woodland, there is grazing also on cropland used only for pasture and, besides, on many cropped areas after harvest. A few crops, such as winter wheat, are grazed during growth without interference with their harvest.

Pine Forests of Georgia," Vol. 8, No. 1 (January 1955); L. F. Smith, R. S. Campbell and Clyde F. Blunt, "Forage Production and Utilization in Longleaf Pine Forests of South Mississippi," Vol. 8, No. 2 (March 1955); Hurlon C. Ray and Marvin Lawson, "Site Characteristics as a Guide to Forest and Grazing Use in the Ozarks," Vol. 8, No. 2 (March 1955); S. Clark Martin, "The Place of Range Livestock in the Missouri Ozarks," Vol. 8, No. 3 (May 1955); Hubert D. Burke, "Game Habitat and the Multiple Use of Southern Forest Ranges," Vol. 9, No. 4 (July 1956); Lowell K. Halls, "Grazing Capacity of Wiregrass-Pine Ranges of Georgia," Vol. 10, No. 1 (January 1957); Robert S. Rummell, "Beef Cattle Production and Range Practices in South Florida," Vol. 10, No. 2 (March 1957).

[3] Hugh H. Wooten and James R. Anderson, *Major Uses of Land in the United States,* Summary for 1954, Agriculture Information Bulletin No. 168, U. S. Department of Agriculture, Agricultural Research Service, Washington, January 1957.

The essential physical characteristics of most grazing areas are aridity, other unfavorable climatic factors, rough topography in many cases, often soils unsuited for cropping even if other factors were favorable, and native vegetation. In each case, almost without exception, one or more physical factors makes cropping uneconomic or impossible.

The chief source of livestock feed on grazing lands are native plants, which grow with comparatively little help from man. These include grasses, shrubs, and forbs (weeds) which are palatable and nutritious to livestock. There has been some invasion of grazing areas by introduced species, those that make good forage, those that do not, and even some that are poisonous. In limited areas man has reseeded the range, usually with grass. His management of livestock has influenced the plant cover in many ways; and his provision of water for livestock, his fencing, and other management practices have also been effective. In the western range region, dependence upon grazing is often relatively heavy for the animals grazed. Sometimes they find all their feed on the range, sometimes they feed for a part of the year. The average per-acre production of forage from grazing land is perhaps one-tenth that from the hay land. Some grazing land is very much more productive than this; other areas, much less so.

The forage available from grazing land is roughage, and often rather coarse roughage. As such, it is suitable for cattle, sheep, and horses; and for them, better adapted to maintaining life, to reproduction, and to growth, than to fattening. In some areas and at some seasons, the grass is sufficiently lush to fatten cattle or lambs, but other areas provide a bare maintenance ration, especially at certain times of the year. The kinds of livestock kept, the specific livestock management practices, and the age and season at which animals are sold, are all greatly influenced by the kind of forage available.

Because the physical output per acre of grazing land is usually relatively small and not highly valuable per unit of weight or volume, the only economical way to harvest the forage is by livestock. Moreover, in order to have a large enough number of livestock to produce a satisfactory income, a rancher must use a large area of grazing land. In the strictly grazing regions, therefore, the density of human population is low. Although output per acre is low, output per man has been relatively high because of the relatively large average size of grazing operation.[4]

[4] Clawson, *op. cit.*, pp. 6-8.

With a low physical and economic output per acre, grazing is obviously not a strong competitor for land suitable for other uses. Its large acreage comes almost by default, and it is a measure of the extent of relatively unproductive lands in the United States that so much is used for grazing. It is true that, were the population pressures much more severe, some of the grazing land could be used for other purposes, notably agriculture, but its net output for this use would be low indeed.

GRAZING AS A MAJOR USE OF LAND IN THE PAST

The area of land now used for grazing and the methods of its use are of necessity influenced by the way the land formerly was grazed. Before discussing present methods and possibilities for the future, therefore, we summarize quickly the major happenings in the history of American grazing.

Grazing in the Colonial Period

Before the white man came to North America, the native forage plants were consumed by big game animals, notably buffalo, antelope, deer, and elk, and the grazing lands were vitally affected by the American Indian. There was some sort of balance between the factors leading to greater numbers of animals, particularly their reproductive capacities, and the factors tending to limit numbers, particularly variable feed supply, the climate, predators, disease, and man. Numbers must have changed from year to year, depending upon one or more of these factors, and it would be a serious mistake to assume a stable balance in numbers of game animals at any time.[5]

The Indian's role in the ecological balance included the use of fire as well as the killing of grazing animals.[6] Accounts of early explorers

[5] William L. Thomas (ed.), *Man's Role in Changing the Face of the Earth* (Chicago: University of Chicago Press, 1956), see particularly Andrew H. Clark, "The Impact of Exotic Invasion on the Remaining New World Mid-Latitude Grasslands."

[6] Carl O. Sauer, "Grassland Climax, Fire, and Man," *Journal of Range Management*, Vol. 3, No. 1 (January 1950).

and anthropological investigations agree that in many areas the Indians deliberately and regularly burned the vegetation.

Into this situation the white man has introduced, sometimes deliberately and sometimes unintentionally, various micro-organisms, larger plants, and animals.[7] The main streams of introduction were two: by the Spanish, through Cuba, Florida, Mexico, and, later, up the Rio Grande and into California; and from North European countries to areas along the Atlantic Coast.[8] The introduced plants and animals often went "wild" or feral—able to reproduce and maintain themselves without further assistance from man, perhaps in spite of his efforts to control or eliminate them. This process took place in a generation or two. In several instances, the new environment being highly favorable to the newcomer, and the diseases, pests and other controls which had formerly held it in check being absent, many species spread rapidly and reached a degree of dominance never attained in the area of their origin.

From the earliest colonial days, the white settlers grazed domestic livestock upon native feeds. Hogs ran in the woods, feeding off the mast or acorns from oaks among other foods; and cattle and horses ran in the woods, the open prairies, and the swampy areas, feeding off grasses and shrubs. Many problems were involved in this type of livestock economy, not the least of which was the escape of the animals or their theft or slaughter by Indians. But the forage was free, while crop production under the techniques of the day was difficult and labor-consuming. The farmers in the settlements took advantage of this natural forage as best they could.

At the same time, there soon developed specialized agriculturalists who raised animals primarily on native forage and marketed their animals in the settlements. This type of production was typically located along the fringes of the established settled areas. Its size fluctuated, depending upon the local situation as to war or peace with the Indians, and in any event its location gradually shifted as permanent settlements spread. There were serious problems of livestock production, growing in part out of adverse climatic factors, in part out of

7 Thomas, *op. cit.* In addition to chapter cited above, see also John T. Curtis, "The Modification of Mid-Latitude Grasslands and Forests by Man"; Edgar Anderson, "Man as a Maker of New Plants and New Plant Communities"; and Marston Bates, "Man as an Agent in the Spread of Organisms."

8 Sampson, *op. cit.*, p. 113.

seasonal variations in forage supply, all of which tended to limit numbers of livestock and their production. At the same time, there were also serious marketing problems. In general, the animals could only be driven to markets in the older towns, but this presented its problems too, and sometimes the chief product marketed became the hides from the animals raised.

Grazing in the First Half of the Nineteenth Century

Grazing of native forage was common over much of the eastern half of the United States during the first half of the nineteenth century. The Indians in this region were gradually subdued or driven westward, and much of the land was settled and developed into farms. Some of the land in the vicinity of farms was grazed by farm animals, and thus supported a type of grazing different from the primary consideration of this chapter. But specialized livestock operations also existed. This was true in the Ohio Valley, for instance, where, until about 1850, operators raised cattle, sheep, and hogs upon prairie areas, and then sold them to drovers who drove the animals to city markets along the Atlantic Seaboard.[9]

The grazing industry of the eastern half of the United States up to 1850 had characteristics which were later to exert a marked influence on the nature of grazing in regions farther west.

(1) The grazing up to 1850 had used native plants in *humid* areas. The productivity of these grazing areas, the ability of the plants to reproduce under varying conditions, and the capacity of native forage to recover after severe use or over-use, were each markedly affected by the relatively high precipitation.

(2) As a frontier fringe industry, it had moved on each few years, never having to face the problems of continued and stable operations on a fixed area of land. There was no need to graze conservatively and to maintain the productive capacity of the grazing plants, because if the land was to be plowed in a few years it did not matter.

(3) Virtually all the land grazed was publicly owned, especially at first; but even toward the end of the period a large part was still pub-

9 Paul C. Henlein, "Cattle Driving from the Ohio Country, 1800-1850" and "Shifting Ranger-Feeder Patterns in the Ohio Valley before 1860," *Agricultural History*, Vol. 28, No. 2 (April 1954) and Vol. 31, No. 1 (January 1957).

lic land. The grazing operator typically had no legal right on the land on which his animals were grazing.

These situations built up or conditioned the attitudes of many people toward grazing land and toward grazing as a use of land—attitudes which were later to raise many serious problems of resources use and public policy.

By 1850 there were some cattle in Texas, in what is now New Mexico, in California, and in Oregon. Cattle had been taken into present-day California from Lower California in 1769, and within a few years had multiplied to the point where they amply supplied local needs for meat.[10] A cattle industry based upon grazing developed. This early industry was centered in southeastern or south central Texas, in the general area from San Antonio to Houston, and northward to Waco, and later northward to Dallas and Fort Worth. By 1842 some cattle were driven to New Orleans, for market there or for shipment by boat to other markets. Cattle were also driven to Ohio as early as 1846.

The settlers of Spanish blood who entered what is now New Mexico took livestock with them, and some livestock were found in the Rio Grande Valley and elsewhere in the seventeenth century or earlier. However, the grazing lands were not extensively used until much later—with the influx of "Anglos" in the last quarter of the nineteenth century.

Settlers in Oregon Territory took livestock with them as early as 1830, but the numbers remained comparatively small until 1860, and were largely confined to western Oregon and western Washington.

The situation at the end of the Civil War was much the same—a few islands of range livestock in the western half of the country and in Texas; but Texan cattle, largely uncared for during the war, had multiplied greatly. Unbranded and often half-wild, they roamed extensive areas. Markets had been almost nonexistent during the war, and few cattle had been killed. At the close of the war Texas, like most of the South, was impoverished; ample cattle available almost for the taking and apparently good markets in the East were powerful forces for the development of a grazing cattle industry.

[10] Robert M. Denhardt, "The Role of the Horse in the Social History of Early California," *Agricultural History*, Vol. 14, No. 1 (January 1940); L. T. Burcham, *California Range Land* (Sacramento: State of California Department of Natural Resources, Division of Forestry, 1957).

Period of Major Grazing Expansion, 1865-1890

"As settlement advanced westward into the prairie regions beyond the Mississippi, this border of herding along its western rim varied somewhat in width but was never very broad. However, it is one of the most remarkable features of American economic history during the nineteenth century that soon after the Civil War this border of grazing, hitherto fairly constant in breadth and area, suddenly shot out into the wilderness and spread with a rapidity that was fairly startling until, in less than two decades, it had come to cover an area greater than all that part of the United States east of the Mississippi devoted to crop raising. Thus was formed the so-called "cow country" or range cattle area, a region in which ranching was carried on for several years upon a scale vastly larger than ever before had been the case until the homesteaders advancing slowly but steadily westward had invaded nearly every portion of it and taken over most of the land suitable for crop growing."[11]

In 1865 the Great Plains had few permanent white settlers, and settlement west of the Plains was largely confined to the Mormons in Utah, the agricultural areas in western Oregon and Washington, mining districts in Nevada, California, and elsewhere, and to some limited agriculture and urban development in California. Travel across the country, or around it, was slow, uncomfortable, and often dangerous. Construction of the first continental railroad began west of Omaha in 1865 and was completed to the Pacific Coast in 1869. A few years later, another line was built as far west as Dodge City, Kansas, and still later other lines were built into or across the Plains much farther north. These railroads opened up the western half of the country to travel, settlement, and shipment of products.

During these same years, the buffalo of the Plains were being slaughtered at a rate which seems today almost unbelievable, and the Indians were being subjugated. Gradually the Indians were forced onto reservations; although they broke away whenever opportunity presented itself, their strength would have been broken in any event by the disappearance of the buffalo, which had been a major source of food, clothing, and housing. As it was, the federal government

[11] Dale, *op. cit.*, p. xiv. Some of the other general works relating to this area and period are E. S. Osgood, *The Day of the Cattleman* (Minneapolis: University of Minnesota Press, 1929); *The Western Range*, cited above.

issued them beef, flour, and other food and clothing, within a few years converting nomadic hunters and fighters into idle paupers.

This combination of expanding railroads, buffalo slaughter, and Indian subjugation set the stage for a rapid expansion of cattle grazing. As we have noted, the cattle numbers built up in Texas during the Civil War were, by its end, ready for market. Some early relatively unsuccessful attempts were made, but, by 1868 and for slightly more than twenty years thereafter, substantial numbers of cattle were regularly driven northward each year to the transcontinental railroads, and shipped eastward. Data on these drives are not complete and perhaps not fully accurate; it appears that between 5 and 10 million cattle were driven northward from Texas during this era. At the same time, some Texas cattle were driven into Arizona and the Southwest to stock the ranges found there.[12]

During these same decades, cattle and sheep spread into the Intermountain country, east of the Cascade-Sierra mountains and west of the Rockies. Some of them came from the same Texas sources, but some also came from Oregon and California.[13]

Sheep were also driven into the Intermountain and northern Plains areas from California. Wentworth estimates that 15 million sheep were driven in this manner in the three decades following the Civil War.[14] Sheep also entered west Texas during this same period, and gradually into the Plains, but not to the same relative extent as cattle.

The markets for the cattle driven from Texas and the Pacific Coast areas into the Plains and Intermountain regions were of three kinds: (1) to the federal government, for beef to give to the Indians cooped up on the reservations; (2) for shipment to eastern markets; and (3) as breeding stock for further production within the respective areas.[15]

[12] Richard J. Morrisey, "The Early Range Cattle Industry in Arizona," *Agricultural History*, Vol. 24, No. 3 (July 1950).

[13] See the following articles from *Agricultural History:* J. Orin Oliphant, "The Cattle Herds and Ranches of the Oregon Country, 1860-1890," Vol. 21, No. 4 (October 1947), and "The Eastward Movement of Cattle from the Oregon Country," Vol. 20, No. 1 (January 1946); Oliphant and Kingston (eds.), "William Emsley Jackson's Diary of a Cattle Drive from La Grange, Oregon, to Cheyenne, Wyoming, in 1876," Vol. 23, No. 4 (October 1949).

[14] Edward N. Wentworth (ed.), "Trailing Sheep from California to Idaho in 1865; The Journal of Gorham Gates Kimball," *Agricultural History*, Vol. 28, No. 2 (April 1954).

[15] Dale, *op. cit.*

It was soon discovered that one excellent method of operation was to hold over winter the animals driven north from Texas, allowing them to graze the following summer, and selling them in the fall, when they would be fatter and much heavier. Several large ranches were established with breeding herds in Texas or Oklahoma and summering or fattening herds in the northern Great Plains.[16]

During the rapid expansion of the grazing industry, it acquired a highly speculative character. Exaggerated stories of high profits led to the investment of much outside capital, some from England and Scotland, by persons who had no direct knowledge of the business. Some high profits were indeed made, but some large losses were also sustained. Climatic and other hazards, such as disease or Indians, added to the economic risks of falling prices to produce instances of very large losses. Credit was extended freely in some areas, sometimes with loss not only to the borrowers but to the lending institutions. Instability was perhaps inevitable in an industry that had grown so rapidly within so short a time. Like the grazing industry of earlier times and more eastern locations, this also was conducted largely upon publicly owned land. While this meant no cost for the land, it also meant that the user had no control over the land. Much of the area was severely grazed, either because one user felt that there was no reason for him to conserve grass for someone else's benefit, or because different users each tried to use the same area without foreknowledge that others planned to do so also.

In nearly every part of the grazing region, this early period of rapid expansion came to grief, either because of climatic disaster, ruinously low prices for cattle, or both combined. The classic example, because of its size, severity, and (at the time) comparative novelty, took place in the northern Great Plains in the winter of 1886-87.[17] When a severe winter followed upon the heels of a hot, dry summer, and prices had already slipped, cattle losses were very high; they may well have averaged 15 per cent, and individual losses of up to 60 per cent, and 100 per cent in extreme cases, occurred.[18]

16 *Ibid.*; J. Evetts Haley, *The XIT Ranch of Texas and the Early Days of the Llano Estacado* (privately printed), Chicago, 1928.

17 Dale, *op. cit.*, p. 108-11; *The Western Range*, cited above, pp. 123-25.

18 W. Turrentine Jackson, "The Wyoming Stock Growers Association, Its Years of Temporary Decline, 1886-1890," *Agricultural History*, Vol. 22, No. 4 (October 1948).

Other areas experienced "hard" winters of various kinds. The southern Plains had had one the preceding winter, but not nearly so severe. The Pacific Northwest had experienced severe winters to varying degrees in 1861-62, 1880-81, and 1889-90.[19] Unusual climatic conditions are a major hazard in the grazing industry.

The situation in 1890 can be briefly summarized as follows:

1. The grazing of domestic livestock on native forage had spread into every *major* part of the nation; many local areas were still unused, perhaps because of lack of water for livestock within the one-to-three-mile grazing radius usually considered necessary, or because the forage was relatively inferior or inaccessible. While the sum total of these areas was large, individually they were comparatively small.

2. The necessity for winter feeding and care of livestock, beyond merely grazing the grass and other forage that grew naturally, was recognized, although the means for so doing had not yet been developed on an adequate scale.

3. Although the grazing industry was almost wholly on public lands, the need for the livestock operator to own waterholes, hay-producing lands to provide winter feed, and other key tracts of range was recognized.

4. The location of markets and the methods of reaching them which prevail today had rather clearly taken form. The railroads were built, at least as to major lines or locations, and the marketing to eastern, and to a lesser extent to western, urban centers had begun. Specialized livestock shipping facilities had begun to be developed by 1890, and the refrigerator car for shipping dressed meat had been developed.

Period of Grazing Readjustment, 1890-1930

The decades from 1890 to 1930 saw major changes in the western grazing region and many economic and social developments which affected grazing use of land. The population increased greatly and the economy as a whole grew rapidly. Means of transportation developed, opening up markets and bringing increased competition.

One important development affecting grazing was the rise of private land ownership. In 1880, and even in 1890, comparatively little of

19 Oliphant, "The Cattle Herds and Ranches of the Oregon Country, 1860-1890," *op. cit.*

the western half of the United States was in private ownership. Land grants to the states and to railroads had been made, or were in process of being carried out, but a major share of the granted land had not yet passed to private ownership.[20]

In 1890 more than half of most western states was still open public domain, except for islands of more intensive settlement. In the tier of states from North Dakota through Kansas, there were large areas where between 10 and 50 per cent was open public land. Comparatively modest changes took place in this general situation by 1900; much larger ones by 1910; and by 1920 open public domain had largely disappeared from the Plains, although there were still large areas of it in the Intermountain areas. Most, but not all, of the land that was not public domain and not in permanent federal reservations of various types, was privately owned.

Statistics on land ownership and transfer are so confused and incomplete that it is impossible to say just how much land within the present grazing region changed hands during these years. Something of the order of 500 million acres of land passed from public or railroad to other private ownership during the 1890-1930 period, and something of the order of 600 to 700 million acres for the 1880-1930 period. These are to be compared with the total land area of 1,163 million acres in the seventeen westernmost states.

Fuzzy as are the over-all statistics, they are better than the almost nonexistent statistics showing how much of this land went to grazing use. The initial disposition from federal to private ownership was often to a would-be farmer; after his failure, the land often got into grazing use and ownership, sometimes by circuitous routes. By and large, ranchers tried to avoid the costs of land ownership if they would obtain and maintain control over public land through other means. In many instances, the values of the associated public lands were all

[20] For histories of the federal lands, see: Samuel T. Dana, *Forest and Range Policy—Its Development in the United States* (New York: McGraw-Hill Book Co., 1956); Benjamin H. Hibbard, *A History of Public Land Policies* (New York: The Macmillan Co., 1924); Roy M. Robbins, *Our Landed Heritage* (Princeton: Princeton University Press, 1942); E. Louise Peffer, *The Closing of the Public Domain* (Stanford: Stanford University Press, 1951); Marion Clawson, *Uncle Sam's Acres* (New York: Dodd, Mead and Co., 1951). Also one article on a specific area and specific piece of legislation is: Arthur R. Reynolds, "The Kinkaid Act and Its Effects on Western Nebraska," *Agricultural History*, Vol. 23, No. 1 (January 1949).

concentrated onto the private lands; and when further land acquisition was necessary in order to maintain control over land use, some shrinkage in private land values was sometimes necessary.

During these same decades, the role of federal land used for grazing also changed. At first, its use was without controls, on a first-come first-served basis, with many extra-legal or local governmental controls gradually developing.[21] Indian reservations had been established at a comparatively early date, under the mistaken notion that the Indian could rather quickly be taught to farm. The areas of the reservations were sometimes larger than necessary, the Indians were incapable or unwilling to use them in white man's fashion, and there was no legal authority by which they could be leased to white men for their use.[22] As a result, pressures quickly developed, and were effective, leading to the drastic reduction in area of Indian reservations, the total area being halved between 1880 and 1900.

National forests were established after 1891, and by 1910 had reached to about the same gross area as today (although there have been some important boundary changes since then). Grazing had not been specifically mentioned as a use of the national forests, in the Act of 1897 which provided for the management of the national forests. The power to regulate grazing and to charge fees for it was challenged by livestockmen in a series of court cases, and it was not until 1911 that the authority to regulate grazing and charge fees was upheld by the Supreme Court.[23] New methods of grazing livestock on the range were developed, and controls were exercised over the numbers of livestock and the seasons of range use. Much pioneering work was needed, both to develop sound practices and to obtain their acceptance by ranchers.

Unfortunately, during the decade after 1910 numbers of livestock on the national forests were increased nearly one-third.[24] Numbers of livestock grazed within the national forests have declined steadily from their peak in 1918 until today they are about one-third of that level.

21 See Dana, *op. cit.;* Peffer, *op. cit.;* and Clawson, *Uncle Sam's Acres,* cited above, with further references in each.

22 J. Orin Oliphant, "Encroachments of Cattlemen on Indian Reservations in the Pacific Northwest, 1870-1890," *Agricultural History,* Vol. 24, No. 1 (January 1950).

23 Dana, *op. cit.,* pp. 144-47.

24 Marion Clawson and Burnell Held, *The Federal Lands, Their Use and Management* (Baltimore: The Johns Hopkins Press, 1957).

During these decades there was substantial development of farming, as distinct from grazing, in the western half of the country, where grazing is a dominant land use. Throughout the grazing region, the best grazing lands were taken into private ownership and plowed for crop production, thereby effectively destroying their capacity to produce native grass, at least until reseeded. In the Great Plains and in the Pacific Northwest, most of the land plowed was used for wheat or other small grain production. In California and the Southwest, much of it was irrigated and used to grow a variety of crops. The process of alienation of the public domain was headlong and without restraint; there was no classification of the land, and much land was disposed of under ill-suited laws and for purposes for which it was physically unsuited.[25] But for a large part of the land plowed, crop production was a higher and more profitable use than grazing; and thousands of farms have been developed on what would otherwise have been ranches.

Where farming was unsuccessful and the land abandoned, it was later often sold to ranchers or to government. In the 1930's some 14 million acres in the grazing region were purchased by the federal government. It has been estimated that there are still 10 million acres of land in the northern Plains that produce wheat crops in some years but that are incapable of continuous profitable wheat production, and that instead should be returned to grass and kept there.[26]

During the early part of this general period, conflicts between ranchers and farmers, and between cattlemen and sheepmen, were fairly common. One frequent form of conflict was over the fencing of federal land.[27] Since this land was used for grazing, the most natural action was to fence it to facilitate its use. Ranchers, especially cattlemen, did fence it in many areas and at many different times. Would-be homesteaders were often beguiled or intimidated into leaving fenced land alone. At various times, action was taken by the federal government; then fences were removed and sometimes the fence-builders

25 Clawson, *The Western Range Livestock Industry,* pp. 86-87, and *Uncle Sam's Acres,* pp. 89-92, both cited above.

26 Mont H. Saunderson, "Range Problems of Marginal Farm Lands," *Journal of Range Management,* Vol. 5, No. 1 (January 1952).

27 Many references could be cited. Peffer, *op. cit.,* Chap. IV, and Dana, *op. cit.,* Chap. 3, are good and give further references. Interesting and specific is Arthur R. Reynolds, "Land Frauds and Illegal Fencing in Western Nebraska," *Agricultural History,* Vol. 23, No. 3 (July 1949).

were punished; but fences had a way of getting back up again. After the passage of the Taylor Grazing Act of 1934, the erection of fences on federal land to facilitate private grazing use was authorized and financially assisted.

During these decades the methods of operating ranches changed also. As more land was owned or brought under the control of the rancher, planned methods of land management gradually replaced the haphazard and unplanned operations of the earlier days. The rancher became, in part, a farmer, at least to the extent of growing and harvesting hay and other feed crops, and sometimes to the extent of some general or grain farming as well. There were also major improvements in the quality of the animals grown; quicker maturing, larger, and better fleshed animals gradually replaced the original stock.

The development of grazing during the decades 1890-1930 may be summarized as follows:[28]

1. Almost everyone concerned failed to understand fully the hazards of an arid and variable climate. The differences, in climate and in plant growth, between a humid and an arid or semi-arid region are very great, and the settlers who entered the drier areas from the wetter ones failed to understand these differences for many years. This is the thesis of Webb's book *The Great Plains;*[29] while he has exaggerated and oversimplified, the situation is nonetheless real. When settlers entered many areas of the grazing region, they found comparatively lush stands of grass—"the sea of grass" is a common legend. While this, too, was often exaggerated, especially in later memory, the fact is that it often represented an accumulation of growth and vigor (perhaps not reduced by the grazing of wild animals because of lack of water or for other reasons) which was far more than annual growth would have been under conditions of annual grazing. The grazing potentials of most western grazing areas were originally grossly overestimated. It should be added that the cropping potentials of many farming areas and the growth potentials of many forested areas were also initially grossly overestimated.

2. Overstocking and improper seasonal use of ranges was common, not to say universal. This was partly due to the lack of understanding

[28] General references for the next few paragraphs are: Sampson, *op. cit.*, especially Chap. 23; Stoddart and Smith, *op. cit.*, especially Chap. XVI; and *The Western Range,* cited above, especially pp. 81-116.

[29] Walter Prescott Webb, *The Great Plains* (New York: Ginn and Co., 1931).

of natural conditions, described above, and partly to the lack of land use control, especially on the public lands. It should also be pointed out that during these same decades, cropping of farm land in this and other regions was also often far more intensive than the land could permanently sustain.

3. As a result of the foregoing, the native plant cover deteriorated considerably, at least from the viewpoint of usability by domestic livestock. The more palatable species were grazed first most thoroughly, so that they were often eliminated or greatly reduced in vigor and volume; woody and unpalatable brush and shrubs often took their place, as well as grasses and annual plants of lesser feed value.

4. Soil erosion was accelerated. It is difficult to know the extent of acceleration, and there has been much exaggeration on this point. It seems clear that many areas were unstable before the white man entered the grazing region, and that erosion was proceeding actively in them. It is also clear that much acceleration has occurred. This is evident in many stream valleys or bottoms, especially the smaller ones. Deep alluvial soils were originally protected against erosion by dense grass cover; when this was weakened or broken—by irrigation ditches and roads as well as by grazing—the soils quickly washed away, leaving great gullies where once had been excellent grasslands. Some acceleration on slopes and hills occurred also, although the results are less dramatic.[30]

5. By the end of the period here considered, the grazing lands of the United States were possibly at their lowest ebb. There had been extensive physical depletion of the natural resources; there was little public understanding of the problems of use of this resource; the users had not developed good techniques for using the land in a way to sustain its productive capacity; and a large part of the publicly owned grazing land was still used badly and without any constructive policy.[31]

6. But it was also true that at the end of this period, in spite of much abuse of the grazing resource, a large and productive resource still existed, and that it was the basis of a large and generally profitable business. "The impact of culture on nature by grazing of un-

30 For a discussion of this situation in one area, see E. R. Smith, "Grazing Industry and Range Conservation Developments in the Rio Grande Basin," *Journal of Range Management,* Vol. 6, No. 6 (November 1953).

31 This point has been well developed for one area in G. John Chohlis, *Journal of Range Management,* Vol. 5, No. 3 (May 1952).

plowed mid-latitude grasslands has been far less than, and on the whole different from, what has generally been implied in conservational writing. These grasslands *have* changed, but they are still essentially grasslands, and, assuming that they were to be used for grazing, they are perhaps not seriously impaired, as yet, for that purpose."[32]

Birth and Rise of Conservation in Grazing, 1930 to Date

Major economic and social movements in these three decades have directly affected grazing as a major use of land. The conservation use of resources has become much more widespread in practice. The conservation movement, as a philosophy and group of ideas, arose earlier—in the first decade of the twentieth century as far as any date can be put upon it. But in large measure it remained an ideal, not a practice, for several decades. Since 1930, however, the nation has been aroused about soil erosion and soil conservation to the point where major public and private efforts are being devoted to this problem. Good forestry, in the forester's sense of the term, has come to be adopted by many large private forest operators, although the character of forestry on the small forest holdings is still often very poor. Our ideas about resources, their use, and their maintenance for future use have undergone major transformations in these recent decades.

This change in attitude and program is no less marked for grazing than for the other major land uses. Prior to 1930 there were comparatively few trained range management specialists, the profession was not well recognized as a separate group, and the typical rancher was ignorant of the grasses upon which his animals grazed. Major changes have occurred in each of these matters. There is today a comparatively large body of competent and recognized range management specialists in the various federal agencies, in the agricultural colleges, including the agricultural extension work, and in private employment. The range management specialists have formed their own professional society, the American Society for Range Management, which publishes its own *Journal of Range Management*. The Society is unique among professional societies in that its membership is drawn about half from ranchers and other practitioners, as well as from the strictly profes-

[32] Clark, in Thomas, *op. cit.*, p. 756.

sional groups. The quality of its journal articles is high. Most striking of all has been the growth in knowledge of grazing by the typical ranchers. Today many ranchers know the chief grasses and other plants in their area, are able to identify them at various stages in their growth, and know their capacities and limitations for grazing. They are able to recognize trends in range condition (discussed below) and have sound ideas as to how ranges may be used for maximum sustained forage production.[33]

One should not infer that all was black before 1930 and that all is rosy today. Some ranchers have long practiced intelligent and farsighted range management.[34] There still remain many uninformed and shortsighted operators in grazing, as in crop farming, forestry, or any other land use. But the dominant attitudes have changed greatly.

One basic factor in the improvement of range use has been the development of range research.[35] The scale of research on range problems was very small until 1928. Forest Service regional experiment stations were established then, one part of which was for range management research. In addition, work on range plants was undertaken by other units of the federal Department of Agriculture and by the agricultural college experiment stations.

Another major development during this period was the passage of the Taylor Grazing Act in 1934.[36] This brought under definite management the remaining federal lands suited to livestock grazing. Numbers of livestock and seasons of use were regulated, and range improvements, such as watering places, fences, reseeding, and others, were constructed. While, especially in the early years, the administration of the Act was hampered by inadequate appropriations, substantial improvements were made. Possibly the educational effect of regu-

[33] For two good articles on this matter, see *Journal of Range Management:* Campbell, "Milestones in Range Management," Vol. 1, No. 1 (October 1948); and F. G. Renner, "The Future of Our Range Resources," Vol. 7, No. 2 (March 1954).

[34] See, for instance, the interesting series of accounts of rancher management of grazing resources in *Journal of Range Management,* Vol. 9, No. 6 (November 1956).

[35] U. S. Forest Service, Division of Range Research, "The History of Western Range Research," *Agricultural History,* Vol. 18, No. 3 (July 1954).

[36] J. Russell Penny and Marion Clawson, "Administration of Grazing Districts," *Land Economics,* Vol. 29, No. 1 (February 1953). See also Dana, *op. cit.,* Chap. 10; and Peffer, *op. cit.,* Chaps. XII-XVI.

lated and controlled grazing, upon the ranchers concerned, was as important a result as any.

Another major influence during this period was the educational work of the Soil Conservation Service and the Agricultural Extension Service. The latter had worked primarily with farmers, rather than with ranchers, in the western states in earlier periods. Gradually, as more research results were available, and as working relationships with ranchers developed, Extension provided information and technical guidance on grazing land use. The Soil Conservation Service, with its programs primarily in the soil conservation districts, co-operated with ranchers and other users of grazing lands. A major part of its program was educational, and no small part of the rancher's greatly increased knowledge of grazing resources and their use, is attributable to the work of this agency. The range field day, when groups of technical men and ranchers toured grazing areas together, comparing ideas and information, has become an established institution in many parts of the West.[37]

The results of this change in attitude toward use of grazing resources, and the various public programs aimed at better use of those resources, have been described by one informed specialist as follows:

> I believe most people who are widely familiar with western ranges will agree that during the past 25 years there has been marked improvement in range conditions throughout almost all of this area. The startling fact is that this improvement has taken place in the face of almost continuously increasing livestock numbers, instead of the reductions once thought necessary.
>
> Since 1928, the livestock population in the seventeen states of the range area has increased 41 per cent, from 32,807,000 to 46,510,000 animal units. Moreover, both the improvement of the range and the increase in livestock have taken place in the face of substantial reductions in the area devoted to range use. In the Plains states, many millions of acres of range lands have been plowed and put into cereals, beans and other crops. In the mountains the encroachment of timber reproduction, and in the Southwest the thickening up of other woody species, has substantially reduced the range area.[38]

[37] Horace L. Leithead, "Field Methods Used to Demonstrate Range Conservation," *Journal of Range Management,* Vol. 3, No. 2 (April 1950). See also the accounts in most recent issues of the *Journal,* of field days by the various sections of the Society.

[38] Renner, *op. cit.*

GRAZING AS A USE OF LAND TODAY

Grazing, as we described it at the beginning of this chapter, is characterized by the use of native grasses and other forage as feed for livestock. It is highly dependent upon the natural growing conditions; there is a close relation between the type of forage and the kind of land use made.

Physical Characteristics of Western Grazing Lands

The various species of plants which made up the natural grazing areas were obviously well adapted to those areas, or they would not have survived. When man does not intervene directly, by plowing, seeding introduced species, or in other ways, he may intervene indirectly, through his control over numbers of domestic livestock and their season of grazing, or through the use of fire, or in other ways. But the native forage plants that persist, and other native plants intermingled among them but not suitable as forage, obviously have a sum total of characteristics that enable them to survive, when in fact they do survive over long periods.

Range management specialists classify natural grazing areas of the West into broad grazing types of vegetative regions, depending upon the dominant species and groups of species. One such classification is shown in Figure 51.[39] Nine major vegetation regions are recognized. These fall into three general groups: (1) four regions in which grasses of various kinds are dominant, (2) two regions dominated by shrubs of different kinds, and (3) three regions of forest or woodland. In our classification of land use, most of the latter group are probably classed as forest, although they may include grasses and other forage plants which provide considerable feed for livestock. In each vegetation region, there are typically many different kinds of plants, a comparatively few of which provide the greater part of the edible plant growth. The growth requirements for different plants in each region differ, as does also the complex of plants in one region as contrasted with that in another. The amount of forage produced, the season at which it is available, the resistance of the plants to grazing, and other

39 Taken from Stoddart and Smith, *op. cit.*, Chap. V. Generally similar classifications are found in Sampson, *op. cit.*, and in *The Western Range*, cited above, and in references cited therein.

Figure 51. Natural range vegetation regions of the range states. (Chart from Stoddart and Smith, *Range Management,* McGraw-Hill Book Co., copyright 1943.)

characteristics of the plants differ considerably, and present the user of grazing land with his more difficult management problems.

In appraising and managing grazing lands, range management specialists have developed the concept of range condition.[40] Briefly, application of this concept involves an attempt to estimate the productive potential of each range site, and then to rate the present condition of the range in relation to that potential.[41] The driest, steepest, hottest sites in the deserts of the Southwest cannot possibly produce as much forage for livestock as moister sites in cooler climates and with deeper soils, no matter how well each might be managed. The productive potential of a site depends upon such climatic factors as annual precipitation, its seasonal distribution, the variability in precipitation from year to year; upon average annual temperatures and variations in them; and upon soil depth and profile, which in turn is related to the climate; and upon topographic exposure (south or north facing slopes, etc.) and slope. Range condition is expressed as the general health of the range in relation to the attainable ideal. Health is measured by such factors as the density of the plant cover, its species composition, the vigor of individual plants, accumulations of plant litter from growth in previous years, soil erosion or stability, and other factors—all in relation to what is considered attainable for this particular site.

Range condition is rarely static; ranges are usually improving or declining in condition. However, range condition (in the sense used here) changes rather slowly from year to year. This approach puts great stress on the measurement of range condition trend; if the trend

40 "Range condition" as used by the range management specialists, and the same term as used by the agricultural statistician, mean two very different things, leading to some confusion. The range management man is trying to measure the general health of the range; the agricultural statistician is trying to obtain a measure of annual or seasonal forage production from the grazing lands. The Department of Agriculture has collected range condition figures, by states, since 1923. These are closely comparable to crop condition figures collected by the same agency. For an analysis of these range condition figures see Marion Clawson, "Range Forage Conditions in Relation to Annual Precipitation," *Land Economics,* Vol. 24, No. 3 (August 1948).

41 The concept of range condition, in the range specialist's sense, is discussed and defined in several writings. See for example, two articles in *Journal of Range Management:* E. J. Dyksterhuis, "Condition and Management of Range Land Based on Quantitative Ecology," Vol. 2, No. 3 (July 1949); and Kenneth W. Parker, "Application of Ecology in the Determination of Range Condition and Trend," Vol. 7, No. 1 (January 1954).

is upward, the management practices may be good, even though the range is yet in poor or fair condition. Range condition and trend are based upon normal or average weather conditions for that location; but it is often difficult in practice to isolate the influence of short-term conditions.

Change in species composition of plants in a particular site is one of the measures of range condition and range trend. The change may be progression or retrogression, in the ecologist's terms; toward a plant association which is higher in the ecological scale, ultimately reaching the climax state, or to one which is lower in the ecological scale. Removal of the circumstances which lead to deterioration of plant composition will be likely to lead to its improvement, but the improvement may not restore the conditions from which the initial deterioration began. Retrogression and progression are not necessarily symmetrical in their paths; one may not be the obverse of the other. The changes, not only in the plant association but sometimes in the soil itself, may have been so great as to make impossible a return to the original condition.

As changes in plant composition take place, certain species of plants often serve as indicators or indexes of the total change taking place. Plants have been classified into (1) decreasers, often the most valuable forage species, which decrease in volume as the range condition deteriorates; (2) increasers, species which were originally found in the plant association but which are able, at first, to increase because of the reduced competition from the declining decreasers; and (3) invaders, or plant species not found originally in the plant association, or at least not common there. In general, increasers are less valuable forage plants than decreasers, and invaders least valuable of all. The relationships among these three groups is shown in Figure 52, taken from the Dyksterhuis article. As range conditions decline, the decreaser plants decline steadily in relative volume, and ultimately disappear; the increasers partially take over the released soil and moisture, increasing in volume at first, but then as range depletion becomes serious, they, too, decline; invaders at first increase slowly as range condition falls, but with subsequent declines they dominate the plant association.

In some circumstances, the original plant association has been so changed over the decades that management to restore it is impossible or practically unattainable. Much of the foothill range in California

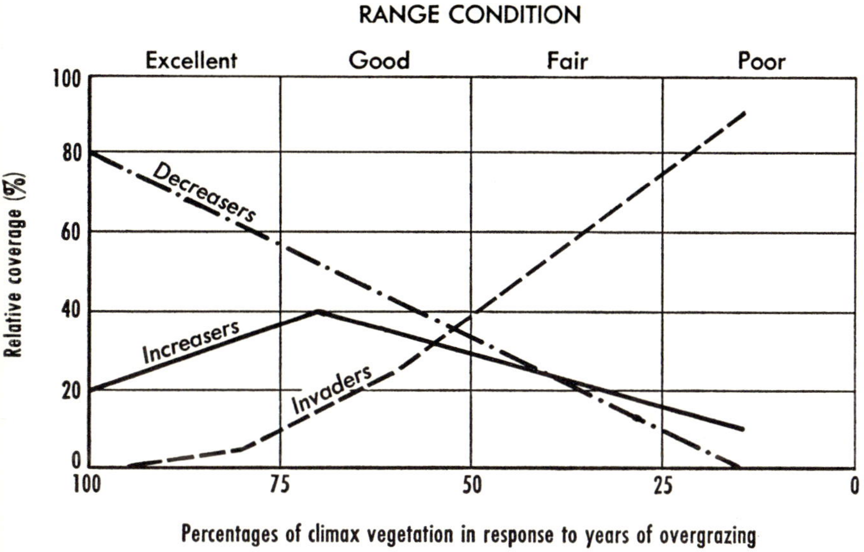

Figure 52. Diagram illustrating a quantitative basis for determining range condition. (From E. J. Dyksterhuis, in *Journal of Range Management,* July 1949.)

was once perennial grasses, but has since become almost wholly annual grasses of introduced species; practical range management in this area must now be based upon the annual grasses, and range condition determined on their basis. A similar situation has arisen in southern Idaho, where the introduced annual grass *Bromus tectorum* has largely replaced the original perennial bunch grasses, to the extent that it is doubtful if any management would now restore the latter. In practice, therefore, range condition must be in terms of what is now a reasonably attainable degree of health of the range, not an ideal once attainable but now lost.

An application of this concept to a specific situation is illustrated in Table 48. This application is in use by the Soil Conservation Service under a wide range of physical conditions in Texas and Oklahoma. Some twelve species of desirable grasses are listed as decreasers; not all of these would normally be found on a single site, but, wherever found, any of these species can be considered as valuable grasses. Another group of plants is listed as increasers; their maximum plant

Table 48. Part I of "Technician's guide to condition and management of ranges in District Group 39, Soil Conservation Service, San Angelo, Texas, September 1947"

Decreasers (all sites)	Increasers (max. in climax)	Range sites[1] 1	2	3	4	5	Invaders (all sites)
		(%)	(%)	(%)	(%)	(%)	
Indiangrass	Texas wintergrass	[2]d	d	d	10	15	All annuals
Big bluestem	Perennial threeawns	5	5	0	0	0	Red grama
Little bluestem	Fall witchgrass	5	5	5	0	0	Hairy triodia
Pinhold bluestem	Silver bluestem	d	5	5	5	10	Tumblegrass
Sideoats grama	Tobosa	0	0	10	20	5	Windmill grass
Neally grama	Sand dropseed	d	5	5	5	5	Ear muhly
Green sprangletop	Texas grama	5	5	5	5	0	Nightshade
Vine-mesquite	Buffalo grass and						
Wild ryes	curly mesquite[3]	d	20	30	40	10	Broom snakeweed
Tall dropseed	Hairy grama	d	d	10	5	0	Mealycup sage
White triodia	Forb increasers	10	5	5	5	5	Western ragweed
Texas cupgrass	Wood increasers	5	30	10	0	20	Woody invaders

[1] Site 1 = *Very shallow upland* (Soil group 24v); Site 2 = *Scrub oak upland* (Soils on which shin oaks are part of climax); Site 3 = *Ordinary upland* (Soil groups 24d, 17); Site 4 = *Deep upland* (Soil groups 1 and 2; heavy clays); Site 5 = *Draws and bottomlands* (Soil group 4; overflow land).

[2] "d" indicates that on this site the species is a decreaser rather than an increaser.

[3] Consider the two species together in estimating coverage. For sites 3, 4, and 5, near 19-inch isohyet use 35%, 50%, and 15%, respectively, and near 29-inch isohyet use 25%, 30%, and 5%, respectively.

SOURCE: E. J. Dyksterhuis, "Condition and Management of Range Land Based on Quantitative Ecology," *Journal of Range Management,* Vol. 2, No. 3 (July 1949), p. 110.

density as increasers depends upon the site characteristics, particularly soil depth. And still another group of plants is listed as invaders on all sites within this general region.

The only comprehensive estimate of range depletion was made by the Forest Service more than twenty years ago before the concepts of range condition had reached their present refinement; but the same general idea was employed.[42] Depletion from virgin range conditions for the entire western grazing region was estimated at 52 per cent,

[42] *The Western Range,* cited above, especially pp. 3-8, 81-116.

and depletion was expressly defined as reduction in grazing capacity. The degree of depletion was estimated by forage types and by classes of land ownership; the various shrub and desert grass vegetative types, and the public domain (then outside of grazing districts), state and county, and private lands showed depletion above average. The tall grass and the open forest vegetative types and the national forests showed depletion much below average. It was further estimated that 75 per cent or more of the grazing lands was still trending downward in condition and only 16 per cent had improved appreciably in the preceding thirty years. The remedy proposed by this study was a major and immediate cut in numbers of livestock grazed, to the estimated productive capacity of the grazing areas.

In retrospect, it appears that the authors of this report were influenced more by the severe droughts of 1934 and 1936, and the accumulated deficiencies of rainfall in the several preceding years, than they realized. The estimates of productive capacity were supposed to be independent of annual fluctuations in forage growth, but instead to be based upon average productivity under conditions of normal rainfall. But it is almost impossible not to be swayed by current conditions, even when one is trying to estimate the long-term trend. This report was admittedly based upon sample information, of less detail and intensity of sampling than ideally would have been desirable.

Events since the issuance of this report have taken a greatly different turn than it anticipated. The western range and grazing areas have, in general, improved considerably in their basic productive condition, although no more recent comprehensive estimates of the degree of that improvement are available; and this improvement in range conditions has taken place at the same time that numbers of livestock grazing on them have increased rather substantially.[43] While more favorable weather conditions than prevailed in the early 1930's have existed, at least up until the early 1950's, the major improvement has been possible primarily because of better management. This was a factor not taken into account, or not weighed heavily, in 1936.

Range survey methodology is now applied by range specialists and by ranchers with much more skill to local areas. Range condition often varies greatly within comparatively short distances; it may be very low around a watering place, or in the comparatively level bottom land of a mountain valley, and within a mile or less be excellent.

[43] Renner, *op. cit.*

For this reason, general averages are in less repute among range specialists today than once was the case, and the approach is more in terms of localized range management situations and programs.

Some localized studies throw considerable light on the matter of the present condition of range health.[44] Allred has shown that in Texas some 150 million acres are now partially or wholly covered with woody plants of various kinds; two chief species are mesquite and cedar (juniper) (Figure 53). The area of these species when the white man settled in Texas is unknown, but it was very much less than at present, and so was the density, even where the plants were originally growing. Heavy grazing has weakened the grass in its competition with the small trees for the available moisture supply. The shrubs or trees have been able to spread, and the degree of their spread and the density of their stands is a direct measure of the extent of the grazing depletion today. Extensive efforts have been made to control or eradicate the brush and trees, often at considerable cost, but the effectiveness of these methods has sometimes been lost because the range management methods followed after clearing were such that the grasses had a hard time reproducing themselves and the shrubs and trees reinvaded the areas.

Parker[45] has summarized the situation in extensive areas of the West, by showing that juniper, sagebrush, chapparal, and lower value annual grasses have either invaded areas where they were not previously found, or have greatly thickened up where they were previously known, in either case at the expense of the perennial grass cover and growth. The balance between grass and sagebrush was apparently always rather close on millions of acres in the Intermountain area, and grazing can easily tip the scales so as to reduce the grasses and allow the sagebrush to increase to the extent of the available moisture supply.

Fisher has also considered the mesquite problem on southwestern ranges, which he considers to be of major concern to livestock raisers. Once established, its eradication or even reduction on a permanent basis is neither easy nor cheap.

[44] See the following articles in *Journal of Range Management:* B. W. Allred, "Distribution and Control of Several Woody Plants in Oklahoma and Texas," Vol. 2, No. 1 (January 1949); Kenneth W. Parker, "Control of Noxious Range Plants in a Range Management Program," Vol. 2, No. 2 (July 1949); and C. E. Fisher, "The Mesquite Problem in the Southwest," Vol. 3, No. 1 (January 1950).

[45] Parker, *op. cit.*

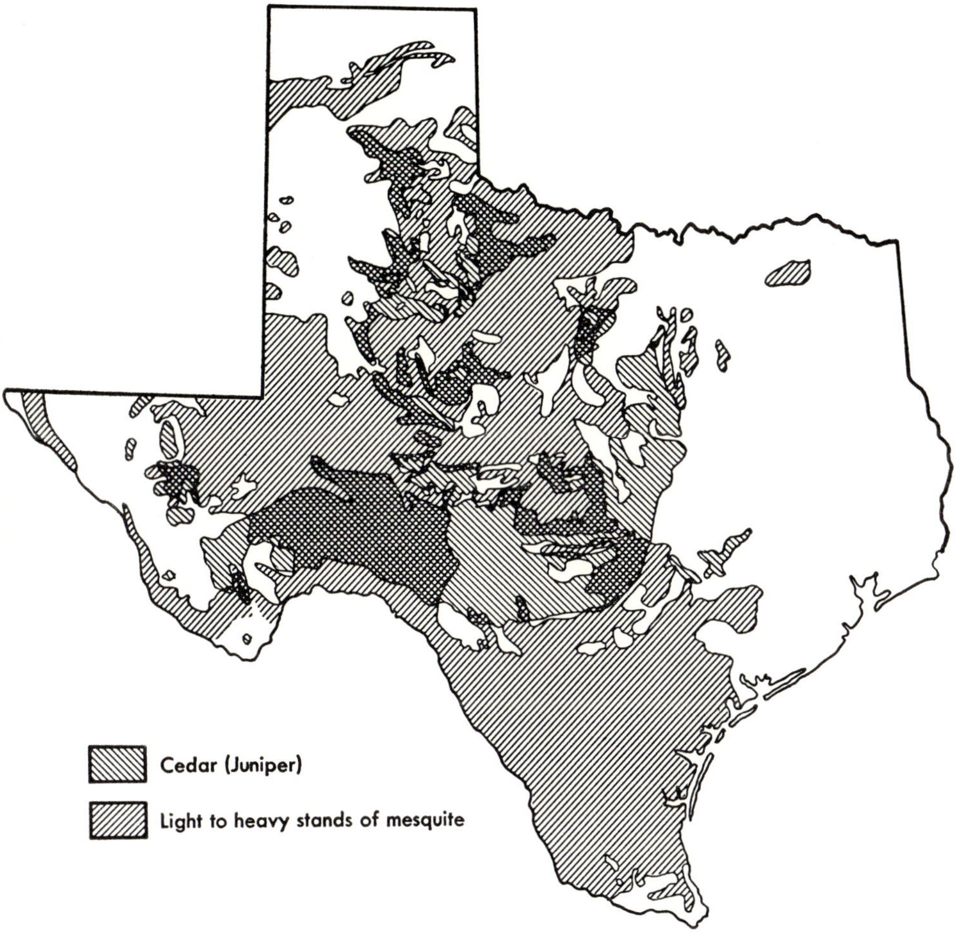

Figure 53. Major mesquite and juniper areas in Texas. (From B. W. Allred, in *Journal of Range Management,* January 1949.)

The Soil Conservation Service has classified most of the range area of western Texas and western Oklahoma. Only comparatively small and scattered areas are classified as in excellent range condition; larger areas, but still only a small part of the total, are classified as in good condition. Substantial areas, some of two or three counties in extent, are classified as in fair condition; but by far the greater part of the whole is classified as in poor condition. This is particularly true in the southern and southwestern edges of the state, and on the Edwards Plateau. The latter area has had heavy use since some time in the 1880's, and in recent decades has been a prime example of common use by cattle, sheep, and goats. This method of grazing undoubtedly produces more total usable forage from a given area, since the forage preferences of the different types of animals are different; but it also nearly always produces heavy overuse, because each class of stock is grazed with inadequate consideration for the requirements of the other kinds. It is also notable that areas of excellent and good range condition are often intermingled with much larger areas of predominantly poor-condition range. This is further evidence of the fact that the character of the management of the range is important, perhaps dominant, in determining range condition.

The Bureau of Land Management has instituted a system of range condition measurement in the grazing districts. It will be recalled that these areas had been subject to uncontrolled use, and hence to overuse and improper seasonal use, for several decades prior to their being placed in grazing districts in the 1930's. These are predominantly shrub vegetative types, many of desert character, and it was necessary to devise new methods of rating range condition that would be applicable to such vegetative types and that could readily be applied.

In 1956 about 2 per cent of the entire area was classed as in excellent condition, 18 per cent in good condition, 52 per cent in fair condition, about 25 per cent in poor condition, and 3 per cent in critical condition. About 20 per cent of the entire area was improving, about 54 per cent was stabilized as far as range condition was concerned, and 26 per cent was still degenerating. While this represents a picture that is far from good, it is generally agreed that these areas are in better condition today than when first put into grazing districts. The 1936 report by the Forest Service estimated that 88 per cent of these lands were on the downgrade at that time.[46] Unfortunately, complete comparability does not exist between these estimates; if it did, the change from 88 per cent to 26 per cent of the area on the downward trend would clearly represent a major improvement in twenty years.

We may summarize this consideration of range condition as follows: (1) the western grazing areas are unquestionably depleted from their original or virgin condition, and also from the condition that

[46] *The Western Range*, cited above.

continued but well-managed grazing would make possible; (2) the extent of the depletion varies greatly, often within very short distances; (3) the present exact situation is not known precisely; but (4) it is generally believed among range specialists and other informed observers that grazing conditions on the whole are considerably better today than they were twenty-five years ago.

The grazing lands are often highly vulnerable to erosion. The limited rainfall that is responsible for their use as grazing lands is also responsible for a limited or scanty vegetative cover, even at the best. When heavy rains came or winds blew with unusual velocity, erosion occurred even before white man came. The earliest explorers were struck with the silt content of many western streams. With grazing by domestic livestock, the grass and other plant cover was often weakened and became less able to protect the soil. Acceleration of erosion was a consequence, and in many situations this so altered the original soil-moisture-plant relationships that a return to the previous condition was impossible. This has often been true in alluvial valleys, small and large, where erosion carried away so much of the soil as to change the ground water relationship and the possibility of plant growth.

With the great increase in public interest in soil conservation of the early 1930's, previously noted, various estimates of the extent and character of the soil erosion problem were made.[47] These often used as their basis reconnaissance information or rough estimates and, while they perhaps served an important need at the time, much more detailed information has become available in recent years. However, no broad regional or national summaries of the over-all erosion problem have been prepared in recent years comparable to these earlier reports.

Unfortunately, no detailed and reliable estimates exist of the total physical volume of output from the western grazing lands. Some information may be pieced together from the various analyses that have been made,[48] from which we may conclude that the western grazing

47 Hugh H. Bennett, *Soil Conservation* (New York: McGraw-Hill Book Co., 1939), especially pp. 60-65.

48 *The Western Range,* cited above; R. D. Davidson, *Federal and State Rural Lands, 1950, with Special Reference to Grazing,* Circular No. 909, U. S. Department of Agriculture, Washington, May 1952; H. E. Selby and Donald T. Griffith, *Livestock Production in Relation to Land Use and Irrigation in the Eleven Western States,* Bureau of Agricultural Economics, U. S. Department of Agriculture, March 1946.

lands normally produce at least 60 million, and perhaps 80 million or more tons of hay-equivalent. This is to be compared with a national hay production of roughly 100 million tons in average years. In 1949, according to the Census of Agriculture, the value of all crops harvested in the United States was $16.3 billion, of which hay was $1.88 billion; the value of all farm products sold in that year was $22.05 million. Farm prices have since risen above those of 1949, but the general relationships are perhaps unchanged. If the range forage had been valued at the same price per ton as hay in 1949, its gross value would have been about $1.1 to $1.5 billion. While hay is often more costly than grazing, simply because more money can be spent to produce hay to complement grazing seasonally, these figures give some idea of the general relationship between grazing and other crops. The value of range forage does enter into agricultural income indirectly, through the value of the animals produced by it. One must conclude that data on grazing are inadequate, compared to its value in the general agricultural scheme.

Site factors, such as soil and climate, greatly affect the general level of forage production from grazing lands, but the influence of many site factors is relatively constant from year to year. Weather factors, on the other hand, greatly influence forage production from year to year. The chief weather factor is total annual precipitation, because of the generally low level of precipitation in the grazing regions generally, but its seasonal distribution, temperatures, and other weather factors may also vary considerably from year to year. The effect of favorable or unfavorable weather is principally in the year in which the weather condition is experienced, but there is also some carry-over effect into the following and perhaps even into later years.

For the northern Great Plains, the correlation between precipitation and forage production from grazing lands is high (Figure 54).[49] The influence of precipitation is almost as great in the following as in the current year; the combined effect, in statistical terms, is a correlation in excess of .80. In the southern Plains the correlation is not quite as close, but the effect of the preceding year is almost equally great. In the Intermountain areas the correlation is somewhat lower, with more years that do not follow the pattern closely, and the influence of the preceding year is nearly negligible. Plant vigor does depend upon

[49] Marion Clawson, "Range Forage Conditions in Relation to Annual Precipitation," *Land Economics*, Vol. 24, No. 3 (August 1948).

growing conditions in earlier years, but many other factors enter also, so that the average relationship is not close. Along the Pacific Coast the correlation between precipitation and range forage production is still less direct; temperature and other factors enter here to a great extent.

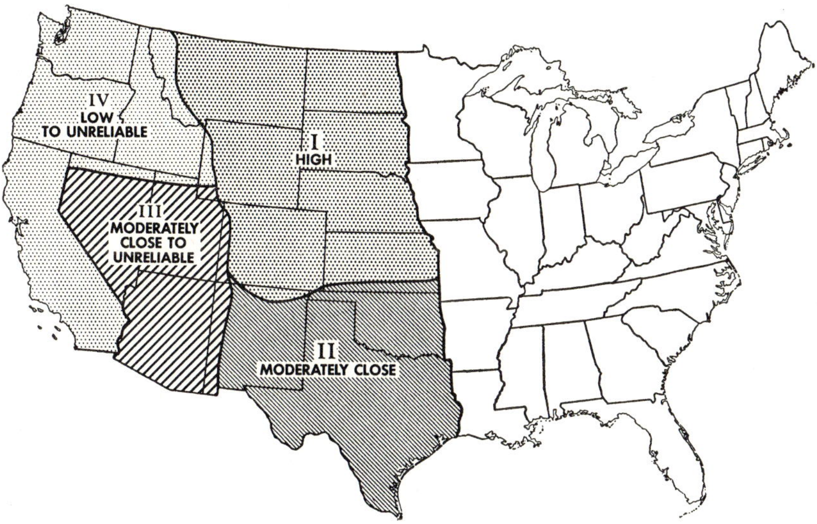

Figure 54. Correlation between precipitation and current range forage yield in four western regions. (Chart from A. W. Sampson, *Range Management, Principles and Practices,* John Wiley and Sons, copyright 1952.)

Ownership of Grazing Land

Land ownership is more closely related to vegetative cover and land use for grazing lands than perhaps for any other major land use. Partly because of their low productivity, and partly because of their history, much of the land on which grazing is the dominant use are in public ownership.[50] Of the 700 million acres where, we estimate,

50 In addition to references previously cited, see: M. M. Kelso, "Current Issues in Federal Land Management in the Western United States," *Journal of Farm Economics,* Vol. 29, No. 4 (1947), Part II; Mont H. Saunderson, *Western Land and Water Use* (Norman: University of Oklahoma Press, 1950).

grazing is dominant, about 200 million are in public ownership. Nearly all the grazing district land, other public domain outside of districts, and much of the national forest area are used primarily for grazing. Conversely, a major part of publicly owned land is used primarily for grazing; of the nearly 400 million acres in federal ownership, about 200 million have grazing as their major use. In addition, grazing is one of the uses on much of the rest of the public land, where forestry or some other use is dominant.

Within the grazing region itself there are considerable differences in grazing land ownership. Nearly all the Great Plains grazing areas —the tall grass and short grass vegetative regions shown on Figure 51 —are in private ownership. West of this area, however, over 60 per cent of all the grazing land is publicly owned.

The forage from the publicly owned land is harvested by privately owned livestock, as part of ranch operations under private management. The private operator uses the publicly owned grazing lands either exclusively or jointly with other ranchers, under various leasing, licensing, and permit systems. Much of the decision-making about the grazing land is in the hands of the public agency—how many animals to graze, at what seasons, and with what range management practices. Private use of publicly owned grazing lands is a distinctive form of land tenure in the United States, and the literature dealing with it has been fairly extensive.[51] In general, the decisions to use public land for grazing and the choices among potential applicants are not made on the basis of competitive bidding. Grazing privileges are awarded on other grounds, and the prices paid are lower than those for which comparable privately owned lands could be rented.

Use Patterns of Grazing Land

Much of the western grazing area can be used only seasonally. The high mountain areas are covered with snow during most of the year, and only during the summer is there plant growth that can be grazed; but such land provides excellent summer range, especially for sheep. There are extensive desert or semi-desert areas, especially in Utah and Nevada, where grazing is possible only in winter, when light

51 Several of the references previously cited deal with this subject.

snow-cover in the valleys and foothills provides a source of water for livestock; these are used almost exclusively by sheep for winter range. In some localities of these desert areas wells or springs provide water for livestock, particularly cattle, for grazing at other seasons. For these two general types of situations, and perhaps for others, seasonal use is rather sharply and strictly limited by climatic and other factors.

There are larger areas of grazing land, the seasonal use of which should be limited. Most range plants grow only when temperature and moisture conditions are favorable; the date during the year at which this occurs depends upon elevation, location within climatic zones, and other factors. When growth begins, the plant draws at first upon food reserves stored in the roots, and then begins to produce new food by the usual photosynthesis process. During these early stages of growth the plant is often highly palatable to livestock, in part because it is tender and succulent, in part because its protein content is high. But it is also highly vulnerable to grazing; unless it can attain a size which permits maximum or at least high food formation, it may not survive or at least will be weakened. Moreover, unless some proportion of the plants of each species can reach seed maturity each year or each few years, the species will not survive permanently. If productive capacity of grazing lands is to be preserved permanently, then grazing must be controlled seasonally; it was lack of such control, perhaps more than any other single factor, which was responsible for the depletion of western grazing areas.

Range management specialists have developed the concept of range readiness.[52] This has been defined as "the date in any one year when the range first reaches the condition in which there is sufficient feed to keep livestock in thrifty condition and when the stock may be admitted without serious impairment of the growth and reproductive processes of the more important forage plants."[53] Range readiness determines the date when grazing may safely begin; there may also be a problem of determining the date when grazing must end, if the productive vigor of the plants is to be maintained. For the higher mountain seasonal ranges, the coming of cold weather almost certainly provides a closing date for the grazing season, but on the winter desert

[52] Stoddart and Smith, *op. cit.*, Chap. X.

[53] *Ibid.*, p. 250, quoting A. W. Sampson and H. E. Malonsten, *Grazing Periods and Forage Production on the National Forests,* U. S. Department of Agriculture Bulletin 1405, Washington, 1926.

regions it may be necessary arbitrarily to close the grazing season. This is because grazing takes place during the winter, when the plants are more or less dormant, and should stop by spring when the plants begin to grow again.

Grazing areas differ also in the flexibility with which seasonal grazing may take place. For some areas, the forage plants available for grazing must be eaten by livestock at the season they are ready, or not at all. For much of the semi-desert foothill areas, where annual forbs provide a large part of the forage, these plants must be eaten when they are ready, or they dry up and are unavailable. At the other extreme, some types of grasses, especially the short grasses, cure on the stem; although their nutritive value drops somewhat, they remain palatable and nutritious for weeks or months.

In practice, seasonal grazing patterns are a compromise between what would be ideal, ecologically and from a range management viewpoint, and what may be necessary in order to provide feed for livestock at all seasons. If there is a shortage of spring-fall range, for instance, livestock may be kept later on winter range, or go to summer range earlier, or be fed farm-produced feed, or some combination of these adjustments. Grazing use during other than the ideal season means lowered production of the grazing area, either immediately or ultimately.

The grazing areas of the West have been classified according to their seasonal use (Figure 55). Most of the short grass type and much of the southern desert areas have been classed as yearlong. Actually, livestock in these areas may need some supplemental feed, either hay, cottonseed cake, or other, during the winter, but some grazing is practiced throughout the year and the reliance upon native forage plants is almost complete. In the mountainous areas seasonal use of range is common.

Seasonal management of livestock on privately owned grazing and feed-producing lands, or integration seasonally of available grazing from publicly owned lands with that from private lands, presents the rancher with some of his most important management decisions. It is in better seasonal use of grazing resources that some of the greatest gains have been achieved through better range management during the past two or three decades. In many instances, more livestock can be grazed, and yet the ranges improved, by better seasonal management.

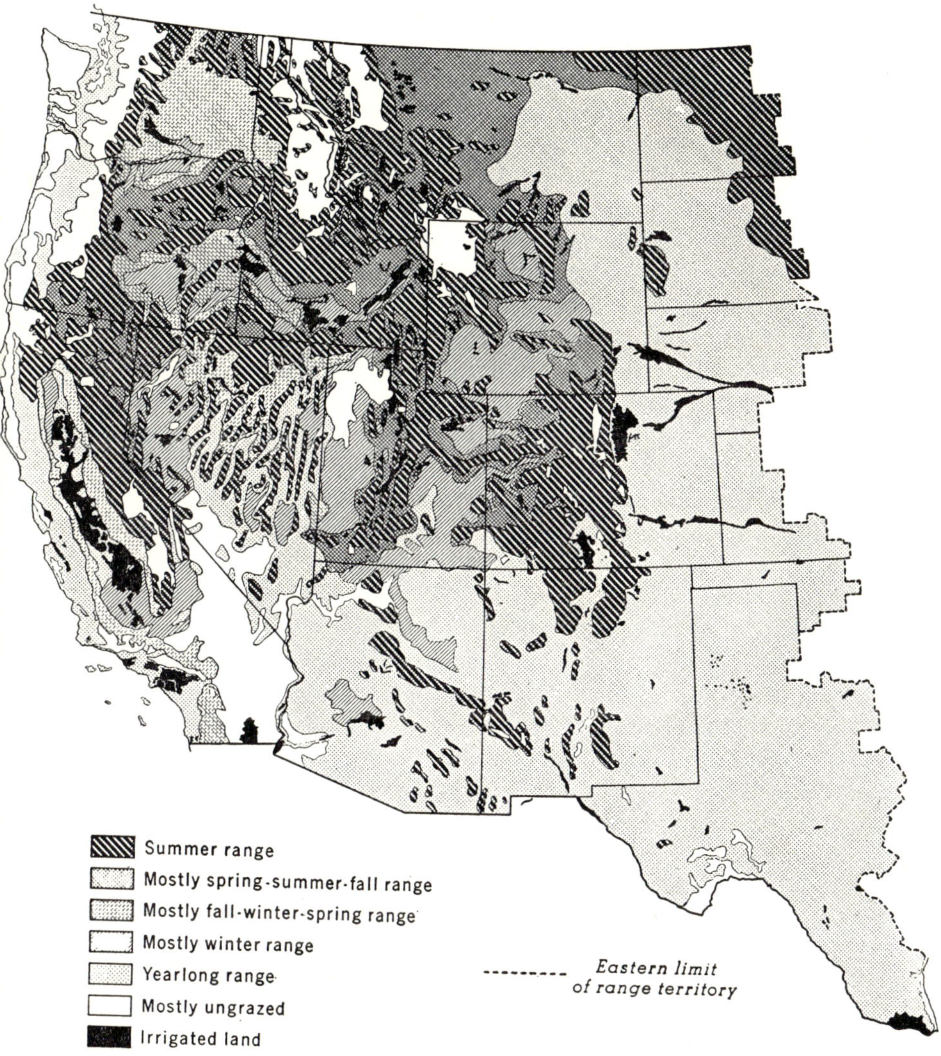

Figure 55. Seasonal use of the western range, 1947. (Chart from *Grass, The Yearbook of Agriculture, 1948.*)

Investment and Improvements on Range Lands

Grazing land is characterized by the low level of capital improvement per acre or unit of area. In the earlier stages of grazing land use, the land and its forage were used as they existed with virtually no improvements—water for livestock was where you found it, for example. As use became more nearly complete, investments were made in order to take advantage of forage which was otherwise unavailable. Some of the earliest and simplest investments were for livestock water: springs were developed, wells dug or bored, dams or "tanks" built to catch runoff and hold it until needed, water troughs installed, and many other devices used to make water available where and when needed. The quantity of water required for livestock is small, compared with the amounts needed for irrigation, and comparatively large expenditures for small volumes of water were sometimes justified. In the mountainous areas roads and trails sometimes had to be built, to enable the movement of livestock into areas with available forage, and to permit their owners to supervise the livestock. Fencing was another early development; at first, comparatively small areas were fenced for special purposes; later, larger areas were fenced for more intensive grazing management. In more recent years considerable investment has been made for vegetative improvement, such as reseeding to grass and removal of brush. In general, grazing lands are not cleared of forest, except occasionally, and by fire; they are not cleared of stone, nor are they levelled; permanent buildings are erected on the range only in unusual instances, although they are placed on ranches; and many other forms of heavy capital investment found on farms are not made on grazing land.

No comprehensive estimates seem to have been made of the total investment made on grazing land—investment in improvements, as distinguished from the unimproved land. Common knowledge of the grazing region suggests that it is very low, on the average—possibly something of the rough magnitude of $1.00 per acre average for the Plains, and of 10¢ per acre for the mountains and deserts. These would amount to something of the rough magnitude of $300 million in total investment on grazing land, including fences, but they are the roughest kinds of guesses. A large part of the total investment—whatever it may be—has been made in the last twenty years, during the period when the conservation attitude toward range has been growing, and a large part of it has been subsidized by the federal government. It is unfortunate that we do not have more reliable data on the amount of capital investment in range land, its distribution as to vegetative types, and other relevant details.

Ranch Organization

A ranch is a farm which raises its livestock primarily by grazing native forage and which ordinarily is highly specialized in livestock production. The same natural conditions that provide native forage often preclude much agricultural production of other kinds.

The 1930 Census of Agriculture, in its "Type of Farming" analysis, was the first to characterize ranches in statistical terms, and to present statistical data concerning them. Livestock specialty farms were those which obtained 40 per cent or more of their gross income from sale of animals, including wool, but excluding dairy products; a ranch was a livestock farm with five times as many acres in pasture as in crops if located in the eastern half of the country, or ten times as many acres in pasture as in crops if located in the western half of the country.[54] At that time there were only about 70,000 ranches in the entire country, well over 90 per cent of which were in the western grazing region. The average western ranch had slightly over 3,000 acres of land, of which only 126 acres were cropland; the value of its land and buildings was nearly $35,000, but only $11.00 per acre; it had 154 cattle of all ages and classes and 347 sheep (excluding lambs); gross income was nearly $7,500, compared with an average for all farms in the United States that year of less than $2,000. These averages conceal great variation among ranches. For instance, while a few ranches had both cattle and sheep, most of them had one class or the other, not both, so that the average cattle ranch had more than the 154 cattle and the average sheep ranch more than the 347 sheep. A few ranches were very large, running into many thousands of acres, a few thousand livestock, and employing many men; but there was also a much larger number of small ranches—too small to provide full-time employment and reasonable income to their operators. Nearly half of them had a gross income of less than $2,500, and more than three-fourths of them were smaller than the average.

[54] See Clawson, *The Western Range Livestock Industry*, cited above, Chap. 13, for a general discussion of ranches.

The 1954 Census of Agriculture separated commercial from other farms, and classified the commercial farms by type. One type was "other livestock," meaning livestock other than dairy and poultry farms. These "other livestock" farms include fattening farms and farms which raise livestock by means other than native forage, hence they include more than ranches.

Livestock farms are the most single numerous type in the West. In acreage of cropland, they are roughly average in size; in terms of total land per farm, they are three times as large as other farms. Investment in land and buildings per acre is from a third to a half as large as the average of all farms, but in total is somewhat larger than average. Livestock farms are about average in terms of off-farm work of the operator. "Other livestock" farms have somewhat fewer home facilities than do average farms; this is not due to lower average incomes, but to the more remote locations of many ranches, which makes the provision of electricity, television and other facilities more difficult and more expensive. On the other hand, it is notable that the ranches have somewhat more home freezers; a home freezer is often a distinct economy on a ranch, since it facilitates ranch slaughter and consumption of livestock at some saving. Ranchers use trucks, automobiles, and tractors to about the same extent as do all other farms. Livestock farms nationally and in the West have somewhat fewer workers than the average of all farms, primarily because they hire fewer workers.

As might be expected, expenditures for feed on livestock farms are comparatively large, partly because numbers of livestock are large. Notable are the considerable numbers of horses; riding horses are very much work animals on such ranches and, besides, some horses are still used for motive power. The average number of cattle per western livestock farm in 1954 was 161, or about the same as the average number of cattle per ranch in 1930. Information on numbers of sheep on these farms is not available for 1954. Numbers of pigs and chickens are low on ranches; the native forage is not adapted to these livestock, and home-grown grain is not available.

A number of studies have been made of cattle and sheep ranches by various western areas in recent years.[55] In the Intermountain

[55] See especially the following, and the references cited in them: H. R. Hochmuth and Wylie D. Goodsell, *Commercial Family-Operated Cattle Ranches, Intermountain Region, 1930-47*, Bureau of Agricultural Economics, U. S. Department of Agriculture, Washington, November 1948; James R. Gray, *Southwestern*

region a ranch that is family operated (essentially by one man, with a little hired labor at peak season) has about 1,500 acres of land and 200 cattle, on the average; in the Southwest, a ranch of similar labor requirements has about 10,000 acres and also about 200 cattle; and one of similar labor requirements in the northern Plains has nearly 4,000 acres of land and about 140 cattle. In the Intermountain area, large additional areas of public land are used. In each area, forage conditions are closely dependent upon weather, including not only total precipitation but its seasonal distribution and other factors; and livestock productivity and hence ranch income are also closely dependent upon forage conditions. Ranch income is also highly dependent upon the relation between prices of livestock products sold and of the various items purchased. There are few alternatives to soften either adverse weather or adverse price conditions. As a result, ranch income is highly variable, from negative in bad years to comparatively high figures in the best years. Net ranch income on the average cattle ranch in each of these areas exceeded $9,000 in one or more years during the period studied.

The typical family-operated sheep ranches studied in these areas were somewhat larger, in that on the average they employed two men throughout the year. In the northern Plains, such sheep ranches had about 5,000 acres of land and roughly 1,000 sheep and about 20 cattle; in the Southwest, the average ranch had about 12,000 acres of land and about 1,200 sheep, and 40 cattle. In each area, dependence upon weather and price was the same as has been described for cattle ranches in the same area.

Investment in these ranches runs high: for the cattle ranches, about $65,000 for the Intermountain region as early as 1947, nearly $150,000 in the Southwest in the early 1950's, and about $85,000 in the northern Plains at the same time. For the sheep ranches, comparable figures are nearly $200,000 in the Southwest and about $75,000 in the northern

Cattle Ranches, Bulletin 403, New Mexico A & M College, Agricultural Experiment Station, February 1956; James R. Gray, *Southwestern Sheep Ranches, Organization Costs, and Returns, 1940-54,* Research Report 7, New Mexico A & M College, Agricultural Experiment Station, May 1956; James R. Gray and Chester B. Baker, *Commercial Family-Operated Sheep Ranches, Range Livestock Area, Northern Great Plains, 1930-50, Organization, Production Practices, Costs, and Returns,* Bulletin 478, Montana State College, Agricultural Experiment Station, November 1951; James R. Gray and Chester B. Baker, *Organization, Costs, and Returns on Cattle Ranches in the Northern Great Plains, 1930-52,* Bulletin 495, Montana State College, Agricultural Experiment Station, December 1953.

Plains, both during the early 1950's. Nearly all the investment is in land and livestock; it is obvious that this is no type of farming for someone short of capital.

It would be difficult to measure the relative productivity of labor and management in grazing and in other parts of agriculture. The results would depend in part upon the precise definitions of each type of farming, and the extent to which part-time and substandard-size units were excluded from the comparison. There would also be difficult problems of measurement of inputs other than labor and management; in particular, valuation of the land input would present special problems. To our knowledge, no comprehensive and carefully comparable evaluation of the relative earnings of labor and management in grazing as compared with other types of agriculture has, in fact, been made.

The foregoing data and general knowledge permit only broad generalizations: (1) that output per acre from grazing is very low; (2) that capital used per man in grazing is very high, but per acre is very low; and (3) that the productivity of labor and management as applied to grazing is not worse than the average for comparable labor and management applied to other agriculture in the United States, and may be slightly higher. Ranching in the United States is a highly commercialized type of agriculture; it produces comparatively high incomes and definitely is not subsistence either in method of operation or in result.

GRAZING AS A USE OF LAND IN THE FUTURE

In the light of the general conditions of population, income, and other factors we have assumed at the start of this book (pages 3-16), and in light of the history and present use of grazing lands in the United States, what seems the most probable use of such lands in the next forty years? By and large, we may say that the grazing lands are more likely to continue to be used much as they are at present than are the other lands we have been discussing. The future use of grazing lands can best be considered under three headings: (1) possible additions to grazing area, (2) possible subtractions from grazing area, and (3) possibilities for more intensive management of grazing lands.

Possible Additions to Grazing Area

The record reviewed in this chapter indicates clearly that no land is used for grazing if it is physically suited for other uses and is in active demand for them. Any possible additions to grazing area will come from land not actively sought for other uses. There is no significant area of land in the United States today (excluding Alaska) that is now not used for grazing or other uses but that is capable of supporting grazing on an economic basis. Some of the sheerest deserts are not used for grazing, because there is nothing there for domestic livestock to graze, and no water for them to drink. There are no indications that such lands can be used in the future for grazing. Additions to the grazing area, if any, must come from land now used for some purpose, not from essentially unused lands, as was possible in the past.

A considerable area of the Great Plains, now used regularly or intermittently for crops, might be regrassed and used for grazing. In the settlement of the Plains substantial areas have been plowed which cannot profitably and continuously produce wheat or other crops.

Some have suffered considerable erosion, and from the viewpoint of maintaining them physically, they should be put back into grass. Moreover, as was pointed out in Chapter IV, even with the prospective population increases and increased demands for food due to higher incomes per capita, the demand for wheat will rise little by the year 2000—certainly not to the level where the product of these lands will be needed. These marginal crop lands usually can produce no crop other than wheat. On the other hand, their beef and other livestock products, if the lands were in grass, would be in active demand. From many points of view, therefore, it would seem that such lands should be regrassed rather than used for crops.

But there are substantial difficulties to such a shift in land use. Wheat is often an attractive crop in spite of its long-run average unprofitability. In years of unusually favorable weather conditions, these lands produce moderately heavy yields of wheat; and since costs are often low, such crops can be highly profitable. There have been many instances in which the profit made on one crop was equal to the cost of the land. Those profits, and often much capital in addition, are sunk in a series of crop failures or poor crops. But the attractiveness of the high crop yield as a gamble is very great. Several million acres of land once in wheat were consciously restored to grass or allowed to drift back toward it during the 1930's, only to be plowed out again during the 1940's, with the lure of highly favorable weather and prices. There is a considerable turnover of farm owners in the Plains. The man who experiences crop failures is often gone when the good weather returns, and the new man has not had the sobering experience of dry years.

Still another obstacle to the shift in use of this land from crops to grass has been the unusually high prices maintained for wheat. The whole parity price concept and its effectuating mechanisms have tended toward prices for wheat higher than the market price would otherwise have been. This has, perhaps, been particularly important for marginal lands that produce large crops of wheat only in occasional years. In the absence of price supports, wheat prices in years of unusually large wheat crops would fall considerably, thus in part offsetting the favorable effect of the high yields. When prices are supported, on the other hand, this decline in prices because of the larger crops is partly or wholly stopped, thus mitigating the effect of the large supplies.

There are other obstacles to a shift in use of the marginal wheat lands. The techniques for regrassing such lands have not been well developed, at least as yet, and failures in grass seedings are common. Conditions for successful grass seeding are at about their worst in the years when wheat yields are low, and hence when there would be the greatest desire to regrass the land. The same dry weather which reduces wheat yields also retards the growing of grass. But, even at the best, the cost of the reseeding is often as great as the successfully reseeded land is worth for livestock grazing. Land values for wheat production are inflated because of the price support effect mentioned above. As a result, grass reseeding has usually required a subsidy in the past, and it is likely that any large conversion of low-grade wheat land to grass in the future will require a major subsidy.

There are still other obstacles to a shift in the use of land from wheat production to grazing. If the shift is made and if the grass is to be used profitably, then the type of farming, usually the size of the farm, and many of the methods of farm operation must be changed. Many wheat farms have no livestock now, and the farmer does not want livestock, for this would tie him to the farm throughout the entire year, whereas now he may live there only temporarily during the crop season. The farm machinery necessary for wheat production is not needed for livestock grazing; on the other hand, a major investment in livestock is necessary. Much larger farms are often necessary if grazing is to provide an adequate farm income. All of these adjustments take capital, and they may involve changes the farmer is unwilling or unable to make in his methods of farming.

Even if the shift in land use from wheat to grass were made, there is no reason to believe that it would not be reversed under temporarily favorable weather conditions. The conversion from grass to wheat can be made in a single season, when the weather is favorable: the livestock can be sold, the land has only to be plowed after machinery has been purchased, and the first crop of wheat is often the best one. Conversion in the opposite direction is slow, uncertain, expensive, and involves greater changes. Returns from the wheat crop are immediate, from the livestock slow in coming. With a process working quickly and relatively easily in one direction, and slowly and haltingly in the opposite direction, there is good reason to expect that the margin between wheat and grass will be shifted somewhat toward the former.

If the shift from wheat to grass is to proceed, and to stay, as far as calculations of average returns from each, under unsubsidized conditions, would warrant, then a different institutional approach may be needed.[56] It may be necessary to make substantial changes in land ownership, in order to make the change in land use and to keep the land in grass. Private landownership, similar to that of the past, is likely to lead to plowing up of grass whenever weather conditions are temporarily favorable. The alternative need not be federal landownership, although this is one alternative. It might be possible to develop a form of mixed private-public ownership, through a mixed corporation, that would have some of the advantages of each. For four decades substantial federal subsidies have been put into the Plains, a large part of it on marginal or submarginal wheat lands; pragmatically, a public interest in the use of these lands has been evident, whether logically there should be such interest or not. More effective methods of expressing this public interest may be possible; the subsidies for the benefit of the farmers and for the land might be channelled into land use changes that would be permanent. Unless some major changes in public programs are made, there seems little

56 Marion Clawson, "An Institutional Innovation to Facilitate Land Use Changes in the Great Plains," *Journal of Land Economics,* Vol. 34, No. 1 (February 1958).

prospect that any major shift in land use will be made permanently.

At the maximum, as much as 25 million acres now in crop, or plowed relatively recently and subject to replowing when conditions are favorable, might be regrassed permanently. The Soil Conservation Service has estimated that "about 14 million acres not suitable for permanent cultivation are now being cultivated on the Great Plains."[57] These lands are naturally productive for grazing. At the largest area and highest productivity, they might add 6 per cent to the output of all grazing lands. Thus, shifts in land use from crops to grass might mean an important but not major addition to the total supply of grazing forage in the nation.

Practically, the shift in land use will be smaller than this. We may thus conclude that additions to grazing land use, and hence to grazing output because of this one factor, will be comparatively small.

Possible Subtractions from Grazing Area

If the possible additions to the grazing area are small, what about the possible subtractions? Although some losses are probable, which locally may be important, in total no major change seems likely. The reasons behind this conclusion are evident if one examines the shifts that are possible.

There will be few shifts out of grazing and into agriculture, as we have defined these two terms, except for some minor shifts for new irrigation development. All the grazing lands suitable for farming without irrigation have long since been shifted; as we have pointed out above, a reverse shift seems desirable. The extent of further irrigation development is limited by water supply, if not by considerations of cost and economic feasibility. Some of the land that might be irrigated is now in farms, often in dryland crops. At the most, only a few million acres of grazing land will be taken from grazing use by irrigation development.

Some land now used primarily for grazing may in the future be used primarily for forestry, but it seems probable that the extent of this shift will be small. Fire control, grazing, and other land use practices under man's control have resulted in the conversion of considerable grazing areas to forests or to woodland, with a reduc-

57 *Facts About Wind Erosion and Dust Storms on the Great Plains,* Leaflet No. 394, U. S. Department of Agriculture, Washington, June 1955.

tion or elimination of grazing. However, it seems probable that most of the conversion that could be made already has been made, and no further major shifts seem probable. As more intensive forestry is practiced in the South, grazing may gradually come to be excluded from some forested lands now grazed. But these lands are mostly now classed as forest, not as grazing land. Brush invasion of grazing lands will continue to be a problem; it seems probable that ways and means of dealing with this problem will become more effective in the future. In any case, the lands invaded by brush may be lost to grazing, and may become woodland, but can hardly become forest in any economic sense.

Recreation will take some land out of grazing, especially in the high mountains. As the recreation use of land increases, its demands will almost surely have an impact upon grazing use in the areas most sought after. These are also areas of high watershed values; and to the extent that grazing interferes with maximum watershed values, it may have to yield. But it seems probable that the combined effect of these two classes of land use changes will not be great.

Urban expansion will take a little land now used for grazing, especially in the Southwest. But most of this land is desert in character, of low productivity for grazing. Elsewhere, most of the land likely to pass into urban use may be pasture land, or cropland, or woods, but little is primarily grazing land, as we have defined the term. Even an extremely large shift of land into urban uses would be a small shift out of grazing, because of the greatly different areas of land now used for the two purposes. Some grazing land will be used for highways and other miscellaneous uses, however.

We thus may summarize by saying that the area of land likely to be lost to grazing will be comparatively small.

Possibilities for More Intensive Management of Grazing Lands

If the foreseeable additions to and losses from grazing use are small, what are the prospects for changes in intensity of use in the present areas of grazing land?

First of all, the demand will exist for the products of an increased intensity of grazing use. The chief product of more grazing output will be beef, with some lamb and wool, and other miscellaneous livestock products. Changing dietary habits have already led and will continue to lead to a greater demand for these products; and the prospective increase in population will mean a greatly increased total demand for them as we saw in Chapter IV. This does not prove that more intensive grazing is the cheapest way to increase the output of such products, for increases in output could come from other sources. But it does mean that if intensified grazing land use is possible and economical, the demand for the resulting increased output will permit its absorption with little or no reduction in price. This is in sharp contrast to the wheat that might be grown on some of these lands, for instance.

The physical possibilities of increased output from grazing lands are somewhat limited. Large areas might be reseeded; Renner estimates that as much as 80 to 125 million acres might some day be reseeded.[58] Reseeding on this scale is impossible at present, partly because desirable species and reseeding practices are not available for some lands, and partly because sufficient seed does not exist even for those species and areas where methods have been tested. Spreading of flood waters onto nearly level bottom lands along streams or watercourses, especially the smaller ones, is also highly practical in some areas. The total area for which topography and other natural factors are favorable is limited, however, to several million acres. Brush control, through fire or mechanical means, and other revegetation practices would increase usable forage output from several million more acres. By fencing, water developments, and other means, livestock can be forced to make a more nearly complete utilization of all forage which grows, but within the safe limits of conservative grazing to protect the grazing capacity of the area. There are thus several kinds of practices, suitable to different natural conditions, that are physically capable of increasing the amount of forage economically usable or used from the grazing region.

Renner has estimated that the output of forage from the western grazing lands could be somewhat more than doubled (Table 49).[59] To us, this seems perhaps the physical limit of increase in grazing output from these lands, assuming no major change in climate comes about, either naturally or through man's efforts at weather control.

It has generally been considered that grazing lands could not be fertilized successfully, because of the lack of moisture. However, in recent years some range lands have been fertilized to yield large

[58] Renner, *op. cit.*
[59] *Ibid.*

Table 49. Estimates of potential improvement of grazing lands in seventeen western states

Range and pasture land	Acres (thousands)	Improvement possible as per cent of present	Equivalent acres (thousands)
Privately owned			
Range and pasture	365,000	200	730,000
Forest range	74,000	150	111,000
Publicly owned			
Range	215,000	300	645,000
Forest range	105,000	200	210,000
Total	759,000	—	1,696,000

SOURCE: F. G. Renner, "The Future of Our Range Resources," *Journal of Range Management,* Vol. 7, No. 2 (March 1954).

increases in forage.[60] These experiments were conducted in areas normally receiving about 16 inches of precipitation. Thirty pounds of nitrogen per acre about doubled forage production, compared with unfertilized areas, and 90 pounds about trebled it. While forage growth was greater in years of high than in years of low total precipitation, the response to fertilizer was almost equally great in each type of year. The response was somewhat greater on areas that had been severely grazed and hence somewhat depleted, than on areas only moderately grazed. There was a considerable carry-over effect of the fertilizer applied, into the second, third, and even later years. It is not clear that this degree of fertilization would be profitable at present price relationships, and it seems probable that no such yield increases could be obtained on areas of much smaller total precipitation. However, the possibility of increased forage production on grazing lands through fertilization should not be dismissed.

The economic possibilities of increasing output from the present area of grazing lands through reseeding or other devices seem less encouraging. Range reseeding costs are still comparatively high. The economic limit to investment in reseeding, water spreading, and other

60 George A. Rogler and Russell J. Lorenz, "Nitrogen Fertilization of Northern Great Plains Rangelands," *Journal of Range Management,* Vol. 10, No. 4 (July 1957).

devices to increase range forage cannot be judged without consideration of the need for forage at the season when it will be made available from these measures and its relation to the supply of forage available at other seasons from grazing areas. Moreover, improvement in forage production at one season sometimes makes possible shifts in seasonal use of other range areas, that, in turn, greatly increases their productivity. For all of these reasons, one cannot say that the average costs of range reseeding are always too high, but one can say that there is grave danger that they will be higher than the returns will justify.

Stoddart has summarized the situation as follows: "It seems unlikely that any but the most favorable sites can ever be seeded at a profit in the arid sections of the West, though certain sections of the northern plains and the Northwest are notable exceptions."[61]

Water spreading has sometimes shown much more favorable financial returns. Where there are broad alluvial valleys alongside relatively small streams, with moderately small watersheds above, the necessary works to divert and spread the water and to withstand floods of considerable magnitude can often be built at moderate cost, especially with the new earth-moving equipment. These water spreading works are automatic in their operation, although they do need maintenance and repair occasionally. They are really a form of cheap irrigation, and in some instances provide relatively cheap livestock feed.

If reseeding, water spreading, and the various other devices to increase grazing output to or approaching its physical limit are to be adopted, it may well be necessary to provide public subsidies. Such subsidies have been available for reseeding and some other range improvement practices in the past, under the federal agricultural programs. The subsidies were designed to cover roughly half of the cost of the reseeding. They were granted on the theory that reseeding and other practices were a form of soil conservation, in which there was a general public interest. Obviously, if half of the cost is met, the margin of economic investment for the rancher has shifted greatly. If subsidies of this kind continue in the future, the amount of investment made for grazing land improvement will be larger than it otherwise would be.

Grazing land has a low capacity to use additional inputs of labor and capital. Such evidence as there is suggests that grazing land makes as large a return upon the labor and capital now used on it, as

61 Stoddart and Smith, *op. cit.*, p. 379.

does other forms of agriculture. The efficiency of grazing land is thus moderately high at the present level of inputs; but the very physical conditions that determine the land shall be used for grazing also limit its capacity to absorb profitable additional labor and capital.

Even if Renner's estimate of the increased forage production is to be realized by the year 2000, the grazing lands in total would just about be expanding their output proportionately to the need for expansion of agricultural output as a whole, as a result of the growing population, increased per capita income, and other changes. To the extent that these physically possible changes are uneconomic or not made, then the output from grazing lands will expand less than proportionately to the need for increased total agricultural output, and the output from other types of land will need to be increased more than proportionately.

VII

Miscellaneous uses of land

The big three in land use, so far as area is concerned, are agriculture, forestry, and grazing; important in terms of the values involved and of their growing importance in modern life, are urban and recreation uses. To these major five land uses, each of which we have considered in a separate chapter, must now be added a number of other land uses. Lacking a single easily described characteristic, they can only be classified as "miscellaneous."

Treatment of the various miscellaneous land uses is plagued by a lack of dependable and consistent data. By and large, they have been studied only by those interested in the particular land use, not as part of a broad scale study of all land uses. To no small degree, these uses are the wastebasket into which accumulate the errors in estimating other land uses. For instance, data on agriculture and forestry may sometimes include the land in local roads serving those who use the land. Yet this same area may also be included in estimates of area in roads. What is a relatively small error for a major land use may become large for a minor use. And what is considered "miscellaneous" often depends upon the detail of the preceding itemization.

We shall consider here a few miscellaneous land uses on which there is moderately good information or which we have reason to believe are important. These include land for transportation, for mineral production, for reservoirs and other water control purposes, and for specialized wildlife refuges. For these, as for the major land uses already examined, we consider only those lands primarily used for the purpose in question. Thus, for wildlife we take account only of wildlife refuges, although there is obviously some wildlife on lands primarily used for forestry, grazing, and other purposes. A number of other land uses we consider not at all or only in passing—military reservations, small towns and places too small to be included under urban use, rural cemeteries, rural school areas, and others. Then, there are some lands that can be described in physical terms relatively easily, but that are harder to identify in terms of use. This is true of some of the swamps, deserts, and mountain tops, as well as some of the dense brush areas that are noncommercial forest.

LAND FOR TRANSPORTATION

Land used for roads or railways is taken out of crops and out of forestry, and thus in one sense is out of production; but the development of the roads or railways so greatly increases the economic productivity of the land left for crop or forestry production, that it may fairly be said that the land used for transportation is the most productive of all. The productivity of land used for transport depends in part upon the markets that are thus opened up, and in part upon the physical productivity of the land through which the road or railroad passes. Lacking either or both of these two essentials, it is possible to waste land for transport purposes; but where they exist, land for transport purposes is highly productive in an economic sense.

In this section, we shall consider the land used for railways, roads or highways, and airports or landing strips. Most land for water transport, in the form of terminals and locks, is included in the data on urban land uses. In general, to anticipate the analysis which follows, the history of land use for transport over the past several decades shows modest expansions in areas so used, great increases in use or output of such areas, and hence marked rises in the efficient use of the land set aside for transport purposes.

Railroads

Between 1830, when railroads began to be established, and 1890, over 160,000 miles of rail line, with 200,000 miles of track, spread across the nation.[1] (Appendix D, Table D-12 and Figure 56.) The

[1] The difference represents double and treble track on the same right of way.

mileage of right of way expanded modestly until the First World War, and the mileage of track continued to expand somewhat until the beginning of the Great Depression in the early 1930's—these were the decades when railroads were dominant in American transportation. In more recent years few new lines or new track along old lines have been constructed, and some lines have, in fact, been abandoned. Rail technology has found ways of greatly increasing the flow of traffic over the same trackage, and new methods of transportation are providing ever-increasing competition. Abandonment of trackage would probably have been greater, were the railroads not regulated by federal and state governments. These have considered the protests of local communities anxious to retain the railroads even though they might be unprofitable to the operators.

Railroads were built across federal land, by permit; the adjoining land was patented to private individuals with the railroad right of way as an easement. If the rail line was abandoned later, the adjoining landowners secured the land. In addition, the railroads were granted extensive areas of public land, which they generally sold, to aid in financing the construction of the railroad. In areas already in private ownership, the railroads were required to buy the land from the individual owners; in this case, on abandonment of the line, the land could be sold. Railroad rights of way varied in width, but 300 feet was common. If one takes this width as an arbitrary basis for calculation, at the railroads' peak about 9¼ million acres of land were within railroad rights of way, and about 8 million acres are still there. In a few instances, hay is cut or other crops grown within rights of way, but, in general, this land is withdrawn from other uses even if not actually occupied by the rail tracks.

Figure 56 shows how greatly the volume of freight moving over the railroads is affected by depression and war. The volume rose rapidly and rather steadily from 1890 to about 1920, then remained relatively steady until 1930. In the depression, freight movement declined almost one-half; but then in World War II it rose again to new peaks, from which it has declined somewhat. Passenger movement also rose from 1890 to 1920—a date which in many ways marks the peak of railroad importance in the United States—and declined through the 1920's as better roads and more automobiles led people to travel by private car. It declined again sharply in the depression, but rose dramatically during World War II as more people sought to travel and as auto travel was restricted. It has since declined again.

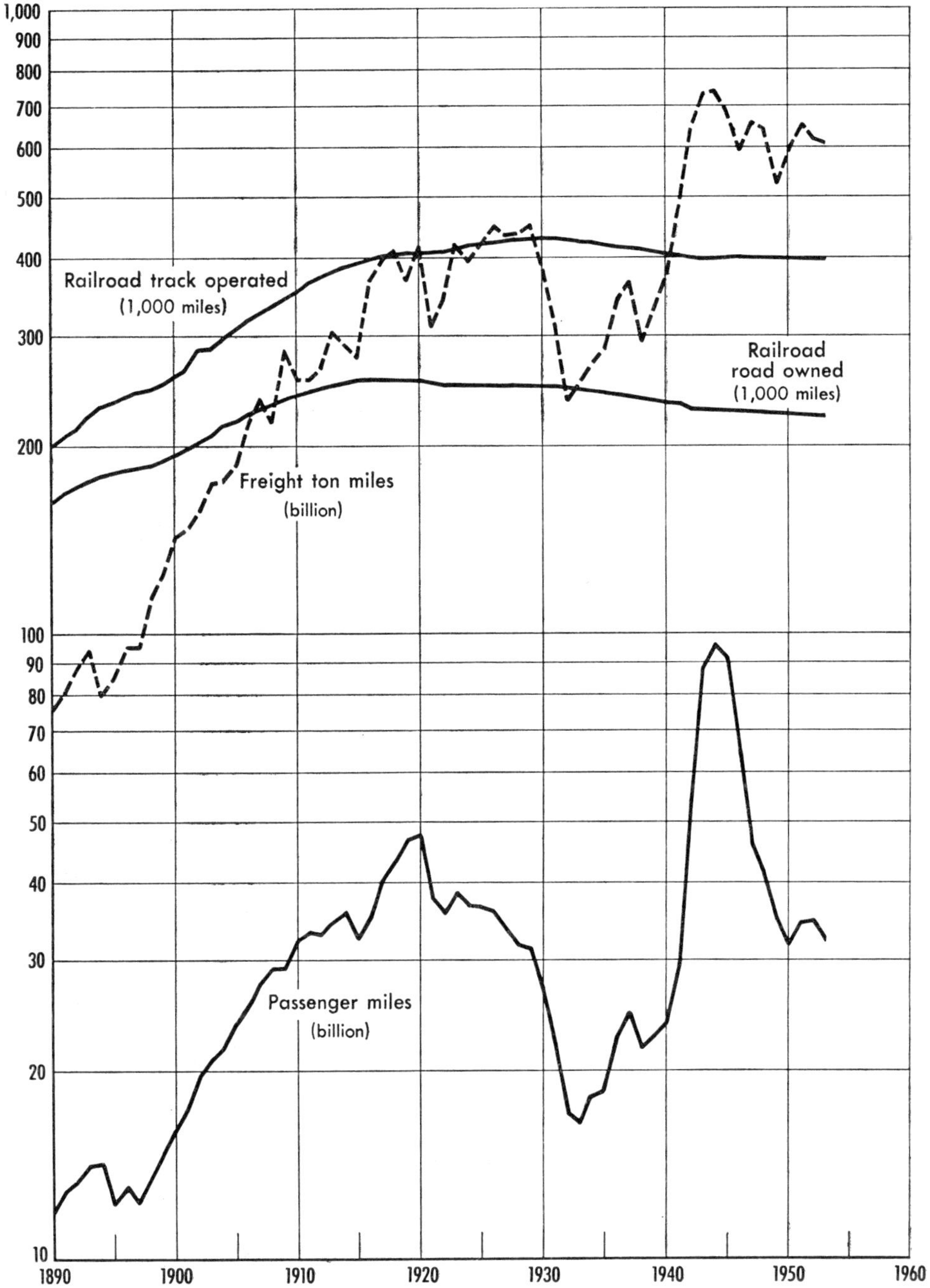

Figure 56. Railroad mileage and freight and passenger movement, 1890–1953.

The area of land within railroad rights of way is used with greater efficiency today than twenty or fifty years ago—the volume of traffic is heavier upon a smaller area. During the war peak, however, it was used with even greater efficiency, and there is still much unused capacity in the present railroad rights of way. If the demand for rail traffic justified it, new equipment of various kinds could materially increase the capacity of the present tracks, and large mileage of additional trackage could be added within the present rights of way or upon very modest extensions. The volume of rail traffic could be multiplied by three or four times, or possibly more, with no substantial additions to rights of way. However, it is unlikely that any large increase in rail traffic will occur; instead, passenger traffic will probably continue to decline sharply, and freight traffic will increase slowly if at all. The increase in total freight movement as the economy expands will be offset by an increasing proportion moving by truck.

Roads and Highways

In the original colonies in the East, public roads were laid out under laws affecting each colony or state; in the public domain states, the states and counties were allowed to lay out public roads over the public domain. The land along the roads in the latter states was patented to individuals, with an easement for the road so that, if the road was later abandoned, it reverted to the adjoining landowners. If the land was already in private ownership when a road was proposed, the adjoining landowners often donated the land as partial inducement to the construction of the road. In cases where the land was purchased, it could be sold by the state or county if the road was later abandoned or relocated. The rights of way for roads varied considerably, usually following state laws. While in a few states they were as wide as 400 feet, the most common width for local rural roads was two rods, or 33 feet. Until the automobile became common, most roads were unimproved. As transportation arteries, they were poor in quality and expensive in terms of the time and effort required to move freight or people over them. But their establishment did serve to withdraw the land from other uses, although in earlier days it was common to graze livestock along the rights of way.

The area of the private roads that exist, for example those within farms and forests, is ordinarily included in the statistics on the land use they serve. Thus, some of the "other" land in agriculture, shown in Chapter VIII, Table 51 and elsewhere, is for roads. The acreage in private roads in intensively managed forests may run as high as 5 per cent of the total. But these private roads are not considered here. Neither are the city streets, which are considered as part of urban land in Chapter II.

The mileage of public roads in this country prior to 1904 is not tabulated. From that date until 1921 the statistics show that total mileage increased about 40 per cent.[2] Since 1920 there has been remarkably little change in the total mileage of roads, but a vast change in their quality and use (Appendix D, Table D-13 and Figure 57). The total mileage of all public roads outside of cities has been within 5 per cent of 3 million miles since 1921. This, on the average, is almost exactly one mile of road per square mile, or per 640 acres for the entire country. In the more densely settled farming areas, where the roads often follow each section line a mile apart, the average can be a mile of road for each half square mile, or for each 320 acres. More roads than this are maintained in many areas; but in considering the average figure one should also have in mind the comparatively large areas that are essentially roadless—the mountainous and desert areas, and others that are forested or used for grazing.

The area of land within road rights of way has nearly doubled since 1904; but most of this increase took place before 1921. The U. S. Bureau of Public Roads estimates that since then the total area within rights of way has increased by less than 20 per cent. This conclusion is perhaps surprising in view of the amount of road construction that has been taking place, and it is indeed possible that the increase has been somewhat greater—road realignments may have required new rights of way without equal relinquishment of older rights of way now unused. Any additions here would be minor, however. The essential facts are that the 20 per cent increase applies overwhelmingly to a relatively small proportion of the total public roads. In 1921, 83 per cent of the total rights of way was in local roads; by 1954, through upgrading of the local rural roads and some abandonment, the proportion had dropped to 68 per cent of the total—still large enough,

[2] U. S. Department of Commerce, Bureau of Public Roads, *Highway Statistics, Summary to 1955*, Washington, 1957.

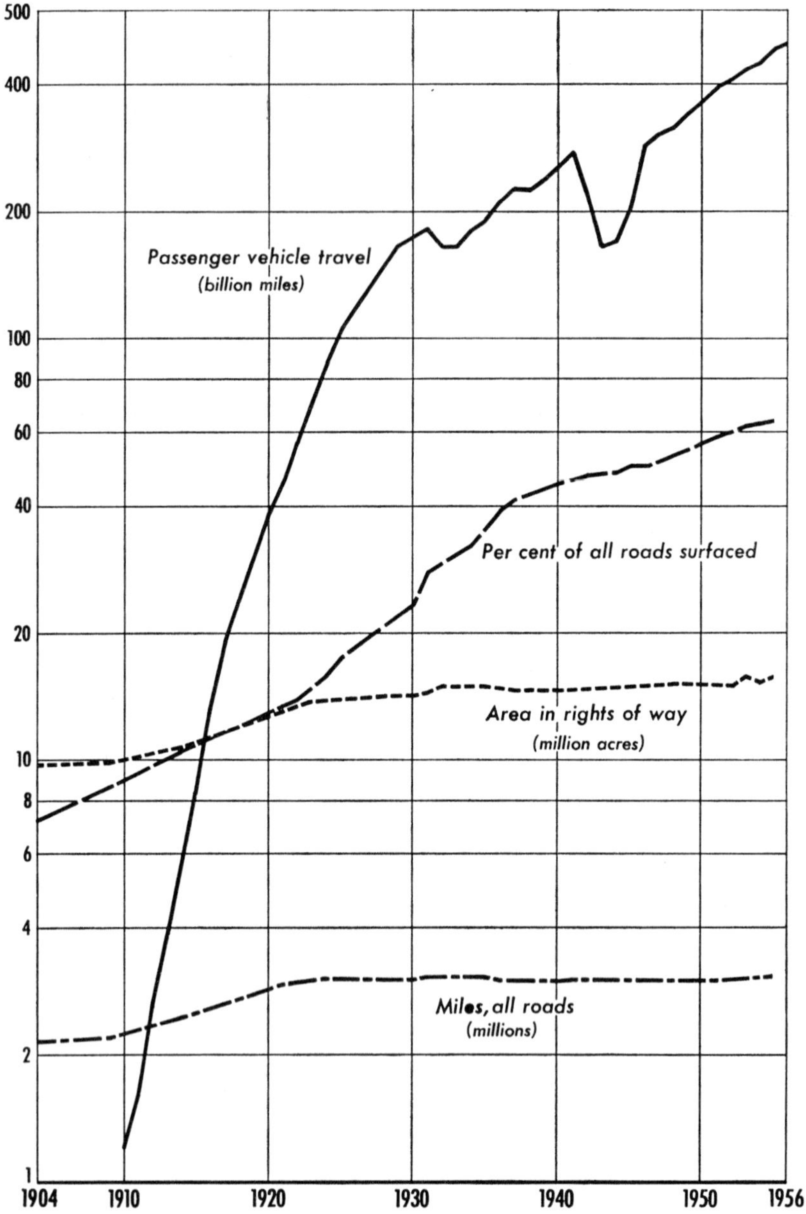

Figure 57. Mileage of all roads outside of cities, area of road rights of way, proportion of all roads surfaced, and passenger vehicle travel, 1904–56.

however, to modify the statistical effects of increases in total road rights of way.

A large proportion of the county and local roads, the major users of land for rights of way, is lightly used, but the roads, nevertheless, are important to the local economy. Sometimes in rural areas the land between the road surface and the boundary of the legal right of way is utilized for grazing. Clearly, many of the local roads could be abandoned and the land in them reconverted to forest and grazing if our need for land were acute.

The most noteworthy trend in roads has been their steadily improving quality. Statistical evidence of this may be seen in the percentage that is surfaced; the 7 per cent surfaced in 1904 had increased to over 60 per cent by 1953. All of the major highways and about half of the strictly local roads are now surfaced. Surfaced roads are wider, as to actual road surface if not as to right of way, than unsurfaced roads. They permit all-weather use, heavier loads, faster speeds, and greater comfort. Only the older person today, remembering how frequently he used to be stuck tightly in the mud in the midst of the best highways the countryside then afforded, can really appreciate the degree of improvement in rural roads over the past forty years.

With changes in road quality, equally impressive changes in road use have come about. In 1920 passenger vehicles averaged about 13,000 miles of travel along each mile of road; by 1955, this had risen to about 160,000 miles, a twelvefold increase in thirty-five years. The increase in truck traffic has probably been even greater. The improvement in automobiles and trucks and the improvement in roads are classic examples of complementary changes in technology; without the other, each would have had limited usefulness.

The United States is embarked upon a new program of superhighway construction, to build a new and vastly improved national net of major roads. These new highways take land, and many observers have been disturbed by the amounts of land involved. Each 8¼ feet in width of right of way results in an additional acre of land taken up for each mile of highway. Thus, a 200-foot right of way takes about 25 acres per mile, and a 400-foot right of way doubles the acreage. The new highway program includes 41,000 miles of road in the interstate system and additional large mileage of other roads; most but not all of this will require new rights of way. If the average width is 200 feet, this will amount to slightly more than a million acres; if the average width is 400 feet, slightly more than 2 million acres; and proportionately for other widths. This is indeed a significant area of land, even for a country as large as ours; but for comparative purposes it should be recalled that there were over 15 million acres of land within road rights of way before this new and enlarged road program began.

Airports and Landing Strips

Since 1927 there has been a nearly sevenfold increase in the number of airfields and landing strips (Figure 58 and Appendix D, Table D-14). Detailed data are lacking, but it appears that small landing fields and strips, rather than the larger airports have been responsible for a large part of the increase. At present, only about 40 per cent of all airports and landing strips are for public use; many of the other privately used places are situated in forest, grazing, or agricultural areas. The acreage in all landing fields and strips in 1945 was about 1¼ million acres; and in 1954, although the number of fields had increased materially, the total area was only slightly greater.[3] Estimates for other years have been made, but possibly are not very accurate; they indicate that the total area in all airports and landing fields has increased from less than ¼ million acres in 1927 to a peak of about 1¾ million acres in 1948.

As with the other forms of transportation, increase in movement of people and goods, or in use of facilities, has been enormously greater than increase in area of land devoted to them. Total miles flown have multiplied by about fifty, and total passenger miles by more than 200 since 1927, while area of land in airports and landing fields was increasing less than sevenfold.

Transportation Area in the Future

Future land requirements for transport are difficult to estimate with accuracy. Little or no increase in railroad rights of way seems probable; the increase in airport acreage, while important locally, seems unlikely to involve a considerable acreage of land; and the increase in rights of way for roads, while large for the new super-highways, will probably remain a small proportion of the total area now in road

[3] Data provided by Civil Aeronautics Administration.

rights of way. At present there are about 8 million acres in railroad rights of way, about 15½ million acres in road rights of way, and less than 1½ million acres in airports and landing strips—about 25 million acres in total. Even if the area in primary road rights of way and in airports each doubled, and the area in other forms of transportation use showed little or no change, the total area required in the year

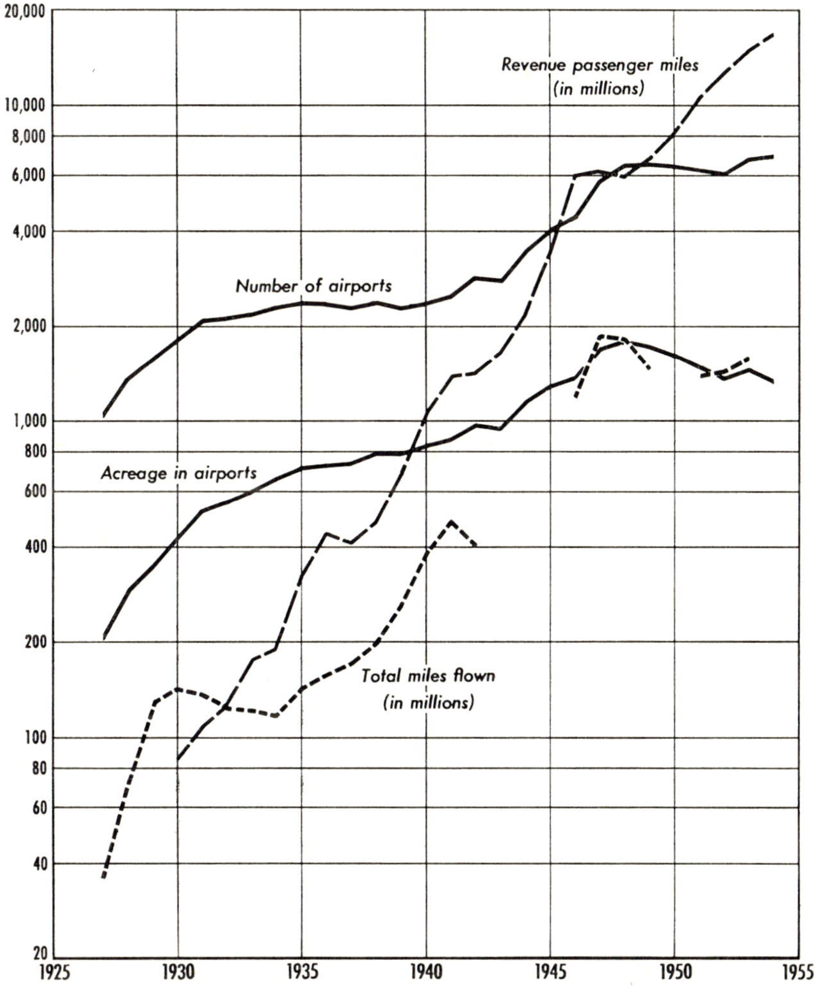

Figure 58. Number and estimated acreage of airports, and passenger movement, 1927–54.

2000 would be only some 30 million acres. With the large capacity shown for heavier travel without proportionate increases in rights of way, this relationship seems not unreasonable. Compared with increases in urban area, which seem highly probable, and in recreation area, which are desirable but perhaps not so probable, the increases in land needed for transportation purposes will be small. This does not deny that changes in land use due to expansion of transportation use may not be significant to the local area. In particular, new roads are often a major factor affecting the use of land adjoining and tributary to the roads. If any level of government seeks to influence private land use, construction of roads and their location is perhaps the most effective tool. But the area actually used by the roads is comparatively small.

Some local roads and branch line railroads will be abandoned. Their rights of way can in some cases be used for agriculture, forestry, or grazing. In areas of farm abandonment or consolidation, fewer local roads will be needed. The roads and railroads going out of use are usually those that are least improved, and trees or grass will rather quickly cover the unused rights of way and, eventually, the travelled areas. On the other hand, more intensive forestry will require more roads; part of their area will show up in the statistics on roads, part in the statistics on forests.

LAND FOR RESERVOIRS AND WATER REGULATION

A considerable acreage of land has been inundated by the construction of artificial impoundments or large reservoirs. About 1,300 larger ones occupy an estimated land area of 11 million acres, or an average of nearly 8,500 acres each. Between 1.5 and 2 million small ponds each average about an acre. Together, large and small artificial bodies of water cover a total land area of nearly 13 million acres. The customary statistical practice is to consider as land all natural water bodies of 40 acres or less. If the same definition is carried over to reservoirs and artificial lakes, then part of this area should be included in the land use of the surrounding area. It seems highly probable, for instance, that the area of farm ponds is included in agricultural or grazing use, and some small reservoirs are included in forestry. Recreation land use also includes some reservoir or artificial lake areas used primarily for this purpose.

The larger reservoirs have been built in every portion of the United States on sites where sufficient streamflow and fall of water occur in natural combination. In many parts of the country, especially in the Southeast, impoundments have inundated valley bottoms formerly used for crop production. One type of dam construction which maintains limited permanent pools is that designed for flood control. Much larger areas back of the dams are irregularly flooded and gradually drawn down as the danger of flooding passes. The lands so affected are usually acquired by public agencies but made available under leasing arrangements where suitable for crops or pasture. The lessor takes the risk for losses incurred by flooding.

A much larger area is affected by permanent impoundments for water power generation and for irrigation. Data are unavailable on acreages involving agricultural and other classes of land that are submerged by these impoundments. However, in the eastern United States more agricultural lands have been affected by reservoirs than elsewhere. It is doubtful that for the country as a whole more than 3 or 4 million acres have been so affected.

The much smaller farm ponds, which have been built in great numbers during recent years, are widely scattered over the area east of the Rockies. Few of them actually flooded usable cropland, and probably in a majority of cases they were built on gully areas. They served both to check erosion and siltation and to retain runoff. To the farmer they served as a source of water for livestock and, if stocked with fish, a recreational use as well, often in an area far removed from any natural water areas.

For a considerable part of the eastern United States the monthly balance between precipitation, on the one hand, and evaporation, transpiration, runoff, and other losses, on the other, is such that relatively modest storage will greatly increase the amount of usable water supply.[4] In the East and elsewhere, the demand for water for urban and industrial uses and for recreation will increase greatly as the urban population grows. In spite of the fact that the best reservoir sites have already been used, considerable further dam building

[4] C. W. Thornthwaite, John R. Mather, and Douglas B. Carter, *Three Water Balance Maps of Eastern North America* (Washington: Resources for the Future, Inc., 1958).

seems probable. On the other hand, in much of the West, where water is scarce, present reservoirs are often large enough to hold most of the economically usable water, although some further dam building for irrigation and power will take place there. Elsewhere in the country more dams and reservoirs for flood control seem probable. In total, as much as 10 million acres for reservoir and water control works may be added by the year 2000.

Small ponds will continue to be built in many areas and for a number of purposes—perhaps even double the present total. Stock water as an objective is likely to change to recreation and water retention as the primary justification for development. These acreages will mostly be included with data on use of surrounding land.

Some land is used for water control works, such as levees along river banks. No information appears to be available on the total area so used, but presumably it is small compared with that in reservoirs or in any major land use. However, such land may be strategic for many purposes. In addition to land actually used for water control works, other bottom lands are flooded so frequently or so severely as to have limited or no usefulness for agriculture, forestry, or grazing. Some of these lands may have value for wildlife purposes, either now or potentially. Some may have value for recreation. In total, there may be several million acres of land along watercourses, especially small watercourses, which will not be used intensively or at all for commercial purposes such as agriculture, forestry, or grazing, but which might be developed for wildlife or recreation purposes. To do so would require their transfer to public ownership, or at least the acquisition of certain easements for public use. This is a type of new land use which might well be explored carefully.

LAND USED IN MINERAL PRODUCTION

Both open-surface and subsurface mining are included in this class of land use. The first group is by far the larger user of lands, open-pit mines, quarries, placer and strip mines being the most prominent on the landscape. Practically no data exist to indicate actual acreages that have been converted through surface mining operations from previous uses. Subsurface shaft mining seldom occupies large areas. Sometimes additional lands are covered with overburden and other waste material removed in advance of actual mining.

Since shaft mining is of such minor importance as a land user, major attention must be given to quarrying, open-pit iron ore mining and strip coal mining. Fairly extensive areas have been affected by these types of mineral activities. Open-pit mining is largely limited to the Mesabe and Cayuna ranges of northern Minnesota, whereas strip coal mining has been concentrated in Illinois, Indiana, Ohio, and a few fields in adjacent states.[5] Most of the land affected by these activities has been either in agricultural or forest uses. In all cases, agricultural uses are eliminated by the mining activity which churns up much subsoil and either dumps it on top of the ground or digs great pits which reach deep into the subsurface into bedrock formations. Frequently great piles, ridges and mounds of infertile subsoil are left adjacent to the large holes, thus creating a landscape far different from the original rolling or level land. Quarrying which is carried on over most parts of the country disturbs far smaller areas in total, mainly in rocky outcrop lands where little use has developed for the land.

Placer mining for gold has been common in the West in the past. This method of mining involves washing comparatively large tonnages of soil, gravel, and sand, and extracting the gold by mechanical means. The earliest gold mining in California and other western states was often of this type. Comparatively small-scale and primitive water conveyance and earth-moving methods were used in the early days, and hence the total disturbance to the natural conditions was rather limited, although considerable sediment was washed into streams. This type of relatively simple placering was replaced by large-scale dredging. These machines are able to dig up soil, gravel, and sand to a considerable depth—30 feet or more in some cases. They are also able to work over large tonnages daily, and to operate profitably on gravels yielding only a comparatively few cents worth of gold per cubic yard. Except as restrained by laws against stream pollution—and, by and large, these have been relatively ineffective—this method of mining has already exhausted most of the sites where it could be done profitably. Considerable areas in California, Idaho, and other western states have been worked over in this way. The results are, in general, similar to those described for strip mining. In nearly all cases,

[5] Lee Guernsey, "Strip Coal Mining: A Problem in Conservation," *The Journal of Geography*, Vol. 54, No. 4, April 1955; Lee Guernsey, "A Study of the Economic Impact of Strip Coal Mining in Hopkins County, Kentucky," *Transactions of the Kentucky Academy of Science*, Vol. 17, No. 2 (May 1956), pp. 101-113.

it has been stream or valley bottoms that have yielded gold, and hence it is these areas which have been worked over. While it is possible to question seriously the social gain of past mining operations of this kind, it seems probable that little more of it will be done, largely because few areas offering prospects of profit are left unmined.

Most of the strip and open-pit mining areas are abandoned by their owners after the mineral deposit has been exploited, and are allowed to fill up with water. Generally they are an eyesore of wasteland which contribute nothing but silt to the rivers of the locality. Until recent years, state regulations requiring either revegetation or levelling of these dumps and pits have been only moderately effective in bringing about any improvement. Lately the coal companies, in particular, have made some effort to improve these areas for recreational uses by tree planting and stocking the waters with fish. In areas without lakes —and this is the rule in most coal mining sections—the local populace welcomes the development of these spoil bank areas for hunting, fishing, and other recreational uses. Rock quarries have been developed in the same way, and local "lakes" with some excellent fishing have become available.

It appears uneconomic to bring spoil bank areas back into agricultural uses, but such lands have value as recreational areas, and sometimes for timber growing as well as for grazing. Seeding to desirable pasture grasses or planting of good species of trees can be accomplished without great difficulty on all but the steepest lands. Nevertheless, recreational uses are likely to be the most important uses in many areas. Since considerable work is needed to convert these lands into uses of public benefit, they present an opportunity for public works programs during periods of less than full employment. Some land is used to mine or extract sand and gravel for local construction. Such areas are extremely important as a source of inexpensive building materials, but their total area is probably not great. Often relatively unproductive areas along rivers or streams provide the best deposits.

We find it impossible to present any firm estimate of how much land may be needed for mineral production in the future. Wherever profitable mineral deposits are found, their extraction will almost always take precedence over other forms of land use. In northern Minnesota, for instance, open-pit ore mining has actually led to the moving of towns that stood in the way. Certainly, farming, forestry, and grazing cannot compete with mining if the latter has any real profitability. Oil, gas, and other subsurface mineral development through drilled holes or even by shafts need not interfere seriously with most other land uses to which the surface is adapted. Strip mining for coal is likely to become relatively more important in the future, because it offers great possibilities for lowered costs through large-scale mechanization. This will undoubtedly take some land, but we cannot estimate how much.

In the field of land use in general, there is perhaps less information available on use of land surface for mineral production than for any other use. It would be helpful to the total over-all land use picture if better information could be assembled on this subject.

LAND FOR WILDLIFE

All land except the concrete canyons of our downtown urban centers and the paved roadways is used by some form of wildlife, but rural areas—farms, woods, swamps, and range lands—constitute primary habitat for most kinds of wildlife. Drastic changes in land use coupled with inroads made by unregulated hunting have had major impacts on wildlife populations, particularly big game animals and waterfowl.

In this section, as elsewhere in this book, we are considering only the land whose primary use is for a particular purpose—in this case, for wildlife. Such areas are in the rough order of 15 million acres. The additional areas on which wildlife is one of perhaps several important uses are much larger. Much of the forested, grazing, agricultural, and recreation land is managed so as to encourage various forms of wildlife. However, wildlife on all these lands is subordinate to their primary use.

The process of settling a continent—clearing farms and building cities—has altered the natural environment necessary for survival of many kinds of wildlife. Not all changes have been unfavorable—particularly those species capable of adapting themselves to man's activities on the land. Song birds and many kinds of small upland mammals have found survival possible in many situations, from city parks to suburban housing areas. Some wildlife have even increased in numbers over original populations, as in the case of the white-tail deer.[6]

[6] Walter P. Taylor (ed.), *The Deer of North America* (Harrisburg, Pa.: The Stackpole Co., 1956).

But some kinds of wildlife have been decimated to levels below those existing prior to civilization.

Wildlife, roaming at will over fences and across roadways, is no respecter of property boundaries. Traditionally, it has been considered common property held in trust by the states and subject to control under the police power. Game laws regulating or prohibiting harvest stem from this constitutional authority. Waterfowl, because of the interstate and international nature of their migratory movements, have been subject to federal laws and international treaties. Wildlife conservation programs have centered around three main efforts: regulation of the harvest; reservation of special public areas for breeding, resting, food, and cover undisturbed by man; and improvement of habitat on public and private lands to encourage greater population.

In 1955 nearly 9½ million acres[7] of federal lands had been reserved primarily as wildlife refuges in the first forty-eight states, and an additional 8 million acres in Alaska. The largest portion of this acreage (5¾ million acres in the forty-eight states, 5 million acres in Alaska) was that devoted to big game range (elk, bighorn sheep, etc.). Some of this was also used for grazing of domestic livestock. Waterfowl nesting, resting, and wintering grounds occupied 3.3 million acres (nearly all in the forty-eight states), most of which were strategic wetlands. In addition, the states have acquired or leased a total of 4.9 million acres[8] primarily for wildlife purposes. Portions of these acreages under both state and federal management are opened to regulated hunting, fishing on open waters, and limited recreational uses. It is estimated that 8.8 million acres in public ownership are exclusively dedicated to wildlife protection, and that the remaining areas used primarily for wildlife are also used for grazing, timber production, or other purposes not in conflict with wildlife purposes. Many of the areas exclusively dedicated to wildlife are strategic, either because they are on waterfowl migratory routes and essential links in their life chain, or because they possess certain unique characteristics and combinations of environmental features needed to assure the continued existence of species close to extinction. In addition, all of the national parks and national monuments, as well as many

[7] Marion Clawson, *Statistics on Outdoor Recreation* (Washington: Resources for the Future, Inc., 1958).

[8] H. H. Wooten and James R. Anderson, *Major Land Uses in the United States, 1954,* Agriculture Information Bulletin No. 168, U. S. Department of Agriculture, 1957, Table 7.

state parks, serve as wildlife refuges because they are closed to hunting. In the national forests wildlife management is one purpose of multiple-use management. On many privately owned tracts major consideration is given to wildlife.

Future Wildlife Land Uses

On most rural lands where wildlife is essentially a product of favorable environmental combinations, some attention to the elements of habitat establishment or maintenance can assure significantly larger wildlife populations. Such measures as brushy fencerows, brush or food patches in corners, shrubby borders between fields and woods, unmowed field corners, and similar habitat provisions occupy relatively small patches of idle or unusable border land. No means of measuring these acreages of land are available. In any case, their area is likely to be included in the statistics on agricultural land.

Conflicts in land use of more significant proportions develop between wildlife and commodity production on drainable wetlands which can be converted to field crops and woods pasture.[9] Clear choices must be made by the landowner between cash income for himself and benefits largely of public value.[10] For nearly a century landowners have brought these additional resources into production, through drainage or otherwise. Considerable areas of undrained pothole marshes still remain in the key North Central states, and woodlots still occupy parts of most farms east of the Great Plains. To the extent that these are drained or cleared for farming, wildlife habitat will be destroyed or reduced.

If any significant acreage of private land is to be devoted primarily to wildlife production or recreation uses, an expansion of leasing and easement arrangements which include compensation to owners for lands so used must expand at a far more rapid rate than at present.

9 Some drainage of agricultural lands improves the soil moisture relationship for cultivated crops, yet the land was never "wetland" in the sense of supporting wildlife dependent on water or marshy areas.

10 Charles H. Stoddard, "Wildlife Economics: A Neglected Tool of Management," *Proceedings of 1951 North American Wildlife Conference* (Washington: Wildlife Management Institute, 1951), and Walter L. Slocum and LaMar T. Empey, *The Role of the Farmer in Upland Game Production and Hunting in Whitman County,* Bulletin No. 552 (Pullman, Wash.: State College of Washington [Agricultural Experiment Station], 1954).

Included in this approach might be a similar arrangement to make picnic and camping areas available on private lands—particularly those bordering waterways—for general recreational use. As yet, no satisfactory system for financing such programs has been developed, but they seem to hold large opportunities for making recreational facilities available on private lands.

Additions to public wildlife refuge lands will probably be made during the next several decades from economically undeveloped swamplands. Wildlife biologists would like to see the present acreage of federal lands in such refuges increased to 12 million acres—an addition of 4 more million acres.[11] Whether this goal will be reached depends entirely upon the availability of public funds. Much of the area for this purpose will come from the glacial potholes of the upper Midwest, where more than three-quarters of the annual duck crop in the United States is raised. Other than waterfowl lands, no great acreage shifts in lands primarily dedicated to wildlife appear to be in prospect.

The intensity of management, and hence the wildlife output, of lands used mainly for wildlife purposes may be increased. By better control of the relation of land and water areas—through development of additional water areas for fur bearers and waterfowl in some localities—and by manipulation of vegetative cover, the capacity of areas to supply food and shelter to upland wildlife may be increased. Programs of this kind require capital investment and expenditures for annual management. In this respect, use of land for wildlife purposes is not different from use of land for any other major purpose.

MISCELLANEOUS PUBLIC PURPOSES

Among any miscellaneous grouping of land uses one that is fairly large is land set aside for military or defense purposes. The forts that were once common throughout much of the country were gradually reduced in area, but during and since World War II a major expansion in military and defense area has occurred. About 23 million acres are now used for this purpose, including atomic energy reservations.

[11] Statement of John L. Farley, Director, Fish and Wildlife Service, before the Subcommittee on Interior and Insular Affairs, House of Representatives, on June 30, 1954.

Much of this land was not highly productive for agriculture, forestry, or grazing; but some areas are located close to cities.

Small towns and urban-like places, too small to be included in the urban category of over 2,500 population, are also users of land. Wooten and Anderson estimate that 10 million acres are in this use; however, they conclude that most of this acreage is also included in the usual statistics for forestry, grazing, agriculture, or other major uses.[12] There are certainly rural cemeteries, schools, churches, hospitals, prisons, and other public or semi-public uses which require land—sometimes rather importantly located tracts of land. However, their total area is probably small and certainly is concealed in the statistics for other land uses.

LAND OF SPECIAL PHYSICAL CHARACTERISTICS

In addition to lands of clearly defined uses, there are those which do not fit readily into any land use class and in a sense may be considered residual. They cannot be considered land "uses" in the strict definition of the term. Lands which have had little or no previous use and those which have not been altered in a major way, as well as those which have been abandoned and have no present use are grouped together into this general category. They include swampland, desert, rocky outcrops, rocky mountain peaks above timberline, densely brushed or other noncommercial forests, and many other smaller areas with no current or immediately foreseeable use. They have frequently been described as being of value only in "holding the world together." Military installations occupy large areas of such lands for various types of activities with extensive spatial requirements. Some of the swampland is also forested; these swamp forests are either special types of forest or are relatively nonproductive, or both. "Desert" is a relative term, and some of what most people would call desert is used for grazing, although it produces only a small growth of forage—small even by grazing standards. Some of the desert areas are in military, wildlife, or recreation usage. Within national parks and national forests there are comparatively large areas of nearly bare rocks above timberline; some of these have low grazing values in national forests—grazing is excluded from national parks—and all have some value for recreational purposes. The wilderness areas of the national forests in-

12 Wooten and Anderson, *op. cit.*

clude a considerable acreage of this type of land. It is hard therefore to establish a clean-cut definition for the types of land here considered.

These lands have one trait in common; they are, with few exceptions, publicly owned and almost all federally owned. Their value as space in a world with rapidly expanding populations is beginning to be recognized as a recreational asset; furthermore many desert, swamp, and other lands in this group have been modified by irrigation, drainage, or other activity in order to convert them to higher uses. Examples of such modification are the filled-in swampy wetlands which now make up the downtown sections of such cities as Chicago and Milwaukee. Many swamps have been drained for agricultural developments and desert lands irrigated for the same purpose. Desert areas are increasingly used and valued for residential and industrial purposes; if water could somehow be made available, a much greater use for these purposes would occur.

One estimate, made in 1954, of the acreage of these lands which have not been included in any other use, gives a total of 80 million acres.[13] This, presumably, is a residual after all accounted-for uses have been deducted from the gross land area. Nearly 9 million acres of this estimated area is in California, much of it in the Mohave Desert; nearly 7 million acres is in Nevada, most of it in the southern and Black Rock Desert areas; and approximately an equal area is primarily desert in Utah and Arizona combined, with lesser amounts in the other western dry states. But much of these desert areas are in military or atomic energy withdrawals, some are in wildlife refuges, a considerable area is in Death Valley national monument, and there are other smaller tracts used for definite purposes. But a total of over 11 million acres in the miscellaneous category is reported for the Corn Belt, part of which is clearly swamps. The size of this figure suggests that statistical discrepancies involved in the calculation of residuals may have entered.

Wetlands

Fortunately, we have much more detailed information on wetlands, based upon a recent study by the Fish and Wildlife Service.[14] It is often extremely difficult to determine whether a particular spot is

13 *Ibid.*

14 U. S. Department of the Interior, Fish and Wildlife Service, *Wetlands of the United States*, Circular 39, 1956.

land or water; in some districts what may be land at one season is water at another. The wetlands are of all gradations—from land flooded only occasionally, to land flooded rather regularly; to shallow marshes, to deep marshes; to open but wet meadows, either fresh or salt; to open water, either fresh or salt, but with growing vegetation suitable for wildlife. This study fortunately distinguishes the chief physical types; but it includes areas that may be included elsewhere in our use classification; it also includes some water areas that are excluded from the usual figures on land area.

These low, usually level, and always wet (at least seasonally) areas have an interesting and somewhat unique history. They were almost always by-passed in the early waves of settlement. They were often regarded chiefly as sources of malaria mosquitoes and as obstacles to travel. The utility of poorly drained areas for agriculture was often questioned; but many have been successfully drained and others have been drained, though unsuccessfully. Some wetlands have been valued for the fur-bearing animals and game they contained. The coastal and other lowlands of the original thirteen colonies remained unoccupied and to a large extent unowned for many years. When they became states they retained title to all unsold lands within their boundaries. Not until the passage of the Swamp Land Acts of 1849, 1850, and 1860 were swamplands of the Public Domain states considered worthy of economic development. The original purpose of these swampland grants was to enable the states to reclaim them for agricultural purposes by the construction of levees and drainage ditches. In all, fifteen states were granted 65 million acres of swamplands under these acts—much of which was only partially swamp.

The Fish and Wildlife Service study of wetlands estimated the original wetland area of the United States to be about 127 million acres; about 53 million acres of this has been converted to other uses, mostly agricultural, and mostly by drainage. The area now considered as "wetland" includes some twenty different types of areas, depending upon the nature of the wetness, whether they are fresh or saline, and upon the type of vegetation (Table 50). The largest single category, with about a third of the total acreage, is seasonally flooded basins or flats; wooded swamps, with nearly a fourth of the total acreage, are next largest in area. In all, seven-eighths of the total are inland land and water areas and one-eighth is coastal. But the grouping used in this survey includes water areas (with permanent standing water but less than 10 feet deep) as well as land.

Table 50. Acreages found in different classes of wetlands

Wetlands	Million acres	
1. Wetlands that are not swamps		
Seasonally flooded basins or flats	23.1	
Fresh meadows	7.5	
Irregularly flooded salt marshes	.7	
Subtotal		31.3
2. Water, not land, areas		
Inland areas:		
open fresh water	2.6	
open saline water	.3	
Coastal areas:		
open fresh water	.2	
sounds and bays	1.1	
Subtotal		4.2
3. Genuine swamps		
Classed as wooded:		
wooded swamps	16.8	
mangrove swamps	.5	
Subtotal		17.3
Other swamps:		
shallow fresh marshes	4.0	
deep fresh marshes	2.3	
shrub swamps	3.8	
bogs	3.3	
saline flats	1.1	
saline marshes	.3	
shallow fresh meadows	2.2	
deep fresh marshes	1.6	
salt flats	.4	
salt meadows	1.0	
regularly flooded salt marshes	1.6	
Subtotal		21.6
Total		74.4

Most of these wetlands are not swamps, but are probably included elsewhere in our use classifications—as agricultural, grazing, or forest land. It is unlikely, however, that all of this land would be so classified—some of it is unproductive for those purposes. The water areas would, in general, not be included in statistics on land area. Probably most of the wooded swamps are included in the data on forestry, but this may not be true for all such areas. Possibly some of the meadows in the "other swamps" group are included as agricultural or grazing land, but it seems more likely that they are not. The genuine swamp areas not included in other land use categories probably range between 21 and 39 million acres.

While the economic importance of wetlands in their natural state is limited mainly to wildlife and recreational uses, many ecologists and some hydrologists feel that they have an important place both hydrologically and in biotic relationships. As water storage areas during periods of high rainfall, poorly drained swamps outside of river bottoms undoubtedly have a value for retarding flood waters.

Wetlands are important sources of local income where waterfowl and fur bearers are abundant. The extensive marshes of Louisiana are the largest producers of muskrat hides in the country and provide much of the winter resting and feeding grounds for North American ducks, geese, and other water birds. Duck hunters bring in substantial income to communities located in the Mississippi and Illinois river bottom lands and in the pothole country of the Dakotas and Minnesota. Wild rice is an important crop from shallow lakes and marshes in Minnesota and Wisconsin. Cranberry culture in Massachusetts, New Jersey, and Wisconsin is established on a limited acreage of formerly acid bogs in these states. But the greatest monetary income from wetlands has resulted from drainage where more intensive agricultural development is made possible. In New York and Michigan extensive use has been made of them for truck crops.

Recent Trends in Wetland Areas. In recent years there has been a renewal of activity in drainage of wetlands for agricultural purpose. Most of the potentially arable wetland in the North Central states has been drained and brought into cultivation, with important exceptions in the northern forested parts of Wisconsin, Michigan, and Minnesota, and some of the prairie "pothole" areas of western Minnesota and the eastern Dakotas. Drainage of lowlands in other parts of the country has been less consistent. Large parts of the Mississippi delta have been drained and cleared for cultivation: much Florida lowland is being drained and diked for agricultural and residential usage.

Recent investigations of drainage in the pothole areas of the upper Middle West have both inventoried marshy lands and made an estimate of the annual drainage rate. In the prairie pothole areas of Minnesota, North Dakota, and South Dakota, the United States Fish and Wildlife Service in co-operation with the state departments of conservation found 1,210,000 potholes covering 4,450,000 acres (1954).[15] These wetland areas in three states produce a larger quantity of ducks than all the other states in the United States—estimated at 4 to 5 million per year—and are critical in this respect. Because many of the shallower potholes contain rich agricultural soil, their drainage has been profitable. Northern Iowa, which once abounded with wetlands, has few left and is no longer important as a duck producing area. Unpublished studies in Minnesota during the past six years show that 460,000 acres of prime waterfowl wetland in 1951 were reduced by drainage to 366 thousand acres by 1957.[16] Shrinkage of wetlands is estimated at 5 per cent annually.

Although no economic studies are available, many wildlife conservationists feel that wetland drainage in recent years has been accelerated by subsidy payments of the Agricultural Conservation Program and by engineering encouragement of the Soil Conservation Service. In 1955 the Agricultural Conservation Program for the nation as a whole shows $3,943,000 paid out for drainage works, and $1,335,000 for water retention projects.[17] However, drainage would probably take place in time anyway without subsidy, as is already the case in Iowa.

Future Developments in Wetlands. The lack of data on past trends in wetland use and development makes any attempt at future projection extremely hazardous and must be qualitative at best.

Much of the present area of wetlands which occupy good agricultural soils and which lend themselves to intensive soil management are likely to be drained for crop production. These are confined

15 Thomas A. Schrader, "Wetlands, Waterfowl and Welfare," paper presented at annual meeting of Minnesota Conservation Federation.

16 Information furnished by David B. Vesall, assistant supervisor in Bureau of Game and Fish, Minnesota Department of Conservation.

17 U. S. Department of Agriculture, *1955 Summary—Agricultural Conservation Program,* Agricultural Conservation Program Service (Washington: U. S. Government Printing Office, December 1955).

mainly in the upper Midwest and in some of the Mississippi and other river bottoms. But since the largest potential of such lands have already been drained and developed, the actual area involved in future development may be no more than 3 to 5 million acres. Some of the peat swamplands of the Lake states were once thought of as being highly valuable for agriculture, but after tremendous expenditures turned out not to be so. Some of the highly productive pothole areas will be acquired for waterfowl and other wildlife uses. Minnesota hopes to acquire 200,000 acres and the Dakotas perhaps 300,000 in all. Thus, there is a limit in the area which can be so developed.

More wetlands which have small water courses flowing through them are likely to be developed as impoundments for a variety of purposes. The shortage of open water areas and an expanding population will probably bring about a significant movement in the direction of flooding them for recreational purposes—summer homes, duck hunting and fishing. Commercial fish culture in these areas, except for bait minnow production, has yet to be developed, but is thought by some to have important possibilities. Forecasting of acreage estimates is difficult because no survey data are available on the acreage of swamplands with water flowing through them.

In spite of some probable development, there still is every prospect that the largest portion of the wetland area will continue in its undeveloped state for many years. Increasing pressures upon it will intensify uses so far as the lands are capable of or have a potential for conversion. Swamplands represent one of the important undisturbed wildlife habitats in the country; there is every likelihood that additional large acreages will be maintained in their present state through incorporation into public hunting grounds and wildlife refuges, not only for waterfowl but for other game as well.

Deserts, Mountain Tops, Rock Outcrops, and Sand Dunes

The area of these residual lands is not accurately known. One estimate is that they run roughly from 40 to 60 million acres in extent.[18] That estimate is taken to exclude all wildlife refuges, all parks, and all land used for commercial forestry or grazing; however, as a residual figure it may be subject to error, possibly substantial error relative to this estimate of its size.

[18] Based on Wooten and Anderson, *op. cit.*

As we have pointed out, "desert" is a relative term, at least in popular usage, although geographers and climatologists have often defined it rather precisely. For our purposes, we are considering as desert any land producing too little forage to be grazed regularly by domestic livestock, and without other major use. The largest such deserts are the Mohave in southeastern California, with possibly 4 to 5 million acres in this category; the southern Nevada deserts, with perhaps as much as 3 million acres; the Black Rock desert of northwestern Nevada, with possibly ½ million acres; the salt flats of northwestern Utah, with roughly 2 million acres; and desert areas in southwestern Arizona with roughly 1 million acres in this category. Including many smaller desert areas, which in total are relatively large, the total desert area, in this definition, may run 15 million acres or possibly more. In addition, other desert areas are used for military, wildlife, or recreation purposes. In recent years people have been pushing out onto the desert, primarily for winter residence but also for year-round living and even for factories. Substantial areas in the general vicinity of Tucson and Phoenix, Arizona; of Twenty-Nine Palms, Victorville, and Barstow, California; Las Vegas, Nevada; and other western towns, have been taken over for these purposes. The great attraction has been sunshine and privacy and clean air; the great drawback has been lack of water. Water is hauled into some of these areas for household use.

The type of mountain-top area included here lacks commercial forest, usually because it is above the timberline; it is ungrazed, either because it lacks forage or because grazing is forbidden; and it is not within an established park. These limitations mean that the primary areas left are within national forests, and these are likely to be included within wilderness areas. There are some areas within grazing districts, Indian reservations, and perhaps other areas that will fall in this category also. Such lands may have mineral possibilities not yet discovered; and they have the special recreational value of wilderness or near-wilderness areas. But otherwise they are now essentially unused, and their remoteness and lack of commercially exploitable resources are likely to keep them from being used in the future.

The total area of rock outcrops in the entire nation is possibly rather large. However, it is not clear when their area has been excluded from that of forested or grazing land, and when it has not been. In many of the grazing districts substantial areas are rock outcrops, but their area is likely to be included in the gross area of grazing land.

In the more humid parts of the country, some rock outcrops have been used to quarry stone or gravel.

The area of sand dunes is comparatively large. Sand dunes are mostly found along ocean and lake fronts, although there are some sand dunes in desert areas. Those along water fronts have a value, potential if not actual, for recreation.

These various areas, about which, partly because they do not fit well into any land use classification, we really know so little, have had very limited use or usefulness in the past. There will be some encroachment upon them in the future for various uses, as the supply of other land becomes more limited in relation to the demand. But it seems probable that for many decades most of these lands will remain more or less as they are at present.

VIII

Future land use in the United States

We come back now to the question posed in Chapter I: How can this country reconcile its fixed area of land with the increased demand for every land use that an increasing population and an expanding economy make probable?

Having traced in some detail the probable future demand for land for each of a number of purposes, we have found in every instance a trend toward greater demand for the products of land and in nearly all cases a greater demand for the land itself. Regarding each use, we tempered our analysis with the knowledge that the total area is fixed and that other uses are increasing in urgent demands. For instance, we did not suggest increasing the area of forest as a means of meeting the future's greater need for forest products; we knew that the area is far more likely to decrease under pressure of other uses. Here, however, we shall attempt to reconcile more explicitly the competing demands and sum up the prospects for the various uses of the land. In addition, we shall briefly consider some alternative patterns of land use if demand factors should differ greatly from our assumptions. Various regional variations in land use merit brief review, so also do certain policy implications and institutional problems, and evident needs for further research.

The summary analysis and discussion builds upon that in previous chapters, and obviously can be no stronger than the basic structure of our study. At the same time, a summary such as this is perhaps more difficult than the detailed analysis, and hence may have some errors of its own.

PROBABLE PATTERN OF FUTURE LAND USE

The heart of our discussion of future land use in the United States is the statistical summary in Table 51. There, we have made estimates of the area in a number of land uses by census dates from 1900 to 1950, and projections for 1980 and 2000.

Before considering the content and implications of the data in Table 51, let us frankly face the unsatisfactory state of the presently available data. No complete land use inventory, with comprehensive but mutually exclusive land use categories, has ever been made in the United States. Data on land uses come from a variety of sources, chief among them being the agricultural censuses. Some duplication in area is inevitable—the local roads mentioned in Chapter VII are an instance, and so are the agricultural areas found within cities and towns. While the areas involved may be small, they are enough to throw off balance any appraisal of total area and total uses. On the other hand, it is evident that some kinds of uses are at least partially omitted from the data. And one cannot but suspect that many areas are reported with a considerable margin of error, which may be cumulative rather than offsetting. In addition to this, and beyond it, there is the persistent confusion of land use, vegetative cover, and land ownership, which we considered in Chapter I.

In consequence, our comparisons in Table 51 should be considered only as generally indicative of the true situation. We have tried to put all acreages on a land use basis, without regard to ownership or other factors. In general, we have started with the situation for 1950; then have constructed estimates for earlier periods based upon such data as were available; and lastly, have made projections for 1980 and 2000. It has been our intention that the data in this table should conform to, and in some instances supplement, the data on each land use found in the earlier chapters.

The area of land used for urban purposes has grown in the past, and will grow in the future, primarily because of more people living in cities and their extending urbanized arms. Much of this land, perhaps nearly half, is not actually used by the city, but is withdrawn

Table 51. Use of land in the United States, selected years, 1900-1950, and projections for 1980 and 2000[1]

(million acres)

Use of land for	1900	1910	1920	1930	1940	1950	1980	2000
Cities of 2,500 or more population[2]	6	7	10	12	13	17	30	41
Public recreation areas[3]	5	9	12	15	41	46	72	95
Agriculture:								
Crops[4]	319	347	402	413	399	409	388	388
Pasture[5]	77	84	78	73	68	69	70	70
Other[6]	53	57	58	45	44	45	45	45
Subtotal	449	488	538	531	511	523	503	503
Commercial forestry:								
Continuous management[7]	0	30	60	200	300	359	385	405
Little or no management	525	482	440	295	188	125	90	50
Subtotal	525	512	500	495	488	484	475	455
Grazing[8]	808	775	730	735	740	700	700	680
Transportation	17	19	23	24	24	25	28	30
Reservoirs and water management[9]	*	1	2	3	7	10	15	20
Primarily for wildlife	*	*	1	1	12	14	18	20
Mineral production; Deserts, swamps, mountain tops, some noncommercial forest, etc.; Miscellaneous and unaccounted for	94	93	88	88	68	85	63	60
Total	1,904	1,904	1,904	1,904	1,904	1,904	1,904	1,904

*Negligible.

[1] The data in this table are necessarily estimates in several instances, sometimes on a relatively scanty basis of fact. This table emphasizes land use, as separate from land ownership or control or from vegetative cover. See text for discussion of projections.

[2] Includes municipal parks.

[3] Excludes municipal parks. Includes national park system, areas within national forests reserved for recreation, state parks and acreages around TVA and Corps reservoirs reserved for recreation. Excludes all areas used primarily for other purposes even though they provide much recreation. Excludes actual water area of reservoirs, which is shown later under its own heading. Excludes also wildlife areas, which are shown below. We have assumed that only part of the increased potential demand will be met. Data for past taken from *Statistics on Outdoor Recreation* (Washington: Resources for the Future, Inc., 1958); projections based on discussion in Chapter III.

[4] Cropland harvested, crop failure, cultivated summer fallow, and cropland idle or in cover crops. See Tables 11 and 12, *Agriculture Information Bulletin No. 168*. U. S. Department of Agriculture, 1957. See text for discussion of future projections.

[5] Only pasture on land which is considered cropland is included. This corresponds to the 1949 and 1954 Census of Agriculture definition. The 1900 figure is an estimate. The acreages for 1910 through 1940 are the difference between crops, as shown above, and estimates of cropland potential which included cropland pastured, given in Table 1 of *Agriculture Information Bulletin No. 140*, 1955.

[6] Farmsteads, farm roads, feed lots, lanes, ditches, and wasteland. See Tables 11 and 12, *Agriculture Information Bulletin No. 168*.

[7] This is a roughly estimated figure. For 1950, it excludes commercial forest land with no fire protection or poorly stocked, as shown in *Timber Resources for America's Future*, U. S. Forest Service, 1958. For earlier years, it is our estimate of comparable definition area.

[8] See text for source and basis estimate of this figure in 1950. Includes some noncommercial forest land used primarily for grazing.

[9] Excluding land around reservoirs and conservation pools of reservoirs, which are included in recreation areas.

by it from other use. This unused land includes vacant lots, engulfed areas, and surrounding tracts believed to be imminent for urban development. There is a relation between city population and average density, and hence of total city area. Average density has risen in the past because the average city has grown considerably in population. In the future it may increase further, or at least stay relatively constant, in spite of a readily observable tendency for density to be lower in the booming suburbs than in the city centers. However, density is rising in the older city areas, and the suburbs are largely taking the place of the independent small towns where density was even lower.

As nearly as we can estimate, the area in urban use in 1950 was 17 million acres; it has increased to this point rather steadily from about 6 million acres in 1900. If our projections of population materialize, by 1980 there will be 30 million acres in urban use, and by 2000, 41 million acres. For the most part, the regions of the country with the largest urban population now will experience the greatest increases in urban land use. While the area used for this purpose will still be only a little more than 2 per cent of the national total area in 2000, it comprises the most importantly located lands in the country. The possibility of reducing this area by sound advance planning is, therefore, highly important.

Recreation is another intensive land use that is growing in importance. With more leisure, as well as higher incomes and better travel facilities, our growing population makes an increasingly heavy demand upon outdoor recreation areas. Some areas are user-oriented—located near the user group for whom they are designed; others are resource-based, and these include our finest outdoor areas; and still others are intermediate both in location and in physical quality. Demand has increased for each type, and will continue to increase, though at different rates. The demand for land for recreation expresses itself primarily through governmental or political action, rather than on the basis primarily of economic considerations. It is, therefore, harder to estimate to what extent the potential demands for recreation will be met.

We estimate that the specialized outdoor recreation areas, outside of those city parks included in urban land use statistics, have grown from 5 million acres in 1900 to about 46 million acres in 1950. Even assuming that only half of the potential demand for outdoor recreation will be met by increased areas used for this purpose, we estimate the use of 72 million acres by 1980 and 95 million acres by 2000.

Agriculture shows one of the most dramatic changes in land use in the past, when one considers the great changes that have taken place in total population. The area in crops expanded all through the nineteenth century and up to about 1920; since then it has varied little. However, total agricultural output has about doubled since 1910, increasing each year at a remarkably constant rate if unusual weather conditions are discounted; and the number of people engaged in agriculture has decreased materially and at an accelerating rate. Incidentally, the data on agricultural land use are some of the best for any major use. There have been some shifts in area devoted to different crops, and some regional shifts, but the total agricultural land use is one of relatively high stability.

Our best estimate is that further increases in agricultural land use intensity and output per acre will equal or outrun the increases in total population. Higher incomes in the future will have little effect upon the amount of food consumed, although they may affect the kinds of foods most in demand. We estimate that something less than the present area used for agriculture can feed and clothe us, at least up to the year 2000. While some farm land will be taken for city and highway use, this will tend to be replaced by other land cleared, drained, or irrigated. There are strong resistances to retirement of land from farming; but in spite of them, the area used for agriculture will, we estimate, decline by about 20 million acres up to 1980 and will not increase beyond this by 2000.[1]

As nearly as we can judge, the total area used for commercial for-

[1] The projected acreage for agriculture in Table 51 does *not* agree with the projection in Table 38, Chapter IV. The latter projects reasonable possibilities of area adjustment, considering the best estimates of future domestic and export demand for agricultural commodities, and considering probable trends in yields per acre. Table 38 indicates that some 36 million acres *less* of cropland than in 1954-56 will suffice in 1980 (averaging the two projections for 1980) and 23 million acres *less* in 2000 (averaging the two projections) than in 1954-56. The area in 1954-56 was, in turn, 15 million acres *less* than in 1950. However, it seems to the authors that a major reduction of 36 million acres from the 1954-56 level, or of 51 million acres from the 1950 level, is unlikely by 1980. As Chapter IV points out, many factors tend to keep land in crops, once it is there. We think a downward adjustment of 20 million acres from 1950 to 1980 as great as is likely to occur, even when allowing for the shifting of some cropland to highways and cities. We think it probable no further downward adjustment will occur by the year 2000. The agricultural acreages for both 1980 and 2000 in Table 51 are thus somewhat *larger* than a strict analysis shows to be possible and desirable, but are smaller than present acreages.

estry has declined slowly but rather steadily since 1900, due primarily to clearing of some land for farms, urban uses, and highways, or to conversion to recreation uses. The major change, however, has been in the increased intensity of forest land management. While there is little direct data on this point, there is rather general agreement as to the trend. Our estimates are admittedly rough. Better fire control, better harvesting methods, tree planting, and other practices attest to the difference in management intensity.

Although the demand for forest products will be materially higher in the future, the demand for land for other purposes will force some decline in area used for forestry. If the increase in demand is to be met, it will be through higher output per acre.

The area used primarily for grazing has shrunk considerably in the past fifty years; much of the increase in agricultural land came from plowing up the Plains. Other areas have been lost to less extensive uses, including some to such miscellaneous uses as military reservations in recent years. Grazing is a poor competitor for land that is in demand for other uses, and thus both in the past and in the future grazing to a large extent is a residual land use. Some further moderate declines in grazing area seem probable in the decades ahead.

Each of the miscellaneous uses for which we have even moderately good data—transportation, reservoirs and water management, and specialized wildlife areas—have increased relatively a great deal in the past, although the total area is still inconsiderable. Relatively large increases seem probable for the next several decades, but again these are of modest dimensions in terms of total area.

The truly miscellaneous land uses—those for which the data largely reflect inaccuracies in other land use data—have decreased rather steadily in the past. Only in 1950 did they show an increase, mainly because land for military purposes increased. Further decreases in area seem probable. The American idea of "waste" land seems to be a flexible one: more generous when there is much land in relation to demand, less expansive when the pressures on land use strengthen. This is perhaps most noticeable with desert; we have moved out on to lands previously considered desert, and we will again.

No tabulation of data of the scope shown in Table 51 can be fully adequate to its task. Set down in this way, the data imply a regularity and simplicity of change, and a comparability of use from period to period, that do not exist. The nature of the use within each major land use group changed continuously over the decades. In earlier years much land in the United States was used lightly, or not used at all in any purposeful and planned sense; we have included it in the group where it most nearly belongs, even if earlier and later uses are not fully comparable. All of the land use changes shown in the table are net; much larger changes occurred which were partly balanced by other changes.

Past changes in major land use in the United States, at least since 1900, have been on a comparatively modest scale. The changes may be broadly grouped as follows:

marked increases in use—urban,
recreation,
managed forestry,
reservoirs,
wildlife;
modest increases in use—transportation;
increases at first, then stable acreage—agriculture;
continuous decline—unmanaged forestry,
grazing,
miscellaneous and unaccounted for.

While the pattern of land use was different in 1950 from that in 1900, statistically the net changes were not great so far as major use was concerned; and on a geographic or areal basis the gross changes not large either—as an example, most of the land used for agriculture in 1900 was still so used in 1950. Anyone familiar with land use in the United States in 1900 would readily have recognized the 1950 data as belonging to the same country.

If the past changes in major land use have been relatively slight, it seems clear that future changes will be still more slight. The future changes may be grouped as follows:

marked increase in use—urban,
recreation,
reservoirs,
wildlife;
modest increase in use—managed forestry,
transportation;
modest declines in use—agriculture,
unmanaged forestry,
grazing,
miscellaneous and unaccounted for.

Those uses that have increased in the past will, in general, increase further in the future; and those uses that have decreased will, in general, continue to decrease.

MULTIPLE USE: ITS POSSIBILITIES AND LIMITATIONS

The basis of classifying land use throughout this book has been that of primary use. At various places, however, we have referred to other, secondary uses on various types of land. In view of the somewhat greater demands that are in prospect for all land uses the possibilities of using land for two or more purposes, as a deliberate policy, becomes increasingly attractive, as well as increasingly difficult.

The idea of multiple use of land has been widely advocated, particularly for certain types of public land. There seem to be two major possibilities. One is double, treble, or more use of the identical land, with uses co-ordinate in importance, or one use dominant and the others subordinate but still important. The other is dominant use of particular tracts of land, with relatively limited subordinate use for other purposes, but with the different tracts so closely intermingled that the area as a whole can reasonably be considered to be under multiple use management. Under the first concept, the same tract would be used for recreation, forestry, and grazing, for instance; under the second concept, small areas would be used primarily for recreation, while other and adjacent areas would be used for forestry or for grazing primarily, with limited secondary use of each for other purposes. In either case there are physical limitations and possibilities, and also economic and institutional ones.

Let us look first at the physical possibilities and limitations inherent in common use of the same tracts. Some uses are physically compatible with others, while other pairs of uses are not (Table 52). Among the dominant and intolerant uses are urban and transport uses of land; little else is possible on land primarily for these purposes, and they can hardly be tolerated on land primarily dedicated to other uses. Grazing and forestry are much more tolerant of one another, and of recreation, wildlife, and some other uses. But there is great variation among forest and forage types in this regard. Rec-

Table 52. Physical compatibility of various major land uses, in multiple use land management programs[1]

Primary land use	Physical compatibility with secondary use for								
	Urban uses	Recreation	Agriculture	Forestry	Grazing	Transport	Reservoirs and water management	Wildlife	Mineral production
Urban	complete	high for city parks; zero for others	none	none	none	very poor, except city streets	none	very poor	very poor
Recreation	none	complete	none	poor to moderate	very poor to none	very poor	poor	fairly high	very poor
Agriculture	none	very poor	complete	zero	zero	zero	very poor	poor to moderate	poor
Forestry	none	high	none	complete	variable—none to fairly high	zero	zero	high	poor to moderate
Grazing	none	high	none	usually very poor	complete	zero	poor to fairly high	high	poor to high
Transport	none	none directly; incidental on rights of way	none	none	none	complete	none	none	none
Reservoirs and water manage-ment	none	poor to high	very poor	very poor	poor to moderate	none	complete	poor to high	very poor
Wildlife	none	high	very poor	moderate	moderate	none	poor	complete	very poor
Mineral production	none	poor	poor	fair	fair to moderate	fair	poor	poor to fair	complete

[1] These are uses on identical or common areas; the problems of different uses on closely interrelated areas is considered later. The definitions of the various uses are those employed in earlier chapters.

reation and agriculture, as we have defined the latter, have a generally low tolerance for other uses. The table does not include watershed management as a specific or major land use. Where it is the primary use—as for instance in a municipal watershed—recreation, forestry, wildlife, and grazing would have a limited place; as a secondary use, it is compatible in varying degrees with most uses except urban and transport.

When it comes to actual dual use of the same tract, many difficult economic and institutional problems must be overcome. For such pairs of uses as forestry and grazing, the landowner may make an evaluation of his comparative income from each, and choose the one, or the combination, that promises the greatest profit. It has been argued that the forest landowner is likely to underestimate the income possibilities from grazing, just as the rancher may underestimate those from forestry. A much more difficult set of problems arises when one of the uses produces values not readily capturable in the market place —recreation and wildlife, for instance. The private landowner cannot capture the income from such land uses; the public land agency has the problem of evaluating their benefits in comparison with the more commercial benefits that might be derived from other uses of the land. The problem is complicated further by the fact that the benefits accrue to different groups of people. One cannot ask the lumberman: Would you prefer a little less cash income from forestry in return for noncash income from recreation? Some land uses produce their incomes directly on the site; others, such as watershed management, produce their income at relatively distant places; and still others, as recreation and wildlife, produce their income on the land but only or primarily for people not resident there. The benefits of different kinds of land uses in a multiple land use program arise in different forms—really different kinds of "money"—to different groups of people located in different areas.

The technical, economic, and institutional problems in actual joint use of the same identical tracts of land are many, and often greater than the benefits that might result from the necessary expenditure of time, energy, and money.

The situation is more promising for a planned and integrated multiple use program of intermingled tracts of land. For instance, attractive forested areas along streams or lakes may be reserved for recreation, and the back country used for commercial forestry. The need for extensive buffer strips to make recreation areas most usable has been indicated already, as has the generally small area used for intensive recreation even within parks (pages 151-53, 170-73, 178, 185). Grazing may be allowed in the mountain valleys, open grassy areas, and similar types of land among generally forested areas that are used for commercial forestry. The farmer may reserve a strip around his fields, to serve as a turning area for his tractor and machinery, and plant it with species to provide for the wildlife adapted to the area. Wayside tables for picnicking and resting while travelling may be established along highways, even the super-highways where it is desirable not to allow local uses to interfere with the capacity of the highway. In some types of parks forestry, including logging, may be practiced under carefully devised rules—logging during the off-season, complete disposal elsewhere of the slash, minimization of scars and planned efforts at their healing, etc. Perhaps the greatest opportunities for this kind of multiple use management exist for forestry and grazing. For one thing, the areas used for these purposes, in total and in individual management tracts of efficient size, are large enough that small areas for other purposes can readily be spared. Another reason is that the returns per acre from these land uses are not usually so high that a few acres cannot be spared for other uses.

Multiple use management of closely intermingled tracts now exists on sizable areas, especially publicly owned areas. It more nearly applies to management of the national forests in this respect than does multiple use of identical tracts. But this type of land management imposes its problems and limitations also. Careful planning for the entire area is needed. This is easiest when the whole area is in single ownership, whether public or private; then a conscious facing of the opportunities and problems of each land use, and of their interrelations, is not only possible but nearly inevitable. Where larger areas are owned by several persons or groups, however, joint planning is possible through various devices, including soil conservation and other districts. Regardless of ownership, single control is desirable to simplify plan making and action programs. If, for example, part of the area is to be used for recreation, access for vacationists through the forested and grazing areas is easier to obtain where there is single management over the entire area. However, a mixed control and management does not make the situation hopeless. For in addition to voluntary agreements, landowners can be offered payments or be required to give rights of way, easements, and other rights to permit the use of their land as part of a larger land use plan.

We visualize increased importance for such multiple land use management. This may be especially important for recreation and wildlife uses of land. The difficulty of getting enough land to meet the potential demand for each of these land uses will be greater if the management of each insists on exclusive jurisdiction over the necessary buffer areas around the areas of intensive use. In many types of recreation areas commercial forestry on part of the area and under the special safeguards mentioned above would remove much of the opposition to enlargement of the recreation areas.

Use of closely intermingled tracts may take place within an area designated as a park or as a forest; we suspect that often it will be hard to tell which name fits best. Not that *all* park areas should be subjected to this type of management; we are fully aware of the opposition to commercial logging in national parks, and are willing to see this arrangement continue there and elsewhere. But provision of enough land to meet, or reasonably approach, the demands for recreation and wildlife may depend very largely upon how well those interested in such uses are able to fit their needs into the use of closely intermingled lands for other purposes.

ADJUSTMENTS TO UNKNOWN FACTORS

At the beginning of this book we dwelt on the hazards of economic and other forecasting for the longer range future, and in Appendix A we illustrate this with examples from demographic projections. We have distinguished between the forecast, which is a prediction of what is most likely to happen; the projection, or calculation from stated assumptions to derived conclusions; and the estimate, or rough projection. We have chosen to use the latter, because we believe that, despite its unpretentious character, it is likely to be as close to accurate as is the more laborious projection. Nevertheless, in any study of the future some assumptions must be made, and we have made them.

How much confidence can we have in the results? How good are the odds that our estimates will work out? What adjustments are called for, if they are in error?

It would be possible to carry forward a detailed analysis of land use, for each kind of use, under a wide variety of other assumptions or estimates as to future population, income, leisure, travel, other basic factors. We could label such assumptions or estimates "high," "medium," and "low"; or A, B, and C; or otherwise designate them for the reader's choice. Such a method has its advantages; primarily, it does suggest the importance of various factors or assumptions on the final results. But it can also be confusing.

We have made our analysis in light of one set of assumptions as to the future economy and society. Under these assumptions, the demand for virtually every service the land can provide is strongly upward. With a total fixed area, much of the increased demand will take the form of more intensive land use, and some of the possible demand cannot be satisfied. The effects of population, income, leisure, and travel upon demand for land are all positive—the more of the basic factor, the higher the demand for land. This is true for each use, and to a considerable degree proportionately for each. If in the future the United States has more people than we have estimated, more land will be needed for cities, and there will be a demand for more outdoor recreation, more agricultural output, more timber products, and so on. Likewise, if incomes rise higher than we estimate, the demand for each element will be larger. Similarly, if any or all of the basic economic and social factors are less than we have estimated, then demand will be lower.

At the best, we think an area of uncertainty equal to roughly 10 per cent of the estimated figure should attach to all acreages for 1980, and one of 25 per cent for the year 2000. These are by no means the outside limits of probability; one could easily set up future conditions that would require either more or less than these limits. But we hazard the judgment that the true figures will fall within these limits.

Equally important with the estimates of future population, income, and other economic and social variables are the estimates one uses with respect to changing technology—to the future coefficients between volume of land product demanded and area of land required to provide it. Technology is perhaps more readily grasped for agriculture than for other major land uses. There, as we have seen, it has dominated the land use picture for the past forty years or so, and we expect it to do the same for the next forty years. But no land use is immune to changing technology. We have less information than we would like about the effects of past technology; this is especially true for land uses other than agriculture. As to the future, we cannot know with any degree of precision the impact of technical innovations yet to be developed or perfected. Nevertheless, such changes must be taken into account when developing estimates for the future. At various places in our calculations we have assumed that certain land use practices will come into existence, even though we cannot be sure of their precise form or timing.

Table 53. Summary of projected land use changes, 1950-2000

Land use	Projected change, 1950-2000: million acres	Projected change, 1950-2000: % of change	Probability of change approximately to extent and direction indicated	Kinds of adjustment that will be made, if our projections of demand and of supply responses are in error
Urban purposes	+24	+141	very high	Area used will change proportionately
Public recreation	+49	+107	high	Some area adjustment; mostly adjustment in intensity of use
Agriculture	−20	− 4	low	Adjustments in area, intensity of use, foreign trade, and consumption all readily possible
Forestry	−29	− 6	high	No major change in area, but in intensity and consumption
Grazing	−20	− 3	high	No major changes in area, but in intensity of use and output
Transportation	+ 5	+ 20	high	Area will change proportionately
Reservoirs and water management	+10	+100	high	Area will change proportionately
Primarily for wildlife	+ 6	+ 43	moderate	Area will change proportionately
Mineral production; Deserts, swamps, mountain tops; Unaccounted for	−25	− 29	low	

The uncertainty about the future has two dimensions, as far as land use is concerned: the degree of confidence we can place in even our best estimates, and the adjustments necessary if the estimates prove wrong. In Table 53 we have tried to deal with both of these. Given the analysis made in earlier chapters, there seem excellent reasons to expect the area of land for urban use to expand greatly; fairly good reasons to expect changes in recreation, forestry, grazing, transportation, and water management uses for land; but only a little confidence can be placed in any estimates for future agricultural area, because of major uncertainties in the basic factors entering any projection formula. Our estimate of the probable accuracy of our best estimates is necessarily a subjective one—but one which nevertheless we think important to make for the reader.

The nature of the adjustment possible differs considerably according to the land use. For urban use, the area will almost surely be larger or smaller in direct proportion as we have under- or over-estimated the need. The same is likely to be true for transportation. For agriculture, on the other hand, adjustments in area are by no means the only possible adjustments nor perhaps the most feasible ones. Land may be used more or less intensively, foreign trade might be either increased or decreased, and the consumption pattern of the people might respond considerably to relative prices for different foods. With forestry and grazing, the area used for them will to a considerable extent reflect changes in other land uses; if we have been wrong about the estimated demand for their products or for the estimated output of these lands, the adjustments are likely to take place in intensity and form of land use, not in area.

If total future demand is more or less, because of more or fewer people, more or less income, or for other reasons, the direct effect will be upon what might be called the overriding land uses—urban, transportation, perhaps reservoirs, and recreation and agriculture to some extent. The need for land for these purposes would vary more or less directly with demand for their products. The indirect effect upon the other major land uses might be considerably different. Their demand, like that of the overriding uses, would vary directly with the economic and social forces. But the area available for their use might vary in opposite direction. That is, if total future demand were more strongly upward than we have estimated, more forest land might be taken for agriculture, and hence less be available for forestry; and similarly for grazing. On the other hand, if total future demand were much less than we have estimated, then less land would be taken for the overriding uses, and more left for the residual uses such as forestry and grazing. If this relationship is correct, the chief variables for forestry and grazing are intensity of operation within the land use and differences in output to meet demand, rather than in area.

REGIONAL DIMENSIONS OF PROBABLE CHANGES

The foregoing discussion in this chapter, as in this book generally, has been in terms of the national or total United States situation. We have been concerned primarily with forces, trends, and situations in their broad or total aspects. However, a land use change must necessarily occur in a specific location. Without attempting a comprehensive and quantitative analysis, a brief discussion of some of the probable major changes that appear likely in each region may be helpful.

Northeast

We mean by this term, all of New England and the Middle Atlantic states, especially along the coast and the immediate tributary area. This is now an urbanized area containing some of the largest cities of the nation; it seems destined to become even more urbanized. As Figure 18 demonstrated, by the year 2000 a more or less continuous urbanized area may well stretch from Maine to Norfolk, Virginia. The problems of conversion of land from agriculture, forestry, or other rural status, to an urbanized status will be major and continuing in this region. For a good deal of the region, this problem will be interstate, intercounty, and intercity; for this reason alone, the problems of developing co-ordinated plans for development of large areas will be especially difficult. There is now an acute need for more recreation facilities in this general region—a need which will become greater as the total urban population rises. At the same time, servicing the vacationist and the tourist is now an important economic activity in the region, especially along the seacoast and in the mountainous areas. Expansion of parks and other recreation areas, better use of the forested areas for recreation, further development of the vacation-tourist business and information about it, and other related problems will surely require major attention during the next few decades.

This general region is heavily forested, but its forests have somewhat low productivity per acre, are badly split up into a multitude of small holdings, and there is a serious lack of markets for the kinds of wood now actually being produced. A major land use problem for this region is to bring these forested lands into larger and more profitable production, both of wood fiber for different uses, and of recreation, wildlife, and other values.

Southeast

Varied land use adjustments are probable or possible in this region. On the one hand, there are still major readjustments to be made between agriculture and forestry; while some of the poorer farm land has been taken out of this use in recent decades and transformed to forestry, often on a rather intensive basis, not all such adjustments have yet been made. Some of the most productive forest lands in the United States are in the South, and these have been or are rapidly being brought into reasonably full production. At the same time, in the hilly areas and elsewhere, there are extensive acreages where production per acre is low, both in volume and in value. A major forestry problem in this region is a better differentiation of forest sites according to their productivity and their ability to absorb profitably inputs of labor and capital, and a more profitable utilization of some tree species that grow well here but that thus far have had only light use. The problem of small forest holdings is serious in this region also.

The South as a whole is the most deficient of all major regions in provision of recreation opportunity, for its own people and to attract tourists and vacationists; this is true in spite of the outstanding development of the tourist industry in Florida and elsewhere. Many of the less productive forest lands and less productive farm lands might be developed into satisfactory, and in some cases into outstanding, recreation areas.

There will be considerable expansion of urban areas in the Southeast—not in large continuous belts, as in the Northeast, but around present small cities and towns. Favorable climate may lead to residence of retired people in suburban areas over much of the South. The total area thus shifted from rural to urban use may be considerable, in part because the density of land use in cities is relatively low in this region. Opportunities for sound planning of urban expansion are as good here as anywhere.

The Southeast is one region where agriculture might expand materially, should we be wrong in our estimates of the need for agricultural land. Along the southeastern coast, in the coastal plain, there are 35 million acres of land in classes I, II, and III (Soil Conservation Service classification system), which, if drained and cleared, could be farmed indefinitely and productively. At present prices and with present demands for land, the clearing and draining of this land will proceed slowly within present farms and even more slowly where such programs must include larger areas than single farms. But if demand were active, these processes could be greatly accelerated.

Corn Belt and Adjacent Commercial Farming Areas

In this region, the major land use change will probably be a further intensification of agriculture. There are several million acres of good land now in farm woodlots in this region. Their productivity as forest is not high, in part because they are split up into thousands of small tracts; but their productivity for agriculture is good. Clearing and conversion of these lands seems probable and desirable. Urban expansion and new highway construction will take some farm land; these processes could often be guided to reduce the loss of good farm lands. This region will need some additional recreation areas, particularly for ready use by residents of the area. It is hard to see how this region can become one of major attraction for tourists and for vacationists.

South Central States

A diversity of natural conditions and of land use in this region will mean various kinds of land use changes. As far as the better commercial agricultural areas are concerned, the changes here will be much like those in the Corn Belt—intensification of agriculture; limited conversion of farm woodlots, when on good soils, to cropping; and modest expansions of urban and transportation uses of land. In this general region, however, relatively large areas of poor farming land are found, where erosion has been serious, farm incomes low, and rural living standards low. A major part of the excess labor supply in agriculture is in the region and in the Southeast. Many thousands of farmers need assistance to get out of agriculture and into more productive employment. Their land should be used to enlarge the remaining farm units, under much more extensive systems of farming. Major land use changes are desirable—combination of land into much larger farm units, shift to general farming with more emphasis upon livestock and upon pastures and forage crops, and some shift of land from farms to forest. The problems of forestry already noted for the Southeast—better identification of differences in productive capacity of different sites, and programs better adjusted to each—exist here also, and will require similar measures.

There will be modest expansion of urban areas in this region. As in the Southeast, this will mostly take place in comparatively small and separate areas, generally around present towns and cities; and similarly it may require a good bit of land, partly because the urban use of land may not be intensive. Excellent opportunities for expanding recreation areas exist in the South Central States, both for local residents for whom facilities are now often grossly inadequate, and as a means of attracting tourists and vacationists. The development of recreation in the Tennessee Valley Authority area illustrates what can be done. The best quality areas, those most likely to attract outside vacationers, are in the mountainous areas; but acceptable areas could be developed elsewhere, especially through the provision of reservoir areas.

Great Lakes Region

In this region agriculture is likely to continue its retreat from the poorer soils and less adaptable areas, while at the same time perhaps becoming more intensive in the better dairying and other farming areas. This region now has a major recreation or vacation industry; its importance in this regard is likely to grow in the future. With a large and growing urban population along its southern border and with the possibility of an unsatisfied demand for recreation in the Northeast, this region has a large potential clientele. The extent of the actual development of this land use will depend in part upon how wisely it is done—on how well it is possible to satisfy the wants of many more people, when one of those wants is for a degree of privacy.

This was once a great forested region. Cut and burned severely, its productivity was gravely impaired, partly because tree associations and other ecological patterns were destroyed. A substantial comeback has been made but, on the whole, these lands are still producing far below their potential. A land ownership pattern which includes thousands of small forest owners, whose chief interest is something other than growth of wood fiber, is a major factor also. This is one of the regions where there are serious problems in obtaining a more productive use of forest lands.

Great Plains

In this region, all land use changes are minor compared to the really major one—that between agriculture, as we have defined it, and grazing. A great deal of this area is marginal for agriculture. The latter may be highly profitable in a wet year, but marginal in an average one, and submarginal in dry years and over a long period. But the temptation to gamble on favorable moisture and on favorable prices is very great; the problem is complicated by the fact that there has been a rather high turnover of farm owners and operators in this region, so that experience gained from adverse conditions in one period has often been lost by the time the next adverse period comes. This area has been the recipient of public assistance programs almost constantly since the days of World War I—feed and seed loans then and through much of the 1920's; emergency relief assistance, drought loans, drought purchase of livestock, various public works programs, subsidized farm loan programs, and others, during the 1930's; and some measure of these in the 1950's, when drought struck again. Price support on wheat in recent years has also been a major subsidy on these Plains. These programs have softened the blows of unfavorable weather and severe depression; but, in the main, they have also operated against a major and permanent land use adjustment program.

The major land use change needed in this region is to get several million acres out of cropping, to which they are not suited on a long-term basis, and into grass, and *keep them there.* Some of this adjustment was made during the 1930's, only to be lost and worsened when a heavy demand for wheat was created by World War II and its aftermath. Partly because of the federal support of wheat prices, land values are so high that the shift to grass must be subsidized. Large federal subsidies have been poured into this region, some to aid in just this shift in land use; but there is nothing to prevent the opposite shift when rainfall or price conditions offer a gambling chance of a large profit. In addition to changes in land use, considerable changes in farm size and type are needed—larger farms, with more livestock.

Intermountain Areas

This is a large and physically diverse region, and its land use problems are also diverse. Perhaps first among the changes will be a great rise in the recreational use of land—not so much more land used for this purpose, as greater use of land now available for recreation. A considerable part of the nation's publicly owned resource-based types of recreation are found in this region—several great national parks and many national forests. Additional areas of land within them will be required for recreation, thus displacing in some degree other uses from these areas. Much of the increase in recreational use of land will be by people resident elsewhere; in this respect, the Intermountain country will benefit by an interregional flow of income much as countries such as Switzerland benefit from an international flow of income.

A considerable part of the higher mountain areas of this region is forested, yet thus far commercial forestry has lagged. Some of the forests have timber of only moderate quality, and growth rates are generally rather low. A more important factor, however, has been restricted markets for the types of forest products available. Paper mills, increased use of the less desirable species and grades for lumber, and other developments will result in a greater use of the forest resources of this region. Public ownership of the generally low-productivity forests is likely to effect a greater output than if they were in private ownership.

No really major changes in agriculture appear imminent here, although some of the general increase in intensity of operations and output will occur which we expect for agriculture generally. For the most part, the replacement of agriculture by urban land use will be confined to the areas adjacent to present cities and towns, and will be comparatively small in total land area. Grazing will continue to be the major use of most of this large region. On the grazing lands with the highest (but still low) rainfall, improvements such as reseeding can lead to a greater productivity and more intensive use; but on much of the desert ranges large increases in output seem improbable.

Water shortages have limited economic development of this large region, and they will continue to do so. A full consideration of water management problems is beyond the scope of the present study. It is sufficient here to note that purposeful management of watershed lands to the end of affecting the volume and seasonality of stream flow may become a major land use management objective in Intermountain areas.

Pacific Coast

This is the region of fastest population growth. Most of the growth has taken place in cities, and this trend seems likely to continue. Figure 19 shows the nature of this growth. All the problems of urban expansion into rural areas, met in other regions, will arise here; and there is the added difficulty that some of this expansion is on to lands uniquely suited to certain specialty farm crops. Although there has been some talk about trying to limit urban expansion to lands not well-suited to cropping, this seems unrealistic. However, a conscious direction of urban growth, a timing of expansion in different locations, or an orderliness of growth to reduce the disconnected expansion into crop areas together with isolation of crop areas among urban developments, measures to reduce the waste of land idled by urban expansion, and control over the type and character of the urban development itself—these and similar approaches all offer promise for building better urban communities.

The use of land for recreation has risen in this region, and will continue to do so, partly as a result of the urban population growth within the region and partly from tourist travel. Considerable expanses, especially mountain and seashore areas, will be needed for this; and that will displace some other uses of these lands. In perhaps no other region is the co-ordination of user-oriented, resource-based, and intermediate recreation as important as on the Pacific Coast.

Most of the remaining old-growth timber of the nation is here, especially in Oregon, Washington, and northern California; a considerable part but by no means all of it is in federal ownership. Orderly cutting of the remaining old growth, prompt restocking of the cut areas, good management of the young growing areas in order to obtain near maximum rates of growth and consequent earlier harvest, and correlation of commercial forestry with recreation are some of the major forestry problems that arise. It is perhaps significant that in this region intensive forestry is rapidly being developed, perhaps second only to the southern pine for pulpwood. Although the rotations are longer in the Pacific Northwest, the annual growth rates, especially in Douglas fir on the better sites, are high.

Agriculture will experience some relocation from areas required for urban expansion on to areas now farmed less intensively, or hardly at all. This will be limited and directed by the extent of irrigation development, primarily federal, and undoubtedly operations will continue to become more intensive on the lands farmed.

POLICY IMPLICATIONS

Our purpose has been to assemble facts, to discover relationships among the facts, to quantify common knowledge, and otherwise to instruct and inform. Formulation and advocacy of particular policies concerning land have not been a major objective of this study. Yet, our results necessarily have implications for land policy. "Facts are stubborn things." Implicit in any lot of facts are certain future changes in human behavior. One function of the researcher is to make them explicit. The men who have studied a problem closely certainly have ideas as to the significance of their research. They may not fully understand the ultimate meaning of their own work; and they may not have considered all the factors that will be operative. We do not advance a national policy of land use, nor make recommendations; but we do offer some comments upon the policy implications of our findings. Let us examine each of the major land uses in turn.

Urban use of land will expand greatly in the future, under any reasonable estimates of the future society and economy. Large investments will be made in and on it, and the lives and welfare of many persons will be affected directly by the efficiency with which cities use land. This makes the gathering of facts about urban land use, their study, careful planning, and sound action all highly important. Most students and observers of modern urban development understand this as well or better than we do. We venture to suggest, however, that too much attention, at least in the more popular utterances, has been given to the spreading suburbs. The blossoming of row upon row of new houses in subdivision after subdivision is indeed impressive, not to say overwhelming. It deserves comment, critical and otherwise. And yet, from a land use viewpoint, at least three other situations are important and to some extent neglected.

First of all, there are extensive suburban areas that have been passed over, at least for the moment, in the spread of cities. How large are such areas? What are their physical characteristics? Why has subdivision not yet proceeded on them? How might they best be used in a comprehensive land use program for the whole metropolitan area? As we have suggested, these are some of the possibilities—perhaps the only ones—for user-oriented recreation for the new suburbs. How do our taxation and other public policies affect present use of these lands, the timing of their development for urban purposes, and the nature of that ultimate urban use?

Second, the central parts of the presently established cities have had some attention, through urban renewal and various efforts to prevent urban blight—yet we would argue that they have been relatively neglected. Far more study, and far greater efforts to promote wise land use and to prevent unnecessary destruction of capital through needless depreciation, would seem warranted.

Third, the land use problems of the smaller cities and towns have had almost no planning or policy attention, yet it is these smaller urban places that take up most of the land used or withdrawn by the cities. It we wish really to minimize the impact of the city upon the countryside, let us start with the most lavish land users among the cities.

The United States is indeed an urban nation, and is becoming more urban with each year. The efficiency of urban land use should be a matter of national concern. Yet, as a nation, we have most imperfectly seen all the consequences of our urbanization. Do we need at the federal level a "Department of Urbia?"

Recreation as a major land use is growing fast—bursting its seams like an adolescent. The present major recreation areas will in many instances be destroyed if trends toward their greater use continue indefinitely—destroyed in the sense of the satisfactions obtainable from their use, if not in a physical sense. Either we greatly expand the acreage of our outdoor recreation areas, or we change their character greatly. All kinds of outdoor recreation areas are parts of a comprehensive system, and they must be so viewed and planned for. At present there are many agencies at every level of government, and many private groups, all interested in recreation, but usually only in some specialized aspect of it. The recently established National Outdoor Recreation Resources Review Commission is the first major attempt at comprehensive planning. It has immense opportunities and responsibilities.

Planning for recreation must be in terms of the physical requirements and possibilities, the economic values and costs, the governmental interrelationships, and the financial needs and possibilities. Perhaps in no other field of governmental activity are the financial rewards of foresight greater than in acquisition of land for recreation well before the need becomes acute.

In agriculture, it now appears that the agricultural surplus is almost a permanent feature of the American economic scene. The demand for agricultural products is highly inflexible, though growing slowly; the supply of these products is also growing steadily and apparently equally inflexibly although, largely as a result of weather conditions, year-to-year variations in output do occur. If prices were allowed to fluctuate freely under these conditions, they would be highly variable. The major national problem in agriculture is to withdraw the surplus manpower, both as an aid to the national economy and as a necessary step in the improvement of living conditions of the people involved. To meet its future demands with the minimum allocation of resources, agriculture needs large-scale reorganization. The problem of the persistent surplus must be coped with. The futility of simple attempts to reduce acreage, while leaving other factors uncontrolled, has been demonstrated beyond dispute. A bolder use of land adjustments, not only as to whole farms but as to larger areas, and not only within the framework of present ownership patterns, might be more successful. Certainly, major land use adjustments are called for in the Plains, in order to reduce the production of wheat to sensible levels and permanently convert some land to grazing. In the hillier areas of the South and elsewhere there is a like need to take land out of low-grade agriculture, and put it into trees or grass.

In forestry, the small private owner is *the* problem—a problem too serious to be ignored, but with a very low output in relation to the potential of the land. In our view, there are grave doubts if much of this land can produce timber products economically in competition for labor, capital, management, and other inputs. Its economic productivity under any size of unit and any system of management is in doubt; but certainly the very small unit is severely handicapped from the start. The nonmonetary ends of forest land ownership are probably dominant for these owners. We need to know far more about them. In particular, how can the objectives of the small forest land owner be met, and at the same time a moderately large output be obtained from his land? We suspect this is not merely a matter of "education": a new and more fundamental approach seems plainly needed.

In grazing, the chief policy issue is how to attain better conservation in use of the land, and at the same time increase output. Here, the problem seems largely one of research and education; great strides have been made, and more are possible.

In transportation as a use of land, the chief policy problems relate to the expansion of the major and super-highways. One problem is to locate them efficiently, yet disturb present land uses to the minimum. A larger basic problem is how to use the highway as a public means of influencing private land use in socially desirable directions. The major highway does greatly affect land use for other purposes. Social scientists have commonly criticized highway engineers for their neglect of the secondary consequences of their road building; yet, we must ask, how far has anyone gone in seeking to study and analyze this problem? We need to do more than assert the basic fact. Precisely how does the highway affect land use, and how may the good effects be strengthened and the less desirable ones inhibited?

For various reasons, the area used for reservoirs and other water control works is almost sure to increase. The economics of such developments raise major problems, beyond the scope of this study. In large part, the difficulties arise because some groups bear the costs and others get the benefits—the disassociation of costs and benefits, in the economist's words. In part a major factor is the set of traditional pricing policies whereby some uses of water are free, others are subsidized, and others repay most or all their costs. Our concern, as far as land use is concerned, is to get the maximum benefits from the land used for water regulation. In practice, this often means adequate consideration to recreation as a benefit, in comparison with other benefits.

The foregoing suggestions, relating some of the major implications of our study to probable or desirable changes in national land policy, are offered with no sense of finality. Even if it is granted that the factual analysis on which we base our views is complete and accurate, the basis of all judgment in such matters must in some part be conjectural, subject to personal predeterminants or prejudices, and to that extent imprecise. Who, in 1918, could have foretold with any certainty the swerves and changes, the interplay of circumstances and consequences, which brought about the succession of major alterations in our national land policies through the 1920's, 1930's, 1940's, and now? The target date for this study centers on the year 2000, roughly as far in the future as 1918 is in the past. And the whirl and pace of continuing forces of change impelling groundline alterations is not diminishing, to say the least.

SOME FACTS WE LACK

No research of any consequence comes up with final values. Its greatest value is to open ways toward further findings and widen spheres of knowledge for greater use. All may agree that better information is needed on the shift of land from rural to urban uses. Unquestionably, this is the one land use change about which we can be most confident as to direction and degree; and, relatively, it is one of the largest. It is also the land use situation about which we know least.

The first step is to ascertain many more facts about the current situation, at each future census period, than are now at hand. Before information can be collected, it is necessary to define rather exactly what we want, and for what areas. One set of information should pertain to the area in each urban land use; the definitions or classes that Bartholomew[2] sets up might be adequate, or some modification of them might be agreed upon by those particularly interested in this field. In addition, information should be obtained upon withdrawn but unused land. We have suggested that this be divided into vacant lots, engulfed areas, and adjacent but idled areas. Perhaps some better classification could be devised; and, at any rate, precise definitions of broad classes would be necessary. There would also be important decisions as to the areas for which data were to be enumerated. Certainly, data should be available for each incorporated city, and for census tracts within the larger ones.

In addition, suburbs beyond the city limits should be included. The use of the urbanized area concept would not be sufficient, because it excludes the larger idle withdrawn areas. But it would be very difficult to devise an exact definition of the latter. All data should be tabulated and published for cities according to the size and economic type of the city, and for metropolitan areas as a whole. One major problem would be: from whom to get the desired information? For residential, industrial, and other property with an owner or his agent resident, there would be no more problem than for farms. And for the public use areas, such as streets and parks, public officials should be able to give the information. But for the idle property, it might be difficult to locate the person who would have accurate information.

Once the best possible data are in hand, research on urban land use could proceed better than at present. All of the matters discussed in

[2] Harland Bartholomew, *Land Uses in American Cities* (Cambridge: Harvard University Press, 1955).

this book, and many more, need careful study, to ascertain how urban land is used. Particular attention should also focus on the idle withdrawn lands.

The effect of the city upon other land uses in nearby areas needs study also. How can the blighting effect of urban expansion on other land uses, on land not yet actually taken over by the city, be minimized? Research on the process of urban expansion, and on possibilities for direction and control of such processes, is urgently needed—on the physical, economic, political, and institutional problems involved. One major aspect of this problem is provision of recreation areas for the expanding urban populations. The problems of urban expansion do not completely fall within city planning as it is usually defined; certainly, they do not fall within rural land economics, although they are of concern here; but perhaps a broader approach to this general problem is called for. The potential gains to society as a whole from the development of good cities and residential areas are very great, and would seem to justify far more effort in fact-finding and research than thus far has been directed toward them.

We have emphasized the greatly increased potential demand for recreational use of land. While the extent of this change is less certain than for urban expansion, yet comparatively large areas of land are likely to be involved and the effect upon the welfare of many people will be large. There is a need for more complete information on present use of land for recreation—primarily upon who uses it. It would be most helpful if we knew more about recreation land users—their age, their family make-up, their income class, their residence location, their use of other areas for outdoor recreation, the frequency of their use of different kinds of areas, their preferences for different kinds of outdoor recreation, the satisfactions they get from such recreation, and similar items.

Information is also needed about the intensity of land use within established recreation areas; we have suggested a scale of use intensities, which perhaps could be improved. But we need to know more about how different areas are used, how much buffer zone is needed, how other land uses such as forestry might be carried on in the buffer zones, and the like. We have made some rough over-all estimates of future demand for outdoor recreation; but greatly refined methods for estimating such demand, especially on a local and regional basis, are needed. We have also suggested some very crude standards of area requirements, based on ideas of the capacity of different kinds of areas to supply different kinds of outdoor recreation. Our ideas are largely taken from the practical experience of others. But this aspect of outdoor recreation needs much more careful study.

The behavior of people under different kinds or layouts of outdoor areas might be studied, and some actual experiments in layout and design initiated and evaluated. If recreation demand increases at all as we have estimated, the intensity of use of recreation areas must increase greatly, even if our projected area expansions are achieved. This poses a threat to the quality of the recreation experience; in attempting to meet the demand, we may destroy the usefulness of the resource. But it is also true that outdoor recreation has generally been developed to meet peak demands, with the result that the resources and facilities are unused or only partly used a substantial part of the time. The whole supply and demand aspects of outdoor recreation need a far firmer and better factual and technical foundation than they now have, if public funds are to be wisely used in provision of outdoor recreation and if the potential demands are to be met from resources likely to be available.

Agricultural land use has perhaps had the most intensive and long-continued study of any major land use. On many aspects, further detail and greater intensity of study are needed, and we may expect this to be provided in the decades ahead. According to our analysis, the critical factor in agricultural use of land over the next several decades will be the rate of technological development and adoption by farmers. While we have a fair idea of the results of technological change, we have a limited understanding of the processes by which new technologies are adopted and made practical to the farmer. Studies of the processes of technological innovation and change may well have more significance for agricultural land use than any studies of land use as such.

In forestry, several problems call for more study than they have had in the past. Estimation of the demand for forest products has begun and has improved in the last decade or so, but it still lags far behind similar studies for agricultural products or for industrial products. If we could be a little surer of future demand for forest products, we could begin programs affecting future supply with more confidence.

Information on the most profitable level of intensity—of inputs of labor, machinery, and capital—is needed for different forest species and types, and for sites of different productivity within each. If, as we have suggested, there is to be a major "sorting-out" of forest land according to productivity and capacity to absorb inputs profitably, errors in the trial and error process might be materially reduced by research. For reasons previously indicated but deserving of emphasis anew, the greatest need of all is for research on the economics of small private forestry. About a fourth of all commercial forest land is in holdings of less than 100 acres, and much more is in larger but still comparatively small holdings. Few of these small holdings have been economically productive. There seems little chance of increasing their output unless drastically different, more rigorous, more imaginative research, on a larger scale, is devised and applied.

One general type of problem affecting land use for multiple purposes needs research in the next several years. This is a careful examination of the physical requirements of different land uses, and a judgment as to the economic importance of different natural characteristics of land for different land uses. For instance, it has sometimes been suggested that urban expansion be directed toward the poorer farm land. Impediments to this arise from the fact that residence construction costs are lower on level or gently sloping land than on steep land, that the home owner wants good soil for his lawn and garden, and that industrial plants seek favorable topography. A study of the relative importance of each factor for different uses should be helpful in arriving at wise decisions. How far can relatively unproductive forest sites serve for recreational areas, for instance, and what, if any, deficiencies for recreation do such sites have? How far is it practical to route highways away from the best farm lands? Many factors are involved in decisions on such points, yet often there is a serious lack of information as to the physical advantages and costs involved, which research should be able to supply.

INSTITUTIONAL PROBLEMS

Economic forces are powerful in their effect upon land use, but they are operated within a framework of institutions and are modified by them. For many private decisions as to land use, such as to shift a tract of land from agricultural to urban use, economic forces may be dominant. Even in this case, however, the profit possibilities may be modified by laws or regulations or other actions governing the use that may be made of the tract if shifted to urban use. For many public decisions as to land use, such as the shift of a tract of land to recreation or transportation use, economic considerations may be influential, though not perhaps dominant; the cost of the change is likely to be an important factor. The institutional framework sets some limits within which the economic forces may operate, and may at times largely overshadow the economics of the situation. A brief discussion of the institutional forces and factors influencing land use may be helpful in suggesting the forces affecting land use decisions and the range within which strictly economic forces may operate.

One major consideration is the dominance of the locational aspect of land use, and the consequent narrowness of the market for land. Our analysis has been primarily in terms of national totals and broad national forces, although we have given some consideration to regions and locational factors. But specific uses of land and changes in land use are always strictly local and locational in operation. If a farmer wants to increase the size of his farm, there are narrow limits to the distance within which the additional land must lie in relation to his present farm. He may sell his present farm and move elsewhere, but this is a major personal as well as a major economic decision. If a rancher wants additional spring-fall range, there are rather narrow locational restrictions within which it must fall, if it is to be usable from his present ranch headquarters. If a town or community wants a new park, its location is rather narrowly limited, if it is to be usable by the persons for whom it is intended. On the seller's side, too, the market is often limited; he may have the greatest difficulty in interesting any potential buyer.

The market for land is generally not a competitive one, in the classical economists' sense of the term. There are not many buyers, but only a few or one; there are not many sellers, but only a few or one; and the different tracts of land are not a single homogeneous product, but differ in characteristics from one to another. The time factor in land use changes is sometimes highly important. If one must sell a piece of land quickly, often he must take a material reduction in price; but if one is forced to buy at once, he may in turn pay a high price. To find the traditional willing seller and willing buyer takes time. In sum, changes in land use are often slow, hard, and costly, and the final adjustment is often an imperfect one.

Many other problems and limitations enter into private decision making on land use. In general, we may assume that the profit motive dominates private decision making, although sometimes an owner will prefer some course of action not promising the maximum return. An individual, whether a landowner, a tenant, or someone seeking to be either, is often severely limited in his knowledge of land—its productivity, its cost, its income possibilities, and the like. Even when his knowledge is tolerably good, he may lack capital or managerial ability or otherwise be incapable of carrying out land management programs that would be sound from a private profit viewpoint. The typical small forest landowner, for instance, often knows very little about the production capabilities of his forest; he lacks capital to institute a forest management program even if this is economically sound; and he can neither manage the property himself nor find anyone else interested in doing so.

For many land uses, the private owner cannot capture all the benefits from his land use management. For instance, the forest landowner may provide some recreation, or some watershed management, the benefits from which may be considerable, but not capturable by him; or a farmer may provide a habitat for game birds or animals, providing a recreation value to someone but not a capturable income. The private landowner may be primarily interested in one type of return from land, largely ignoring others. We have mentioned how this may be true for forestry and grazing.

The individual landowner may have only a short-run interest in management of his land. Slow growing forests, for instance, may yield a larger income if cut and abandoned than if managed for sustained yield; if the individual's dominant interest is maximum income, he can hardly be expected to practice sustained yield forestry under these conditions. In some instances, laws or governmental regulations may prevent the individual from doing what would be most profitable for him, in their absence. Zoning limits land use, sometimes to the disadvantage of the individual landowner.

For all of the above reasons, the individual landowner may often not make that perfect adjustment of use of his land to economic forces which a purely economic analysis would indicate. If a particular course of action promises to be highly profitable, compared to any other, then it is likely to be done. But more often many inhibiting factors prevent a perfect adjustment. While this is true of economic activities of all kinds, it is possibly more general with land because of the intensely localized nature of land.

These are by no means all the problems attending public decisions as to land use. Many types of use—recreation, transportation, and others—require public decisions. Although such decisions often greatly affect the welfare of the average citizen, it is hard to gain his interest in the proposal. Even when such matters of public decision as the location of a new expressway through a city are hotly debated publicly, a large proportion of the general public remains indifferent to the whole issue. When a problem or situation is somewhat complex, the chance of getting the general public to learn the facts and take an active interest is even less. Most people rest content with what they can learn from a single newspaper account, which in turn was simply written because they would not read anything more complex. And when it comes to a long-run viewpoint on land use, most citizens, on this as on other long-time considerations, refuse to bother. It is, for example, almost always cheaper in the long run to buy land needed for city parks well before it is urgently needed for this purpose; this is true even when interest on the money so invested is included; but it is generally difficult to get enough popular interest in park establishment until all the potential sites have been pre-empted for other uses. By that time, the costs are enormously higher.

All of these matters of public decision making are difficult enough, at best. Organized interest groups of one kind or another help to make decisions and also hinder them. They help, because they arouse the interest of their members, and bring that interest effectively to bear; they hinder, because sometimes their tactic is obfuscation and confusion, on issues that are already unclear to the average citizen. In a democratic form of government, the citizen, through his vote, has the last word on public action, including action relating to land use; we would not have it otherwise. Even so, sound decisions and wise action are hard to come by, and slow.

In saying this, the authors seek simply to raise a small flag of warning. We realize that land use adjustments are affected by many forces, not always quickly and easily responsive to economic motivation alone. Implicitly, throughout this book, we have tried to include in our calculations the obstacles, as well as the incentives, to land use adjustment. The economic forces leading to specific land use changes are somewhat general but often persistent; the obstacles are more localized and immediate.

The general public, acting through government at all levels, can modify the institutional framework so as to facilitate land use changes when they are economically sound. One way is through fact gathering, research, and planning. If people can learn more about particular areas of land rational action will be facilitated. Another major approach is through public actions themselves—the building of roads, water supply systems, sewage lines, the provision of public credit for private lenders, and others. Over the decades, certain fields of activity have come to be accepted for public action, to supplement private economic action. Public action or inaction will greatly affect private land use. Much as we may decry the kind of suburban land development of the past decade or more, we should recognize that to a large extent it is the logical consequence of our public actions affecting land. A third general approach is modification of the laws and other publicly made institutional framework affecting private land use. Some of this is direct, or mandatory; some of it permissive, or influential.

If the demand for various land uses rises in the future as we estimate, and the competition between land uses thus becomes stronger, the public role in aiding, modifying, or controlling private land use undoubtedly will become greater.

BEYOND 2000

The target date for our analysis has been the year 2000. Looking forward, we find no cause for immediate alarm, but a closer reckoning of resource expenditure must replace heedless overconfidence. This nation can meet its own needs for products and services of land adequately, even handsomely; but to do so will require planning, action, and investment. Moreover, some compromises may be needed. We cannot meet every wish of everyone—we never have, and we never can.

But the life of the nation and of the people does not end with the year 2000. Forty years into the future is perhaps as far ahead as rational economic planning can go; costs and values become extremely difficult to estimate for longer time spans. Yet responsible and thoughtful people will surely give some thought to the world of their children and their children's children.

At the rate of population increase that has prevailed since 1950, by about the year 2100 the people living in the United States would number 1.9 billion, or one person per acre of total land surface—mountains, deserts, forests, plains, and all. By about 2500 or 2600, we would number 3,000 *billion,* or one person per square yard of total surface. As human history goes, periods up to five, six, or seven centuries are short; and as planetary history goes, they are mere seconds. Impossible, you say, that any such increases will occur; surely something will intervene to stop them. Maybe so, maybe not. Who, five hundred years ago, could have foreseen today?

The maximum capacity of this country to feed itself is so far beyond the present need, that in this study it seemed to us pointless to attempt a precise calculation. Even under the limitations of present technology and current culture, it is plainly possible to step up food output enormously, especially over a period as long as a century. By shifting diets toward cereals and other vegetable products, we could still further increase our ability to feed hordes of people. Moreover, as the need increased, there is good reason to think that many means could be found to increase output still further.

We could *feed* 1, 2, 3, or more *billions* of people in some future century. But is mere food for bare existence all we want from life? If population should ever reach any such crowding as we have suggested, most other uses of land would vanish, save for the minimum areas needed merely to stand or lie, and for travel. With one person per acre for all land, there would not be much privacy! And with one per square yard, we would be stacked several layers deep, and every ray of the sun would be cherished.

John Stuart Mill has put it very well:

> There is room in the world, no doubt, and even in old countries, for a great increase of population, supposing the arts of life to go on improving, and capital to increase. But even if innocuous, I confess I see very little reason for desiring it. . . . A population may be too crowded, though all be amply supplied with food and raiment. It is not good for man to be kept perforce at all times in the presence of his species. A world from which solitude is extirpated, is a very poor ideal. Solitude, in the sense of being often alone, is essential to any depth of meditation or of character; and solitude in the presence of natural beauty and grandeur, is the cradle of thoughts and aspirations which are not only good for the individual, but which society could ill do without. Nor is there much satisfaction in contemplating the world with nothing left to the spontaneous activity of nature; with every rood of land brought into cultivation, which is capable of growing food for human beings; every flowery

> waste or natural pasture ploughed up, all quadrupeds or birds which are not domesticated for man's use exterminated as his rivals for food, every hedgerow or superfluous tree rooted out, and scarcely a place left where a wild shrub or flower could grow without being eradicated as a weed in the name of improved agriculture.[3]

All of this is to say, we cannot go on forever as we are going now. No crisis appears probable in the lifetime of anyone now in their middle years, nor perhaps in the lifetime of our young. Possibly some forces now completely unforeseen will enter our national life, to make as absurd these fears as time until now has made of Malthus' fears. But surely we cannot now imagine what force it would be. The present population-resource balance in the United States is neither equilibrium nor is it rational dynamic; the trends which grow out of our very culture carry the seeds of their own destruction.

Obviously, we may, and probably shall, materially modify our rates of reproduction, before we come to any such state as suggested. Yet this, too, will present its major problems, social, cultural, economic. But all of this is, as far as we can see, beyond 2000.

A SUMMING UP

At the risk of oversimplification, let us conclude with a few very summary statements:

1. Large shifts in land use, from one major use to another, are unlikely in the future, at least up to 2000. As a nation, we have "matured," as far as land use is concerned, and there will not again be large and rapid shifts in major land uses such as occurred before 1910.

2. But some changes in major uses of land will take place. They will tend to be localized, but there the change may be large or complete—for example, when a city grows and absorbs farm land.

3. Changes in major land use in the future will be made with more difficulty and will be accompanied with more stresses and strains, public and private, than past shifts in land use. As uses have become more firmly entrenched on a given tract, they can be displaced only

[3] J. S. Mill, *Principles of Political Economy* (London: Longmans, Green, Reader, and Dyer, 1871), Book IV, Ch. 6, p. 454.

with more difficulty. This is but a corollary to the idea of maturity in the general land use pattern.

4. Change *within* each land use is likely to be more important than change *between* land uses. The area of land used for agriculture will change comparatively little, but the intensity of its use will change greatly. The same is true, in general, for all land uses. The intensive margin of use will be more important, comparatively, than the extensive margin.

Appendices

Appendix A

Review of population projections

Population projection is basic to any economic forecast. Estimates of future numbers of people, classified according to various age groups, households, states or regions, and other categories are essential to estimating future labor force, may seriously affect estimates of total output, and certainly are a major component of estimates of consumption. Almost every economic analyst who tries to make quantitative estimates for the future starts with population projections.

For reasons stated early in this book, we have used as our own basis for projection rough estimates of total numbers and regional distribution of people in the future. But this decision was made only after we had examined current methods of population projection and had reviewed many past projections and estimates. This appendix is presented in the belief that our review may be of interest to those readers who wish to probe more deeply into demographic methods.

Modern demographic methods for projecting the total population to some future date in essence are comparatively simple.[1] One starts with data on the numbers of persons now in the population, by age groups (usually five-year age groups) and by sex. For some purposes, data on numbers by race are also desirable. Most census counts somewhat underenumerate the total population; a correction for under-

[1] A brief and relatively nontechnical discussion of this process, and its illustration in a specific projection is found in *Illustrative United States Population Projections*, Actuarial Study No. 46, Division of the Actuary, Social Security Administration, U. S. Department of Health, Education, and Welfare, May 1957.

enumeration may be entered, if there is a reasonably satisfactory basis for doing so. The number of persons in each age and sex group can be projected into the future, to any distant date, by applying survival rates appropriate to each group. This assumes that data on survival are available for each age and sex group; this is true in the United States today, but not in all countries. However, the real issue is whether to use present mortality and survival rates or some estimate of future rates. Mortality rates have fallen greatly in the United States over past decades—more for some age classes, some racial groups, and some economic groups, than for others. Further decline in mortality rates seems more probable than that the present rates will continue. But how large a decline is most probable? This is the first major forecasting point, if the projection is to be a forecast.

To the survivors from the original population must be added the survivors at the end of each five-year period of the net immigrants during the five years. The number of immigrants, their age, sex, and family composition, have changed in past decades. Before World War I, there was a net immigration of 3 million in a five-year period; in recent decades, it has been slightly in excess of 1 million per five-year period. While immigration is not a major population factor, it does enter into the calculations, and in the past has been rather generally underestimated. The second major forecasting point is thus net immigration.

The number of births in each five-year period must be estimated by applying age-specific birth rates to the number of females of child-bearing ages at the midpoint of each period, and multiplying the result by five to get the total number of births for the five-year period. The number of births is divided according to sex—one of the most stable of all population ratios—and the number projected into the future in a manner similar to the projections of the original population. One kind of birth rate can be obtained by dividing total births by total numbers of women of child-bearing ages. Variations in such a birth rate may occur, without any changes in age-specific rates, if the age distribution of child-bearing women changes; this is why age-specific rates should be used. Births are commonly underreported, and corrections can be made if an adequate factual basis is available. Underreporting of total population, and hence of women in child-bearing ages, and underreporting of births, tend to offset one another, at least to a degree. Crude birth rates (number of births divided by total population) for the whole of the United States fell from about 25 per 1000 persons in 1915, when complete birth registration data were first available, to about 16 in the early 1930's. Data for some states where birth registrations were available for earlier years indicate that the decline in crude birth rate began much earlier and from a much higher level. Age-specific birth rates are available from 1920; they declined until the early 1930's for all age groups except women under 20 years. Beginning in the latter 1930's, birth rates began to rise slowly; they rose faster during the war, and still faster in the postwar years. Since about 1951 they have been more or less steady, at about 24 to 25 crude birth rate. The age-specific rates rose for every age group of women except those over 40 years. The rise was most marked for women between 20 and 30 years of age.

A substantial part of the rise in birth rates in the past two decades was due to the earlier marriage of women. During the depression of the 1930's a substantial backlog of postponed marriages had been built up; during and since the war, as these people married and had children, they added to the birth rate. The average age of marriage and the average age of birth of the first child have each fallen considerably in these years. Do these children born at the mother's younger age merely replace children who would otherwise be born later in her life; that is, is the number of children born per average woman more or less constant, regardless of the age at which she has her first child, or is it a larger total number, on the average?[2] Only time will tell, for sure.

In the past and still for the most part today, demographers have been firmly convinced that there is a basic long-term downward trend in birth rates. This conviction has dominated their thinking and their estimates, sometimes to the point where it has seemed to overbalance evidence at hand. It is true that there is much evidence in the world that such a long-term downward trend in birth rates exists. Many demographers have gone so far as to assert that no major reversal of this trend will be encountered by any nation; but we have clearly had one in the United States in the past twenty years. Has it been relatively temporary, or is it permanent? What will be the trend in birth rates over the next half century or century?

Thus, the third and most important forecasting point in population projections is birth rate. Assumptions as to future birth rates are the dominant factor in projections as to total population, since the range of reasonable variations in this factor are much greater than the range

[2] Pascal K. Whelpton, *Cohort Fertility, Native White Women in the United States* (Princeton: Princeton University Press, 1954).

of reasonable variation in the other factors that enter the projection formulas. However, the number of people in certain age groups and the number of households can be projected for some years into the future without assumption as to future birth rates, because such projections depend only on the number of persons now living.

Davis has expressed some doubts as to the accuracy of this whole process, as follows:

"Population forecasting is *not* a simple matter. Available techniques do *not* permit reliable prediction to be made for five, ten, twenty, or fifty years ahead. The best may be *far* wrong. Our net reproduction rate is *not* near unity, but has been well above it ever since 1940. It is *not* reliable as a basis for prediction. There is *no* assurance of *any* peak population, at *any* future date. The age structure of the population does *not* 'inherently' point to cessation of growth and eventual population decline. Our major population problem is *not* prevention of such decline. There is *no* adequate basis for expecting the fertility rate, or the crude birth rate, to drop to or below the level of the early 1930's and to remain at that low level. While the long-term trend of our population growth may still be downward, this does *not* necessarily support extrapolation of the curve from the mid-1930's. We do well to recall Raymond Pearl's observation that in this country the word 'extrapolation' is usually mispronounced with the stress on the syllable 'trap.' Let us be on guard against that subtle disease, 'trenditis,' and especially its more dangerous variant, 'short-term trenditis.' Finally, planning for food, agriculture, industry, schools, et cetera, can *not* be safely done on the basis of supposedly expert population forecasts."[3] (Emphasis in the original.)

Literature on population is littered with the wrecks of woefully inaccurate population projections and forecasts.[4] The distinction be-

[3] Joseph S. Davis, "Our Amazing Population Upsurge," *Journal of Farm Economics*, Vol. 31, No. 4 (November 1949) Part 2. Other publications on the same subject by this author include: *The Population Upsurge in the United States*, War-Peace Pamphlet 12, Food Research Institute, December 1949; "Our Changed Population Outlook and Its Significance," *American Economic Review*, Vol. 42 (June 1952).

[4] For a review of the earlier projections and forecasts, we have relied upon Lowell J. Reed, "Population Growth and Forecasts," *The Annals of the American Academy of Political and Social Science*, Vol. 188 (November 1936). For the period from roughly 1925 to 1949, we have relied heavily on Davis, *op. cit.* For the more recent period, we are indebted to our former colleague, Joseph Lerner, who has permitted us to read an unpublished manuscript.

tween projections and forecasts has not been maintained, in these or in any other terms. Yet the competence of many of the persons responsible is beyond question. Today our basic data and perhaps our methodology is better than in the past, but the hazards remain. Knowing how past projections have proved in error may help to guide us.

Reed, writing in 1936, points out that there are essentially two methods of population forecasting: (1) those "based on the idea that a population count is a statistical quantity having a certain orderliness with time, the forecasts being made by projecting the population into the future on the basis of its past orderliness"; and (2) those based upon demographic methods, using birth rates, death rates, immigration, etc. as we have described above.[5] The first method can in turn be divided in two: (1) informally graphic methods, in which the projector uses his judgment as to likely trends but without making explicit his assumptions or bases; and (2) mathematical, using various formulas. The formulas may be a simple straight line on arithmetic scale paper, or a straight line on logarithmic paper (thus with a constant or constantly changing percentage rate of growth), or it may be a more complex one, which Reed calls the logistic. This is an S-shaped curve, with a rapid and accelerating growth at the lower end, a rapidly rising part, then a slower rising part, and an ultimate ceiling.

The very earliest forecasts were necessarily based upon past trends, using roughly Reed's first method, and were mostly rather simple and graphic in application (Table A-1 and Figure A-1). There were no accurate, comprehensive demographic data on births and deaths by age groups, even had more sophisticated demographic methods been known. The formula used were usually simple. In 1815, with only three national population censuses to draw on, Watson made a forecast as far ahead as 1900; it was 33 per cent in error, but its average annual error was comparatively low. A series of forecasts was made by the Bureau of the Census after the 1850 census; the one favored by the Bureau was amazingly accurate, with an error of only 8 per cent in 1900, and of only 17 per cent in 1950. Other forecasts of varying accuracy, using similar simple methods, were made up until about 1910.

[5] Reed, *op. cit.* Reed does not describe the various methods in exactly these terms; he stresses the difference between the graphic and mathematical means of using the first method. But he does describe the latter also.

Table A–1. Summary of various projections of total population[1]

Name of projection	Year made	Total population (millions) projected for									Average annual error to last projected year[2]
		1900	1920	1930	1940	1950	1955	1960	1975	2000	
Enumerated population	—	76	106	123	132	151	—	—	—	—	—
1. Actuarial Study No. 46[3]	1957	—	—	—	—	—	[4]173	184-188	215-238	263-343	[5]
2. Actuarial Study No. 33[3]	1952	—	—	—	—	—	166	173-174	189-201	210-254	—
3. Actuarial Study No. 24	1946	—	—	—	—	—	147-157	148-164	147-191	124-241	−1.31 to −.49
4. Bureau of the Census	1955	—	—	—	—	—	[4]165	176-179	207-228	—	[5]
5. Bureau of the Census	1953	—	—	—	—	—	164-165	174-177	199-221	—	− .09 to +.21
6. Bureau of the Census	1950	—	—	—	—	—	158-167	162-181	—	—	− .77 to +.33
7. Bureau of the Census	1947	—	—	—	—	—	149-156	150-163	152-186	164	−1.16 to −.63
8. Hagood and Siegel	1952	—	—	—	—	—	162-167	167-181	—	—	− .47 to +.55
9. National Resources Planning Board	1943	—	—	—	—	—	146-151	148-157	148-174	129-199	− .93 to −.67
10. National Resources Committee	1937	—	—	—	—	—	138-154	138-160	132-180	—	− .89 to −.35
11. Committee on Economic Security	1934	—	—	—	—	—	144	146	150	151	− .59
12. Thompson and Whelpton	1933	—	—	—	—	—	—	—	[6]145-190	—	—
13. Whelpton	1931	—	—	—	—	—	—	—	[7]145	—	—
14. Dublin	1931	—	—	—	—	—	—	—	[8]148	[9]154	—
15. Pearl and Reed	1920	—	107	122	136	149	—	159	[10]171	185	[11]− .04
16. Woodruff	1909	—	—	115	—	—	—	—	—	—	− .30
17. Tucker	1850	74- 80	—	—	—	—	—	—	—	—	− .05 to +.11
18. Bureau of the Census	1850	70	—	—	—	125	—	—	—	—	[12]− .17
19. Bureau of the Census	1850	99	—	—	—	—	—	—	—	—	+ .60
20. Watson	1815	[13]101	—	—	—	—	—	—	—	—	+ .39

[1] The first 11 lines of this table are based on Table 14, Actuarial Study, No. 46. The actuarial studies listed were made by the Division of the Actuary, Social Security Administration, now in the Department of Health, Education, and Welfare, but previously an independent agency, and published in the years indicated. Those by the Bureau of the Census are described, respectively, as: *Current Population Reports,* Series P-25, No. 123, October 1955; *Current Population Reports,* Series P-25, No. 78, August 1953; *Current Population Reports,* Series P-25, No. 43, August 1950; *Population Special Reports,* Series P-46, No. 7, September 1946. Margaret Jarman Hagood and Jacob S. Siegel presented "Population Projections for Sales Forecasting," *Journal of the American Statistical Association,* September 1952. W. S. Thompson and P. K. Whelpton prepared projections which were published by the National Resources Planning Board in August 1943—*Estimates of Future Population of the United States, 1940-2000.* They had previously prepared the estimates published by the National Resources Committee in October 1937—*Population Statistics, National Data.* When the social security program was under consideration in 1934-35, the Committee on Economic Security prepared estimates of future population growth, published as *Issues in Social Security, A Report to the Committee on Ways and Means of the House of Representatives by the Committee's Social Security Technical Staff, January 1946.* Where several projections were made, the figures shown are those resulting in the widest range.

The remainder of the table is a summarization of the discussion in Lowell J. Reed, "Population Growth and Forecasts," *The Annals of the American Academy of Political and Social Science,* Vol. 188 (November 1936). The specific references cited there are: W. S. Thompson and P. K. Whelpton, "The Recent Population of the Nation," Chapter I of *Recent Social Trends in the United States* (New York: McGraw-Hill, 1933); P. K. Whelpton, "The Future Growth of the Population of the United States," *Problems of Population, Proceedings International Union for Scientific Investigation of Population Problems* (London: George Allen and Unwin, 1932); L. I. Dublin, "The Outlook for the American Birth Rate," *Problems of Population, Proceedings of International Union for Scientific Investigation of Population Problems* (London: George Allen and Unwin, 1932); R. Pearl and L. J. Reed, "On the Rate of Growth of the Population of the United States Since 1790 and Its Mathematical Representation," *Proceedings, National Academy of Science,* Volume 6, 1920, pp. 275-88; Charles E. Woodruff, *Expansion of Races* (New York: Rebman Co., 1909); the Tucker and Bureau of the Census projections made in 1850 are found in *Compendium of the United States Census for 1850*—they employ different methods; and Elkanah Watson, quoted in R. Hunter, *Poverty* (New York: Macmillan Co., 1917). Blanks in the table mean that no figures were available for these dates.

[2] The difference between the projected or forecasted population and the actual population is calculated in millions of people; then expressed in terms of percentage of the actual population; and this percentage is divided by the number of years between the time the forecast was made and the year of enumeration. It may be argued that this method gives an unfair advantage to forecasts for the near future, since population can hardly change greatly in 2 to 5 years, as compared with present population; on the other hand, division by the few number of years involved may mean a high average annual error even though the absolute error is not large.

[3] Includes Alaska, Hawaii, Puerto Rico, and the Virgin Islands.

[4] Current estimate of population.

[5] Inapplicable, because no forecast for a year yet experienced.

[6] 1980, not 1975.

[7] Modal forecast for 1970.

[8] 1970, not 1975; peak population reached at this level.

[9] 1990, not 2000; a different projection than the one for 1970, and believed by its author less probable.

[10] Arithmetic mean of estimates for 1970 and 1980.

[11] Largest average annual percentage error for any forecast in this series was for 1940 (+.16).

[12] The average annual error for the 1900 estimate was about the same as this.

[13] Calculated from statement by Reed that forecast exceeded actual number by 33%.

In 1920, using data available through the 1910 census, Pearl and Reed[6] first developed their logistic curve. These men were biologists and biostatisticians, and they reasoned from known growth data on populations other than man, such as insects, as well as from known data about human population growth in other countries. Their estimates for the census years up to 1950 were highly accurate; the greatest error was for 1940, when they were out by slightly more than 3 per cent. However, their estimate for 1950 was only slightly more than 1 per cent too high. On the basis of average annual error of estimate, their estimates are the most accurate of any, and cover more specific years. Their formula fitted well the period from 1790 to 1950, but it seems to have broken down since then. It would produce a ceiling of about 197 million, and an estimate of 185 million by 2000. However, if their formula were to be recomputed today on the basis of the longer historical record now available, it would yield different figures, and some which would be more in keeping with our present ideas of future population growth.

Beginning in the late 1920's and early 1930's, the demographic approach began to be applied. At that time, data on a complete birth registration basis were available for only a few years, during which there had been a steady and marked downward trend in birth rates. It is apparent now that these early demographic studies were unduly influenced by these data—they fell heir to what Davis calls "short-term trenditis." For the whole period—roughly thirty years—that demographic projections methods have been used, the results produced by them have been seriously greater in error than the forecasts produced by the earlier and less elegant mathematical procedures. One is tempted to conclude that the demographers were carried away by their methodology and by recent events, and lost their discriminating judgment in the process. Perhaps it is more accurate to say that the importance of various factors affecting birth rate have changed, and demographers have been unable to keep up with the changes, much less to explain them.

By the late 1920's and early 1930's, demographers and others, sharply impressed with the regular and relatively steep downward trend in birth rates, were almost without exception projecting—and forecasting—greatly reduced rates of population growth and a rela-

[6] R. Pearl and L. J. Reed, "On the Rate of Growth of the Population of the United States since 1790 and Its Mathematical Representation," *Proceedings, National Academy of Science,* Vol. 6, 1920.

tively early levelling off or even a cessation of population growth. This thinking was highly influential, if not dominant, in the agricultural field. When the social security program was under serious study in 1934, an essentially constant population was forecast for all periods up to 2000. The first studies for the National Resources Committee produced a wide range of estimates, the lowest of which were for a

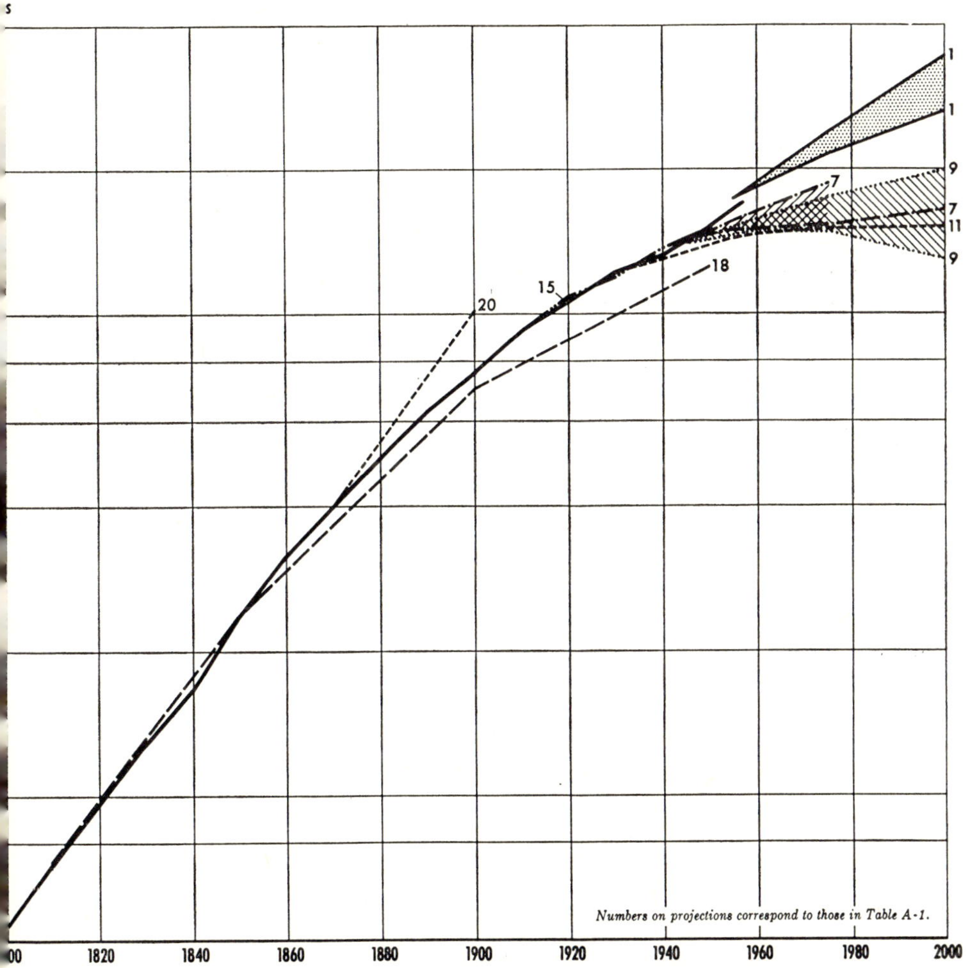

Figure A–1. United States population, 1800–1957, and various projections or forecasts.

material reduction in population by 2000 and the highest of which was substantially below what we have experienced to date. By 1943 these were raised somewhat, especially as to the low estimate, which now included some increase in all periods; but the high estimate was still well below experience to date.

Davis has traced the truly amazing projections of the Bureau of the Census during the 1940's.[7] In 1946 Whelpton made projections for the Bureau for 1950; although over half of the decade since 1940 was over, his estimate was 5 million too low, or underestimated the decade increase by more than 25 per cent. Even in 1949, the Census estimate for 1950 was nearly 1 million too low. The Census estimates of 1947 showed only a slight rise to 2000, for the low estimates; and a larger rise, but well below experience to date, for the high estimates.

Dorn has studied the same history and has made a careful and stimulating review of it.[8] He considers the early Malthusian approach, and how and why it has not applied to the United States. He takes up the logistic curves of Pearl and Reed and shows that, though they applied well during certain periods, they seem to have broken down for the later years. Then he traces in some detail the history of demographic approaches to the problem. He presents a tabular analysis of the various projections and forecasts (Table A-2). He points out that, even when demographers made estimates which time has shown to be accurate, such as Whelpton's estimate of 1928, they had so little confidence in their results that they quickly changed them—and, as it turned out, much for the worse.

Dorn concludes:

> You have heard the charge: I have reviewed the evidence. What is the verdict? I find the demographers guilty of
>
> 1. Giving the impression that projected populations were relatively inevitable and certain.
>
> 2. Underestimating the effects of scientific developments on lowering mortality rates.
>
> 3. Believing that the demographic development of a theoretical stable population must inherently characterize current demographic developments in an actual unstable population.

7 Davis, *op. cit.*

8 Harold F. Dorn, "Pitfalls in Population Forecasts and Projections," *Journal of the American Statistical Association*, Vol. 45, No. 251 (September 1950). His article gives an extensive bibliography.

4. Assuming the observed fertility and mortality of a single year or short period of years is a reliable guide to future trends.
5. Giving undue weight to recent downward cyclical fluctuations in fertility while making long range population projections.
6. Assuming that because birth rates have declined for several generations they must inevitably continue to decline.
7. Forgetting that the voluntary control of fertility can cause the birth rate to rise as well as to fall.
8. Consistently overestimating the rate of decrease in the growth of the population of the United States.
9. Being too uncritical of the work of fellow demographers.

Dorn himself believed that the rate of population growth in the United States is slowing down, but is slowing down more slowly than demographers up to 1950 had assumed to be the case. He pointed out that up to 1950 no conclusive evidence was shown that the size of completed families had increased permanently; all, or most of the rise in births up to that point, he felt, could be explained by the "catching up" on delayed marriages. He argued that a permanent rise in the average size of completed families would be a new experience in the demographic history of the United States. While he made no specific forecasts himself, it seems clear he envisaged a population of less than 200 million by 2000. His critique of the work of others may have been stronger than the suggestions he was able to make for the improvement of demographic projection.

Almost without exception, each later projection using demographic methods has been higher than the earlier ones. The earlier projections of a slowly increasing or stationary population were based heavily upon the experience of the late 1920's and early 1930's, and to some extent upon attitudes toward the future which were dominant then; the later revisions have been based, belatedly, upon the experience of the late 1940's and early 1950's, with their much higher birth rates. In each case, very recent experience (sometimes of only two to five years) has been the basis for revisions of long-term projections; this is probably what Davis had in mind by his reference to "short-term trenditis." Projections which must be drastically revised within a very few years of being made, and before any substantial part of the period to which they relate has passed, cannot command much confidence.

More recent (1955) estimates of the Bureau of the Census are much higher. Even these are based upon mortality rates now existing,

Table A–2. Enumerated and estimated population of the United States*

(in millions)

Year	Census[1]	Pearl[2]-Reed[2] I	Pearl[2]-Reed[2] II	Dublin[3] 1931 High	Dublin[3] 1931 Low	Scripps[4] 1928	Scripps[5] 1931	Scripps 1933[6] High	Scripps 1933[6] Low
						(Jan. 1)	(Jan. 1)		
1900	76.0	—	—	—	—	—	—	—	—
1910	92.0	—	91.4	—	—	—	—	—	—
1920	105.7	107.4	106.1	—	—	—	—	—	—
1930	122.8	122.4	120.1	—	—	123.6	—	—	—
1940	131.7	136.3	132.8	131.0	131.0	138.3	132.5	134.5	132.5
1941	133.2	—	—	—	—	—	—	—	—
1942	134.7	—	—	—	—	—	—	—	—
1943	136.5	—	—	—	—	—	—	—	—
1944	138.1	—	—	—	—	—	—	—	—
1945	139.6	—	—	—	—	—	—	—	—
1946	141.2	—	—	—	—	—	—	—	—
1947	144.0	—	—	—	—	—	—	—	—
1948	146.6	—	—	—	—	—	—	—	—
1949	149.2	—	—	—	—	—	—	—	—
1950	151.2	148.7	143.8	139.0	139.0	151.6	139.8	148.5	140.5
1960	(180)	159.2	153.0	147.0	146.0	162.7	143.9	—	—
1970	—	167.9	160.4	151.0	148.0	171.5	144.6	—	146.0
1980	—	174.9	166.3	154.0	148.0	—	142.9	190.0	145.0
1990	—	180.4	170.8	154.0	145.0	—	decline	—	decline
2000	—	184.7	174.3	152.0	140.0	186.0	there-after	—	there-after
		ultimate 197.3	ultimate 184.0						

* Except as noted in footnotes 1 and 10, this table is copied exactly, including numbering and placing of footnotes, from Harold F. Dorn, "Pitfalls in Population Forecasts and Projections," *Journal of the American Statistical Association*, Vol. 45, No. 251 (September 1950).

[1] Enumerated population, 1900-1940; calculated as of July 1, 1941-1949. (To these, present authors have added 1950 enumerated figure, and a rough estimate of 1960, the latter based on population in the fall of 1958 plus allowance for increases to 1960.)

[2] *Science*, 92, 486-488, 1940.

[3] *Problems of Population*, 1932, pp. 115-125.

[4] *The American Journal of Sociology*, 34, 253-270, 1928.

[5] *Problems of Population*, 1932, pp. 77-86.

[6] *Recent Social Trends*, p. 48.

Table A–2. Continued

Scripps 1935[7]			Scripps 1943[8]			Scripps 1947[9]			Census estimates[10] 1949
High	Medium	Low	High	Medium	Low	High	Medium	Low	
	(Apr. 1)						(July 1)		(July)
—	—	—	—	—	—	—	—	—	—
—	—	—	—	—	—	—	—	—	—
—	—	—	—	—	—	—	—	—	—
—	—	—	—	—	—	—	—	—	—
132.6	132.0	131.2	—	—	—	—	—	—	—
—	—	—	—	—	—	—	—	—	—
—	—	—	—	—	—	—	—	—	—
—	—	—	—	—	—	—	—	—	—
—	—	—	—	—	—	—	—	—	—
139.2	137.0	134.1	138.7	138.5	138.3	—	—	—	—
—	—	—	—	—	—	—	140.8	—	—
—	—	—	—	—	—	—	142.2	—	144.0
—	—	—	—	—	—	—	143.3	—	146.2
—	—	—	—	—	—	—	144.5	—	148.2
146.1	141.6	136.2	145.0	144.4	143.0	148.0	146.0	144.9	149.9
159.5	149.4	137.1	156.5	153.4	147.7	162.0	155.1	149.8	160.0
172.8	155.0	134.0	167.9	160.5	148.7	177.1	162.0	151.6	—
185.8	158.3	127.6	179.4	165.4	145.8	increase	increase	decline	—
—	decline	decline	189.4	167.1	138.9	there-	until	there-	—
—	there-	there-	198.7	166.6	129.1	after	about	after	—
	after	after		decline	decline		2000		
							then		
							decline		

[7] *Journal American Statistical Association,* 31, 457-473, 1936. (High is based on high fertility, low mortality, 200,000 net immigrants annually beginning 1940; medium is based on medium fertility and mortality with 100,000 net immigrants annually beginning 1940; low is based on low fertility, high mortality, no net immigration.)

[8] National Resources Planning Board, 1943, p. 29. (High is based on high fertility, low mortality, no net immigration; medium is based on medium fertility and mortality, 100,000 net immigrants annually, beginning 1945; low is based on low fertility, high mortality, no net immigration.)

[9] Bureau of the Census, 1947. (High is based on high fertility, low mortality, 200,000 net immigrants per year beginning 1945; medium is based on medium fertility and mortality, 100,000 net immigrants annually beginning 1950; low is based on low fertility, high mortality, no net immigration.)

[10] Bureau of the Census, Series P-25, No. 18, 1949. (The present authors have added the word "estimates" to the column heading.)

with no allowance for future reductions in mortality and consequent increase in survival rates. The highest projected figure is based upon 1954 and 1955 birth rates; no recognition is given in the figures to the possibility that future birth rates may be even higher. Their maximum estimate is not, therefore, truly a maximum; it is more nearly a projection of the present birth and death rates. The idea of an ultimately declining birth rate still dominates demographic thinking. The Division of the Actuary, in Acturial Report No. 46,[9] makes estimates using alternative mortality rates for the future, even the higher of which is below present mortality levels, and the lower, considerably lower. While it employs alternative fertility rates, its highest starts with present rates and trends downward slightly only after 2000; but it regards this as "very unlikely to eventuate." The higher one which it regards as probable starts with present rates and by 2000 is down to about the level of 1940. Its lower one starts at about the level of the early 1940's and trends downward to a rate in 2000 sufficient in the long run only to maintain the population.

9 *Illustrative United States Population Projections,* cited above.

Appendix B

Some general urban relationships

Reference was made in Chapter I to the complex relationships between a city and its economic hinterland and to the interrelationships among different parts of the city. In the discussion of that chapter, centered on urban land use, it seemed undesirable to consider those relationships at too great length. The present appendix presents somewhat more information on the subject—not a complete treatment, but one designed to provide a starting place for those students interested in following this subject in greater detail.

Rural-Urban Population Relationships

While the total population of the United States and of each major region has been growing over the decades, there has been a profound redistribution of people as between the country and the city. This population redistribution is a major factor affecting the demand for land for urban purposes.

The concept of "urban" has itself changed over the decades. At one time, it was fairly easy to tell what was town or city, and what was not. The only question was how small a grouping of people was to be considered a town. The Bureau of the Census has used varying bases at different times to define which part of the population of the United

States is urban, and which rural.[1] The first classification of national population into urban and rural territories was made in 1874, and then only as an incident to the preparation of maps showing population density in various parts of the country. The first definition of "urban" applied to cities of 8,000 or more persons; by 1910, the boundary line between urban and rural was placed at 2,500 persons per town or city—a number which has become relatively standard. Since that date the Bureau of the Census has prepared tabulations of numbers of persons in cities of different sizes, for all census dates.

In recent decades, however, a new complication of urban-rural definition has arisen. A substantial number of people now live outside the legal or political boundaries of any town or city, but often in relatively close settlement patterns which in all essential respects are more urban than rural. This presents complications in defining land use, and also makes enumeration and tabulation of population data difficult. The Bureau of the Census has used one definition for 1930 and 1940; for the 1950 census, data are tabulated according to the previous definition and also according to a new definition. The latter is more inclusive of urbanized areas, and includes larger acreages and more people, although it does exclude some areas that were previously classed as urban.[2]

The essential data on numbers of cities of different sizes and on numbers of people according to size of city and other residence location are given in Table B-1; data are given for 1950 on the basis of both the old and the new definitions of "urban." The same data are shown in simpler form in Figure B-1, where only the new definition for 1950 is used and where the data for 1940 and 1930 have been roughly adjusted to the basis of the new 1950 definition.

In 1790, when the United States was almost wholly rural, the first census showed only 5 per cent of the people living in towns of 2,500 or more. As late as 1840, nearly 90 per cent of the people were rural, and even in 1900, 60 per cent were still rural. Up to 1910, the numbers of rural people increased, although at a rate much slower than the rate of increase in urban population. In the United States birth rates, and hence the rate of natural increase in population, have always been

[1] Leon E. Truesell, *The Development of the Urban-Rural Classification in the United States: 1874 to 1949,* Current Population Reports—Population Characteristics, U. S. Bureau of Census, Series P-23, No. 1, August 1949.

[2] Henry D. Sheldon, "Changes in the Rural Population, 1940 to 1950," *Rural Sociology,* Vol. 17, No. 2 (June 1952).

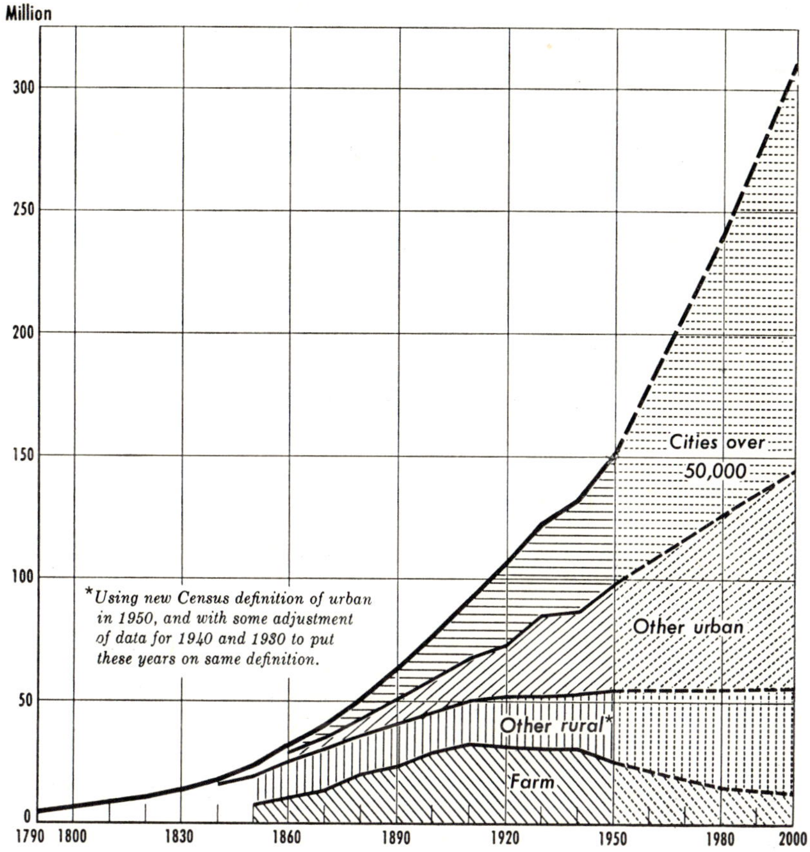

Figure B–1. Population of the United States by urban and rural groupings, 1790–1950, and projections to 2000.

higher in the country than in the city. The faster rate of increase in the cities is thus due in large measure to the heavy and continued migration of people from country to city, as well as to immigration from abroad. The growth of rural population almost ceased in 1910; at that time there were 50 million rural people and by 1950 (new definition) there were but 54 million—an increase of 8 per cent while urban population was more than doubling.

During these decades, also, change has taken place within the rural population. In 1910 (and apparently earlier, although the estimates

Table B–1. Number of urban places and population according to size of city and type of rural location 1790 to 1950, and projections for 1980 and 2000

Year	Urban population											Rural population					
	1,000,000 or more	500,000 to 1,000,000	250,000 to 500,000	100,000 to 250,000	50,000 to 100,000	25,000 to 50,000	10,000 to 25,000	5,000 to 10,000	2,500 to 5,000	Under 2,500	Total	1,000 to 2,500	Under 1,000	Total	Other rural popu-lation[1]	Farm popu-lation	Total
1. *Number of places*																	
1790	—	—	—	—	—	2	3	7	12	—	24	—	—	—	—	—	—
1800	—	—	—	—	1	2	3	15	12	—	33	—	—	—	—	—	—
1810	—	—	—	—	2	2	7	17	18	—	46	—	—	—	—	—	—
1820	—	—	—	1	2	2	8	22	26	—	61	—	—	—	—	—	—
1830	—	—	—	1	3	3	16	33	34	—	90	—	—	—	—	—	—
1840	—	—	1	2	2	7	25	48	46	—	131	—	—	—	—	—	—
1850	—	—	—	[2]6	4	16	36	85	89	—	236	—	—	—	—	—	—
1860	—	2	1	6	7	19	58	136	163	—	392	—	—	—	—	—	—
1870	—	—	—	[2]14	11	27	116	186	309	—	663	—	—	—	—	—	—
1880	1	3	4	12	15	42	146	249	467	—	939	—	—	—	—	—	—
1890	—	—	—	[2]28	30	66	230	340	654	—	1,348	1,603	4,887	6,490	—	—	—
1900	3	3	9	23	40	82	280	465	832	—	1,737	2,128	6,803	8,931	—	—	—
1910	3	5	11	31	59	119	369	605	1,060	—	2,262	2,717	9,113	11,830	—	—	—
1920	3	9	13	43	76	143	465	715	1,255	—	2,722	3,030	9,825	12,855	—	—	—
1930	5	8	24	56	98	185	606	851	1,332	—	3,165	3,087	10,346	13,433	—	—	—
1940	5	9	23	55	107	213	665	965	1,422	—	3,464	3,205	10,083	13,288	—	—	—
1950—old	5	13	23	66	128	271	814	1,133	1,570	—	4,023	3,408	9,827	13,235	—	—	—
1950—new	5	13	23	65	126	252	778	1,176	1,846	457	4,741	4,158	9,649	13,807	—	—	—
1980[3]	12	15	29	100	191	405	1,252	1,930	4,166	—	8,100	—	—	—	—	—	—
2000[3]	19	22	39	150	234	538	1,595	2,975	4,828	—	10,400	—	—	—	—	—	—

2. *Population* (1,00C)

1790	—	—	—	—	—	62	48	48	44	—	202	—	—	—	—	—	3,929
1800	—	—	—	—	61	68	54	94	45	—	322	—	—	—	—	—	5,308
1810	—	—	—	—	150	80	109	116	70	—	525	—	—	—	—	—	7,240
1820	—	—	—	124	127	70	122	155	96	—	693	—	—	—	—	—	9,638
1830	—	—	—	203	222	105	240	231	126	—	1,127	—	—	—	—	—	12,866
1840	—	—	313	205	187	235	405	329	172	—	1,845	—	—	—	—	—	17,069
1850	—	—	—	[2]1,175	284	611	561	596	316	—	3,544	—	—	—	—	[4]7,200	23,192
1860	—	1,379	267	993	452	670	884	976	595	—	6,217	—	—	—	—	[4]10,200	31,443
1870	—	—	—	[2]4,130	768	930	1,710	1,278	1,086	—	9,902	—	—	—	—	[4]13,300	39,818
1880	1,206	1,917	1,301	1,787	948	1,446	2,189	1,717	1,618	—	14,130	—	—	—	—	[4]20,000	50,156
1890	—	—	—	[2]9,698	2,028	2,269	3,451	2,384	2,277	—	22,106	2,509	2,249	4,758	13,284	[4]22,800	62,948
1900	6,429	1,465	2,861	3,272	2,709	2,801	4,338	3,204	2,899	—	30,160	3,298	3,004	6,302	10,833	[4]28,700	75,995
1910	8,501	3,011	3,950	4,840	4,179	4,023	5,549	4,217	3,728	—	41,400	4,234	3,931	8,165	10,330	32,077	91,972
1920	10,146	6,224	4,541	6,519	5,265	5,075	7,035	4,968	4,386	—	54,158	4,712	4,255	8,967	10,612	31,974	105,711
1930	15,065	5,764	7,956	7,541	6,491	6,426	9,097	5,897	4,718	—	68,955	4,821	4,363	9,184	14,107	30,529	122,775
1940	15,911	6,457	7,828	7,793	7,344	7,417	9,967	6,682	5,026	—	74,424	5,027	4,316	9,343	17,355	30,547	131,669
1950—old	17,404	9,187	8,242	9,614	9,073	9,496	12,467	7,879	5,565	—	88,927	5,383	[6]4,200	9,583	27,129	25,058	150,697
1950—new	17,404	9,187	8,242	9,479	8,931	8,808	11,867	8,139	6,490	578	[5]96,468	6,473	[6]4,100	10,573	18,598	25,058	150,697
1980[3]	55,000	12,000	12,000	19,000	16,000	16,000	23,000	16,000	16,000	—	185,000	—	—	40,000		15,000	240,000
2000[3]	90,000	17,000	15,000	26,000	18,000	20,000	29,000	22,000	18,000	—	255,000	—	—	43,000		12,000	310,000

[1] Calculated as difference between other enumerated items and total; ignores fact that some farm population lives in towns and cities.

[2] Including numbers of cities and population of all cities 100,000 and over.

[3] Projections of the authors. See p. 8 for projection of total population. Total number of cities estimated on basis of past trends in average population per city. Number of cities and population for cities of each size class estimated from rank-size distribution discussed in text.

[4] Estimated on basis of 5.0 persons per farm reported by Census.

[5] Including 7,344,026 persons in unincorporated urbanized areas, in addition to those in places of under 2,500.

[6] Estimates, based on numbers of rural places of this size.

SOURCE: U. S. Bureau of the Census. *Statistical Abstract, 1955* and *Historical Statistics of the United States, 1789-1945.*

may not be highly accurate), farm population was about three-fourths of all rural population. In recent years there has been a major decline in the numbers of farm people, while at the same time rural nonfarm population has been increasing. By 1940 about 47 per cent of the rural population was nonfarm, and by 1950, according to the new definition, over half. With the growth of good roads and the increased numbers of automobiles, more and more people live in the open country beyond the suburbs, and commute to work. At the same time, the need for labor in agriculture has been declining.

Urban groups, too, have changed in population distribution. In the early years the cities were relatively small, at least by modern standards. In 1790 there was no city with as many as 50,000 people; as late as 1830 there was none with as many as 250,000; the first city of a million inhabitants shows up in the 1880 census. The average size of city has risen greatly and more or less continuously (Figure B-2). In

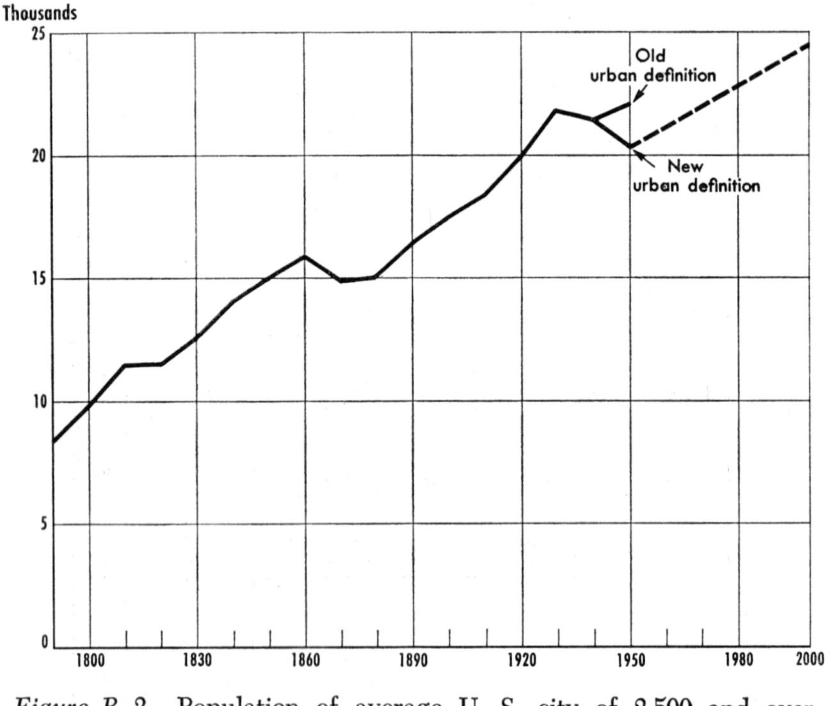

Figure B–2. Population of average U. S. city of 2,500 and over, 1790–1950, and projections to 1980 and 2000.

1790 and 1800 it was less than 10,000; in 1930 and later census periods, over 20,000. Although there has been a very great growth in numbers of cities, especially smaller ones, the increase in population of the larger cities has raised the average. Moreover, the available statistics somewhat understate the situation: they are based upon legal boundaries of cities, whereas in many cases cities that in earlier decades were independent and separate have now nearly grown together, so that the city as an economic entity is now often much larger than the city as a legal unit.

The prospects for the future are for a further urbanization. Farm population will continue to decline; rural nonfarm population will continue to rise moderately, about offsetting the decline in farm population. All the net increase will thus take place in urbanized areas, mostly within cities and towns or closely adjacent to them. The average size of city will continue to rise, to nearly 25,000 by the year 2000; since the amount of land in relation to population required for urban purposes varies with the average size of the city, this has an important effect upon land requirements. The increase in urban population will be greater, we assume, in the larger than in the smaller cities; in the future, as in the past, cities will "graduate" from small to larger sizes.

Population-Distance Relationships[3]

The foregoing discussion is too simple a description of what is essentially a complex set of interrelationships between cities of different sizes and between the people inhabiting them.

In recent years several scholars have attempted to measure and analyze the relationship between the relative size of various cities in the United States and the population each contains.[4] The charts

[3] For a recent article bearing on this and other subjects discussed in this chapter, including some excellent references, see Otis Dudley Duncan, "Population Distribution and Community Structure," *Cold Spring Harbor Symposia on Quantitative Biology*, Vol. 22, Long Island Biological Association, 1957.

[4] Isard, in *Location and Space-Economy* (New York: Technology Press of Massachusetts Institute of Technology and John Wiley and Sons, 1956), discusses and gives full references for various studies of this kind. He leans heavily on G. K. Zipf, *Human Behavior and the Principle of Least Effort* (Cambridge: Addison-Wesley Press, 1949). See also Rutledge Vining, "A Description of Certain Spatial Aspects of an Economic System," *Economic Development and Cultural Change*, Vol. 3, No. 2 (January 1955).

they have devised graphically demonstrate their methodology. If the major cities of the United States were ranked according to their population along the X-axis of a double log chart, and the population of each city shown on the Y-axis, an almost exactly straight line would appear from the dots for the individual cities. This would seem to suggest that the size of a city is in part a function of the size of other cities; this might be due to complex economic interrelationships among cities, or to the fact that the size of each is a function of the total population and economic structure of the nation. Moreover, the line would be at a 45-degree angle to each axis. This means that doubling the number of the city in the ranking from large to small results in halving its population. In 1950, for instance, changing the rank of a city from fifth to tenth place results in cutting its population from 2.4 to 1.2 million; changing the rank again to twentieth place has the effect of cutting the population from 1.2 to 0.6 million, and so on.

Similar calculations have been made for all censuses from 1790 to 1950. The result is a remarkably parallel set of lines indicating the constancy of this general relationship. The line for each successive census lies above and to the right of the previous census, because total population of the country rose. The size of the largest city, and of each successive city in the ranking, is larger at each later census; but the relation between the largest city and the next largest, between it and the next, and so on down the list, remained almost exactly constant. Moreover, the lines connecting the dots for the different cities each year are nearly straight, and the relationship grows more regular in recent years. The position of individual cities in the hierarchy of cities, or the position of the dot for individual cities at different census years, does not stay the same throughout the entire period. There is great stability in the *system* of cities in the United States, but there is equally great mobility of individual cities within this stable general system.

We have used this relationship in Table B-1 to estimate the future number and total population of cities of different sizes. Where there are now five cities with 1 million or more population, it may be estimated that by 1980 there will be twelve, and by the year 2000, nineteen. The total number of urban places of 2,500 or more population will increase to over 8,000 by 1980, and to over 10,000 by 2000. These calculations assume that the hierarchical relations between very large, large, medium, and small cities in the future will remain essentially as in the past.

Bogue has approached this general subject from another angle.[5] He classifies the entire population of the United States into sixty-seven metropolitan regions and hinterlands of these metropolitan centers, according to certain objective standards he has established. Accordingly, there are less than half as many Bogue metropolitan regions as there are SMA's; but every individual in the country, no matter how remotely he is located, is included within one of the metropolitan regions. Bogue then studies the relationships among cities and population groupings within each of these metropolitan regions. His results can be presented in double log charts similar to those described above. As Bogue warns, his summary charts imply a degree of regularity in the data which a detailed examination does not sustain; nevertheless, they do show central tendencies well and strikingly.

Each of the following five factors decreases rather regularly and rather uniformly as distance from the metropolitan center increases: population, value added by manufacture, retail sales, wholesale sales, and receipts from services. That is, as one moves away from the center of the metropolitan region, the number of persons decreases, and so does each of the measures of economic activity. As in the case of city rank and population, the relationship is closely approximated by a line at a 45-degree slope, indicating that doubling the distance results in halving the population or the economic activity.

Vining and Isard have approached this matter with data that is concerned particularly with volumes of products moved and distances of freight movement to and from different areas.[6] Although their analyses are in somewhat different terms, they come to essentially the same conclusions as those indicated above. The great bulk of freight moves comparatively short distances, but some commodities move much longer distances. After all, distance means cost, in movement of freight, although costs—and charges—do not usually increase proportionately with distance. If adequate supplies of a commodity may be had nearby, no additional costs for long-distance freight will be incurred. The contrary is true if adequate, or good-quality, or low-priced supplies are not available from nearby areas. Half of all commodities move less than about 250 miles, and 80 per cent move less than 800 miles. A bulky and low-value product like crushed stone, which is locally available in most parts of the United States, is moved

[5] Donald J. Bogue, *The Structure of the Metropolitan Community: A Study of Dominance and Subdominance* (Ann Arbor: University of Michigan Press, 1949).

[6] Vining, Isard, *op. cit.*

only short distances; products like oranges and grapefruit, on the other hand, limited as they are to definite areas of cultivation, are moved much greater distances. Some states, like California, ship their products over a wide area and also draw on a wider area for supplies than is the national average; other states, Rhode Island for example, ship their products much less than average distances.

These instances from the studies that have been made, although oversimplified, illustrate the common-sense fact that there is a complex interrelation between the economic activity at the center of the metropolis and within its trade territory, and between one area and another, and that in general the closeness of this interrelation diminishes with distance. The United States is a vast trade area with comparatively few barriers to free movement of goods—this is one source of its great material prosperity. But the costs of movement, in money and in time, tend to limit economic activity to more local areas. Social and political interrelationship are less amenable to quantitative and graphic expression.

Population Shifts within Cities

There are also complex interrelations within cities. As a city grows and develops over the decades, the population which first lived, throve, and increased near its center, gradually shifts residence to outlying areas as the city center becomes a focal point for commerce, trade, and perhaps industry. Some of the shift in population takes place within the boundaries of the city as a legal entity; more of the shift will be from the central city to satellite cities or suburbs, or to unincorporated territory—areas which are part of the city only in an economic sense. This shift from city center to city edge or suburb is a land use change which has attracted much attention in recent years. It involves major readjustments, economically, socially, and politically *within* the city itself.

The shift from the center outward within the city (as a legal entity) has been illustrated nicely by McKenzie.[7] Figure B-3 shows this for Cleveland, Boston, and St. Louis since 1830. The boundaries of each city have expanded over the years as population reached out into new areas. In the case of Cleveland, the city as it was defined in 1830

[7] R. D. McKenzie, *The Metropolitan Community* (New York: McGraw-Hill Book Co., 1933).

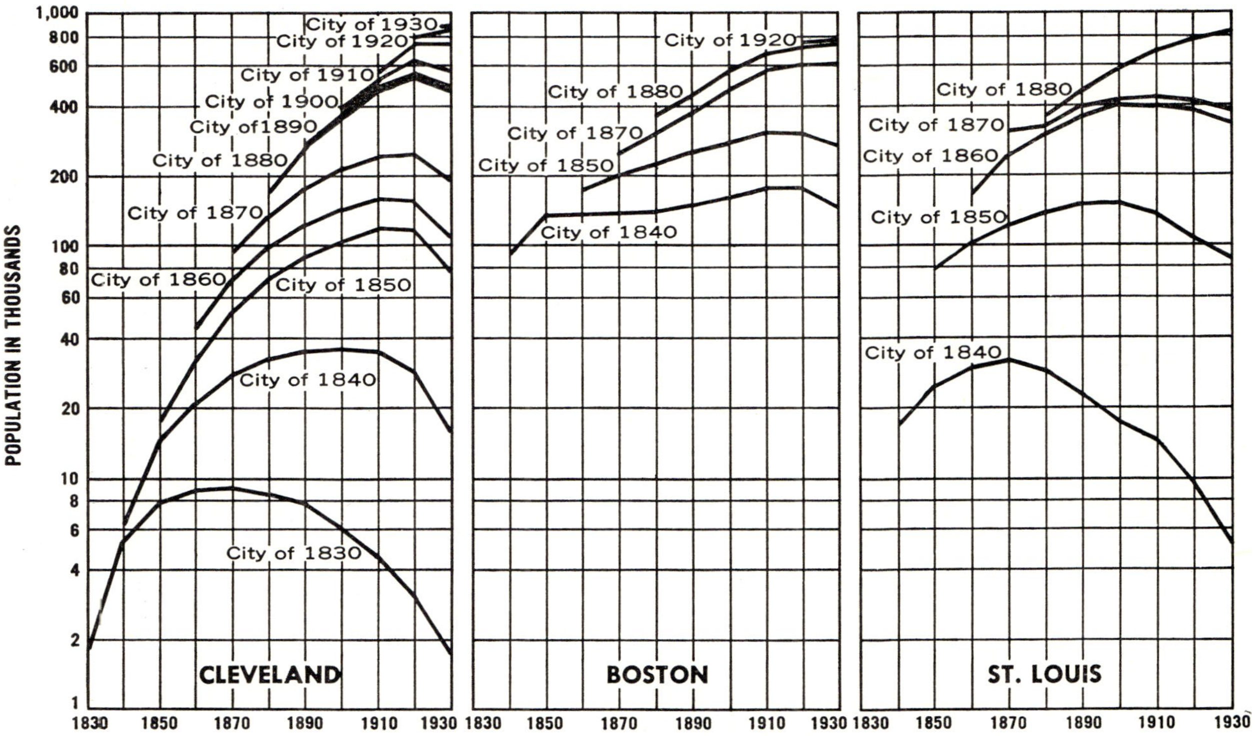

Figure B–3. Population of three major U. S. cities at different dates, according to varying city boundaries. (Chart from R. D. McKenzie, *The Metropolitan Community*, McGraw-Hill Book Co., copyright 1933.)

reached its maximum size in 1870, from which it declined so much that by 1930 the original area had no more residents than had lived there a hundred years before; the city was larger in area in 1840, and as such reached its peak in 1910, after which it sharply declined; even the city of 1910 shows some decline between 1920 and 1930. A similar but not quite so dramatic change has occurred in St. Louis. There, the city of 1840 reached its peak in 1870 and has since declined to a far lower point than in 1840; the larger city of 1850 has also declined from its peak. In Boston the changes have been less marked. The precise picture varies from city to city, but the general relationship is a common one—a decreasing number of residents near the center of the city as the years go by, but probably a rising daytime work population in these areas.

In addition to the changes within the boundaries of the city as a legal entity, there also have been changes in population distribution of the larger metropolitan areas composed of several cities. These have been analyzed in great detail for the 1900-1950 period by Hawley.[8] In general, the population of the metropolitan areas has grown at a rate faster than that for the entire United States, and much faster than that for other areas. From 1900 to 1920 while the central cities were growing somewhat faster than the satellites, there was a slight centralizing tendency. Since 1920, however, the satellites have grown much faster than the central cities, producing a strong degree of decentralization. In most cases the localities of very rapid growth have situated five to fifteen miles from the city center.

In considerable part, these more rapid rates of population increase in satellite than in central cities have been caused by a partial filling up of previously vacant land within cities and by the growth of industry and other land uses tending to displace residences. They are also attributable to a growing desire among urban people to live in lower density residential neighborhoods. The "flight to the suburbs" is a statistical fact. At the same time it is also true that the central cities have themselves grown in population, on the average at fairly high rates.

Political or Governmental Problems of Cities

This discussion of population in cities would not be complete without some brief mention of the political problems of the metropolitan

[8] Amos H. Hawley, *The Changing Shape of Metropolitan America: Deconcentration Since 1920* (Glencoe, Ill.: The Free Press, 1956).

areas. For the purposes of land use study, three major points should be made:

1. The metropolis does not exist as a single political entity, and hence cannot exercise the powers of government. In few cases in the United States does the central city of the metropolis have the necessary legal powers to deal with the economic, social, and political problems of the metropolis in the way that a self-contained city can and should do. Neither do the suburban or satellite cities have these powers, nor does any combination of central and satellite cities have them. Various legal entities (satellite cities) within the metropolitan complex seek to meet their particular problems by zoning or other actions which in effect merely shift the problem to some other unit within the metropolis.[9] The metropolis has grown up as a major economic structure, but political instruments to match the economic growth have not evolved; the problems and responsibilities of the population aggregate are not matched by powers and authorities of equal scope. In many metropolitan areas, some of the necessary functions of government for the whole area are carried on only with the greatest difficulty. Special districts for water supply, sewage disposal, police protection, parks, and other purposes have been created in many instances. The difficulties of creating and managing such districts are great; though some have done very well. Many of the functions of the metropolis which concern land use are carried out poorly.

2. In many metropolitan areas, the central city is becoming the residence of racial minorities and economically disadvantaged groups.[10] In the fourteen metropolitan areas each with more than a million population, white population of the central city increased only 4 per cent between 1940 and 1950, while negro population increased 68 per cent, whereas their rates of growth in the suburbs were nearly equal. There is every reason to believe that this disproportion in growth ratio has continued, and perhaps has been intensified, since 1950. The suburbs are often "protected" by various kinds of arrangements which keep negroes and other racial minorities out; and the higher level of housing and life generally in the suburbs necessarily results in costs which keep the lowest income groups of every race in

9 For an interesting discussion of this point see Charles M. Haar, "Wayne Township: Zoning for Whom?—In Brief Reply," *Harvard Law Review*, Vol. 617, No. 6 (April 1954).

10 Morton Grodzins, "Metropolitan Segregation," *Scientific American*, Vol. 197, No. 4 (October 1957).

the center of the city—often in its slums, or near-slums. The movement to the suburbs has been predominantly of upper and middle class whites. As the central city becomes more and more a special social class, its ability to govern itself may change, and certainly the willingness of the suburbs to have the central city take leadership in government for the whole metropolis will decline. As a result, the outlook for better metropolitan government is unfortunately not bright. In another thirty years, from 25 per cent to 50 per cent of the total population of most of the larger cities will probably be negro.

3. The suburb or satellite cities have their major problems also, which greatly affect land use. The typical suburban city continues to grow rapidly in population. Most such cities lack good political leadership, and the relationship between citizens and political leaders is often weak. If the suburban city were to carry out all the governmental functions it needs, far more revenue would be required than could be raised. An unusual proportion of children, as well as a growing population, necessitates many more schools; expansion into new areas requires more water lines and sewer lines, and more fire stations and police stations. Capital investment is needed, but the city's revenues are often too low to permit the economy of immediate investment. Most satellite cities lack an adequate tax base. Some forms of land use, such as parks, tend to be neglected because they are less urgent than other uses which command the available capital. The suburban or satellite city is often too weak, or too preoccupied with urgent current problems, to plan ahead for several years, and too weak to resist the pressures of "developers" who want to carry out land uses not in conformity with sound, long-range plans. Moreover, the interests of satellite cities are not necessarily those of the entire metropolitan community. A suburban city might solve some of its problems by forcing a through highway somewhere else, for instance; but what does this do to the metropolitan transportation system?

Appendix C

Classification of urban-like areas

There are four major kinds of urban-like areas in the United States for which different kinds and amounts of information are available (Table C-1). In popular discussion these different kinds of areas are not always clearly distinguished; on the contrary, names strictly applicable to one are sometimes applied erroneously to others. For instance, "city" has a definable meaning as a political entity within which considerable numbers of people live; but the term "city" is often applied to any fair-size aggregation of people, whether in fact a legal city exists or not. The various kinds of areas are not separate and distinct but, rather, overlap considerably. All of these terms are used by the Bureau of the Census, and some data are available for each type of area.

Standard metropolitan areas, or SMA's, each include a "county or group of contiguous counties which contain at least one city of 50,000 inhabitants or more." The purpose of the SMA is to provide a common unit of area for data tabulation and analysis, that will include the entire population in and around the city whose activities form an integrated social and economic system.[1] Counties contiguous to the

[1] For discussion of the history of this idea and some of its possible uses, see (1) Stuart A. Rice, *Statistical Programming for Problems of Urban Agglomerations* (Indianapolis: American Association for the Advancement of Science, December 27, 1957); and (2) Henry S. Shryock, Jr., "The Natural History of Standard Metropolitan Areas," *American Journal of Sociology,* Vol. 63, No. 2 (September 1957).

Table C–1. Comparison of standard metropolitan areas, urbanized areas, cities, and urban places, 1950

Item	Standard metropolitan areas[1]	Urbanized areas[1]	Incorporated cities[1]			Urban places over 2,500[1]		
			All[2]	Over 50,000 population	Under 50,000 population[2]	All	Over 50,000 population	Under 50,000 population
Total number of SMA's, areas, cities, urban places	168	157	3,872	232	3,640	4,284	232	4,052
Total population (1,000)	84,501	69,249	86,454	53,261	33,193	88,546	53,261	35,285
Average population per unit area	502,982	441,077	22,350	229,573	9,110	20,650	229,573	8,690
Total acreage in all unit areas (1,000 acres)	132,853	8,139	12,996	4,421	8,575	—	—	—
Average acreage per unit areas—acres	790,792	51,843	3,350	19,056	2,355	—	—	—
Average density of population—persons per sq. mile	407	5,438	4,255	7,590	2,475	—	—	—
in major or central city or cities	7,517	7,788	—	—	—	—	—	—
in urban fringe or outside of central city or cities	175	3,200	—	—	—	—	—	—

[1] See text for discussion of meaning of these terms.

[2] Includes an adjustment in area for cities not reporting area.

SOURCE: U. S. Bureau of the Census, *Census of Population: 1950,* Vol. II, *Characteristics of the Population,* Part I, *U. S. Summary.*

central county are included within the SMA if "essentially metropolitan in character and socially and economically integrated with the central city." Specifically, a contiguous county "must (*a*) contain 10,000 nonagricultural workers, or (*b*) contain 10 per cent of the nonagricultural workers working in the standard metropolitan area, or (*c*) have at least one-half of its population residing in minor civil divisions with a population density of 150 or more per square mile and contiguous to the central city"; and "nonagricultural workers must constitute at least two-thirds of the total number of employed persons of the county." In addition, the degree of economic and social integration of the county with the central city is tested by the following: "(1) Fifteen per cent or more of the workers residing in the contiguous county work in the county containing the largest city in the standard metropolitan area, or (2) Twenty-five per cent or more of the persons working in the contiguous county reside in the county containing the largest city in the standard metropolitan area, or (3) The number of telephone calls per month to the county containing the largest city of the standard metropolitan area from the contiguous county is four or more times the number of subscribers in the contiguous county." In New England, where data are compiled more on a city and town than on a county basis, "a population density criterion of 150 persons or more per square mile, or 100 persons or more per square mile where strong integration was evident," was used.[2]

Each SMA contains one principal central city of 50,000 or more. Any other city of 25,000 or more within a standard metropolitan area having a population amounting to one-third or more of the population of the principal city is also a central city. However, no more than three cities have been defined as central cities of any standard metropolitan area.

In 1950, there were 168 standard metropolitan areas within the United States. They included nearly 85 million people, or about 56 per cent of the total population of the country; they averaged slightly more than a half million persons each. They included nearly 133 million acres of land, or about 7 per cent of the total area of the country. Most of this land is rural, not urban, in use. For the entire average SMA, density of land use was relatively low—slightly more than 400 persons per square mile. This average, however, was made up of a comparatively densely settled central city or cities, with an average of

[2] U. S. Bureau of the Census, *Census of Population: 1950*, Vol. II, *Characteristics of the Population*, Part I, *U. S. Summary*, p. xii.

about 7,500 persons per square mile, and of a thinly settled hinterland, with an average of only 175 persons per square mile. Within each of these two broad categories, moreover, numerous cities or parts of cities exhibited a great range in density of settlement.

Urbanized areas each contain at least one city with 50,000 or more inhabitants; it also includes the "surrounding closely settled incorporated places and unincorporated areas" that comprise its urban fringe. In general, a minimum density of 500 dwelling units (roughly 2,000 persons) per square mile was used to measure urbanized areas outside of central cities. Both contiguous and noncontiguous areas, if the latter were within 1½ miles by shortest road of main contiguous urbanized part, were included. In addition, territory occupied by commercial, industrial, transportational, recreational, and other purposes functionally related to the central city were included. While each urbanized area has at least one central city of 50,000 persons or more, as many as two other cities of 25,000 or more are considered central cities, if each contains as much as one-third of the population of the largest city in the area.

It is evident that the central parts of the SMA's and of the urbanized areas are closely similar in character; and, in fact, they are mostly identical areas. The lists of the two differ somewhat because the urbanized area category had to be established before the actual field enumeration of the 1950 census, whereas the SMA was established on the basis of the data obtained from the census.

In 1950 there were 157 urbanized areas in the United States, on the basis of the foregoing definition; they included slightly more than 69 million people, or about 46 per cent of the total population of the country, or about 82 per cent of the population of the SMA's. They averaged about 440,000 persons each. They included slightly more than 8 million acres of land, or less than half of 1 per cent of the total land area of the nation and about 6 per cent of the area of SMA's. Much less than half of all the land used for urban purposes in the United States was in these urbanized areas. Their average density was more than ten times that of the SMA's; this was made up of a central part with substantially the same density as the central part of the SMA's, and a fringe area with a moderate urban density.

SMA's are based upon legal boundaries of units of local government—those of counties, for the most part, because it is only on a county basis that certain data are available for earlier decades. The urbanized areas have boundaries that are a combination of local governmental units—chiefly cities—and of physical features, such as streams, roads, and the like. Use of land within cities is largely but not wholly for urban purposes; those parts of urbanized areas that lie outside of cities were included because their use of land was urban in character, at least by certain criteria. Both SMA's and urbanized areas include the larger concentrations of people in the United States—those clustered around one or more cities of at least 50,000 people, including adjacent counties—and each excludes independent cities or other concentrations of people which do not include one city of 50,000 or more.

Cities, if incorporated, are units of local government whose boundaries are legally determined, sometimes on rather arbitrary bases, and often do not include all the contiguous area of essentially urban living pattern. They have the virtue of clear and definite existence and bounds, and the importance that their legal position and powers give them. In 1950 there were nearly 4,000 incorporated cities of over 2,500 population each; they had about the same total population as the SMA's—the latter excluded some small cities but included roughly an equal number of persons from outside of incorporated cities. They averaged slightly over 22,000 persons each, had nearly 13 million total acres, and an average density of more than 4,000 persons per square mile.

However, the average of all incorporated cities with 2,500 or more persons included a very wide range in size of city, from small country trading towns to New York City. There were 232 incorporated cities with 50,000 or more persons; thus, more than a third of all these larger cities were included in SMA's or urbanized areas which contained one or more other cities of this relatively large size. Nearly two-thirds of all the people living in incorporated cities of over 2,500 population lived in cities of over 50,000; but only about one-third of the land was in such larger cities. These larger cities had about the same average density as the central cities of the SMA's and urbanized places, because they were very largely the same cities. The very much larger number of incorporated cities between 2,500 and 50,000 in size averaged less than 10,000 persons each; they had an average density slightly less than 2,500 persons per square mile, or substantially lower than the fringe areas of the major urbanized areas. Their total land area was relatively great—twice that of the larger cities and 50 per cent above that of all urbanized areas.

Urban places, as this term is used by the census, includes all incorporated or unincorporated places of 2,500 inhabitants or more.

In 1950 there were nearly 4,300 such places, or about 400 more than the number of incorporated cities of the same size range. Over 50,000 population, all urban places were incorporated; the difference between numbers incorporated and total number of urban places grows larger as the size decreases.

Urban places are based partly upon legal definitions of local units of government, and partly upon actual type of land use. In addition to urban places of 2,500 and over, additional smaller urban places exist. When these are adjacent to larger incorporated or unincorporated urban places, they are included with the totals for urban population. But when they are not (and there are several thousand "places" or groupings of people, both incorporated and unincorporated, within this category) they are included with the rural population. Such small places occupy a substantial area of land in total—some authors estimate as much as 10 million acres.[3] However, as nearly as can be ascertained, this acreage is included in the statistics of land use for surrounding areas—forestry, grazing, agriculture, etc.

From the viewpoint of this study, data on urban places have one fatal defect—they do not include information on the area of land included in such places, as can be seen from Table C-1. Indeed, for the small urban place, unincorporated and often vaguely defined as to boundaries, it would be difficult to ascertain the acreage properly belonging to it. Whether or not it would be desirable to use urban places in analysis of urban land use, in order to get a more nearly totally inclusive figure, is thus not a pertinent question, because the lack of data makes it impossible to do so.

[3] H. H. Wooten and James R. Anderson, *Major Uses of Land in the United States—Summary for 1954,* Agriculture Information Bulletin No. 168, U. S. Department of Agriculture, January 1957.

Appendix D

Statistics

Table D–1. Population, developed area, and density, 53 selected central cities[1]

No.	City and state	Population	Developed area—acres	Density—persons per sq. mile[2]	Total area—acres
1	Atchison, Kansas	12,648	1,430	5,760	1,828
2	Bar Harbor, Maine	3,300	277	7,620	330
3	Baton Rouge, Louisiana	34,719	2,301	9,650	3,392
4	Battle Creek, Michigan	43,453	3,923	7,090	6,856
5	Brookhaven, Mississippi	8,220	919	5,730	1,383
6	Carlsbad, New Mexico	7,116	1,011	4,505	1,543
7	Centralia, Illinois	17,000	1,291	8,550	1,775
8	Frankfort, Kentucky	11,916	879	8,680	1,276
9	Freeport, Illinois	22,467	1,897	7,570	2,670
10	Glasgow, Kentucky	7,040	828	5,430	1,432
11	Jacksonville, Illinois	20,600	2,152	6,125	3,120
12	Jefferson City, Missouri	23,200	2,579	5,750	6,089
13	Kankakee, Illinois	22,241	2,035	7,040	2,560
14	Marshall, Michigan	5,740	1,119	3,280	2,324
15	Mason City, Iowa	25,500	3,012	5,420	7,995
16	Mexico, Missouri	12,500	1,887	4,230	4,287
17	Muskogee, Oklahoma	37,500	4,009	5,980	5,795
18	Naples, Florida	1,740	826	1,370	6,540
19	Petersburg, Virginia	30,631	2,534	7,370	3,608
20	Quincy, Illinois	42,750	3,298	8,280	3,795
21	Rock Island, Illinois	42,000	3,132	8,570	5,298
22	Roswell, New Mexico	20,400	3,036	4,310	4,292
23	Santa Fe, New Mexico	24,320	2,170	7,175	16,551
24	Streator, Illinois	17,268	1,812	6,080	2,192
25	Tuscola, Illinois	2,967	516	3,680	621
26	W. Palm Beach, Florida	44,000	4,097	6,870	9,018
27	Williamsburg, Virginia	3,500	1,130	1,985	1,830
28	Woodward, Oklahoma	6,270	837	4,800	1,095
29	Binghamton, New York	85,397	4,453	12,300	6,919
30	Davenport, Iowa	75,000	6,502	7,370	12,911

Table D-1. Continued

No.	City and state	Population	Developed area—acres	Density—persons per sq. mile[2]	Total area—acres
31	Decatur, Illinois	57,500	4,232	8,700	5,193
32	Greenville, South Carolina	57,932	5,536	6,700	9,994
33	Hamilton, Ohio	54,500	3,742	9,320	4,499
34	Lansing, Michigan	83,500	5,565	9,600	7,660
35	Lincoln, Nebraska	97,423	10,205	6,110	14,535
36	Portsmouth, Virginia	81,957	2,965	17,700	4,168
37	Racine, Wisconsin	72,000	3,916	11,780	5,698
38	St. Petersburg, Florida	60,812	10,933	3,560	33,522
39	Sioux Falls, South Dakota	51,000	5,793	5,630	7,931
40	Schenectady, New York	87,549	5,085	11,030	6,643
41	Topeka, Kansas	67,802	5,803	7,470	7,468
42	Corpus Christi, Texas	110,900	9,218	7,690	81,191
43	Des Moines, Iowa	153,000	17,797	5,500	35,530
44	Kansas City, Kansas	125,000	8,845	9,040	13,008
45	Oklahoma City, Oklahoma	226,000	13,839	10,450	16,852
46	Richmond, Virginia	193,042	14,989	8,240	25,431
47	Utica, New York	106,750	6,599	10,350	10,387
48	Wichita, Kansas	114,966	11,343	6,480	14,326
49	Dallas, Texas	338,000	26,111	8,260	32,163
50	Dayton, Ohio	255,474	14,415	11,350	17,337
51	Memphis, Tennessee	260,000	19,824	8,400	31,133
52	Newark, New Jersey	429,760	12,381	22,200	16,435
53	St. Louis, Missouri	821,960	33,441	15,720	39,836

[1] Data taken from Harland Bartholomew, *Land Uses in American Cities* (Cambridge: Harvard University Press, 1955) Table 1. Data are for various dates, mostly after 1940.

[2] Slide rule division.

Table D–2. Population, developed area, and density, 33 selected satellite cities[1]

No.	City and state	Population	Developed area—acres	Density—persons per sq. mile[2]	Total area—acres
1	Berkeley, Missouri	3,100	408	4,860	2,910
2	Bettendorf, Iowa	4,022	630	4,080	1,997
3	Glendale, Ohio	2,359	621	2,430	1,013
4	Lincolnwood, Illinois	2,450	802	1,955	1,742
5	Morton Grove, Illinois	3,150	1,250	1,615	2,624
6	Northfield, Illinois	900	530	1,087	1,246
7	Olivette, Missouri	1,130	380	1,905	1,251
8	Edwardsville, Illinois	8,008	940	5,450	1,415
9	Falls Church, Virginia	8,343	1,040	5,130	1,306
10	Ferguson, Missouri	7,299	1,290	3,620	1,951
11	Skokie, Illinois	9,700	2,768	2,245	6,528
12	West Vancouver, British Columbia	8,545	3,292	1,665	5,390
13	Wyoming, Ohio	5,231	768	4,360	1,057
14	Brentwood, Missouri	12,500	906	8,825	1,405
15	Clayton, Missouri	13,069	1,238	6,750	1,597
16	Highland Park, Illinois	15,300	3,848	2,550	7,780
17	Kirkwood, Missouri	11,389	2,069	3,520	4,183
18	La Grange, Illinois	10,000	913	7,020	1,240
19	Richmond Heights, Missouri	12,754	911	8,950	1,381
20	University Park, Texas	16,000	1,482	6,910	2,197
21	Webster Groves, Missouri	16,487	2,371	4,550	3,158
22	Wilmette, Illinois	14,900	2,106	4,525	3,408
23	Winnetka, Illinois	13,000	2,079	4,000	2,452
24	Beverly Hills, California	28,669	2,722	6,740	3,219
25	Bloomfield, New Jersey	44,165	2,724	10,390	3,456
26	East Chicago, Indiana	54,637	4,800	7,300	6,848
27	East Orange, New Jersey	68,945	2,265	19,500	2,516
28	East St. Louis, Illinois	74,347	4,989	9,540	7,751
29	Evanston, Illinois	70,900	4,032	11,260	4,870
30	Irvington, New Jersey	60,199	1,778	21,700	1,955
31	Maywood, Illinois	26,648	1,551	11,000	1,777
32	New Westminster, British Columbia	25,000	2,458	6,520	3,788
33	Oak Park, Illinois	70,372	2,901	15,520	3,006

[1] Data taken from Harland Bartholomew, *Land Uses in American Cities* (Cambridge: Harvard University Press, 1955), Table 5. Data are for various dates, mostly after 1940.

[2] Slide rule division.

Table D–3. Population, developed area, and density, central city and adjacent unincorporated area, 11 selected cities[1]

		Central city			Adjacent unincorporated area			Total urban area		
No.	Urban area and state	Population	Developed area—acres	Density—persons per sq. mile[2]	Population	Developed area—acres	Density—persons per sq. mile[2]	Population	Developed area—acres	Density—persons per sq. mile[2]
1	Battle Creek, Michigan	43,453	3,923	7,090	24,272	4,763	3,260	67,725	8,686	4,990
2	Brookhaven, Mississippi	8,220	919	5,730	2,600	655	2,540	10,820	1,574	4,395
3	Corpus Christi, Texas	110,900	9,218	7,690	8,925	3,061	1,865	119,825	12,279	6,245
4	Frankfort, Kentucky	11,916	879	8,680	4,619	603	4,900	16,535	1,482	7,140
5	Freeport, Illinois	22,467	1,897	7,570	2,170	603	2,305	24,637	2,500	6,310
6	Jacksonville, Illinois	20,600	2,152	6,125	2,000	872	1,470	22,600	3,024	4,780
7	Jefferson City, Missouri	23,200	2,579	5,750	2,800	2,077	863	26,000	4,656	3,570
8	Lincoln, Nebraska	97,423	10,205	6,110	6,202	8,270	495	103,625	18,475	3,585
9	Sioux Falls, South Dakota	51,000	5,793	5,630	4,000	1,314	1,950	55,000	7,107	4,950
10	Streator, Illinois	17,268	1,812	6,080	3,452	1,155	1,915	20,720	2,967	4,470
11	Williamsburg, Virginia	3,500	1,130	1,985	3,650	6,560	356	7,150	7,690	595
	Total	409,947	40,507	6,470	64,690	29,933	1,384	474,637	70,440	4,310

[1] Data taken from Harland Bartholomew, *Land Uses in American Cities* (Cambridge: Harvard University Press, 1955), tables in Appendix. Data are for various dates, mostly after 1940.

[2] Slide rule division.

Table D–4. Area and population of cities 2,500 and over, by states and regions, 1950 (Population in 1,000's; area in square miles)

State and geographic division	Cities of 2,500 to 5,000					Cities of 5,000 to 10,000					Cities of 10,000 to 25,000					Cities of 25,000 to 50,000		
	Total		Reporting area			Total		Reporting area			Total		Reporting area					
	No.	Population	No.	Area	Population	No.	Population	No.	Area	Population	No.	Population	No.	Area	Population	No.	Area	Population
Maine	4	15	4	120	15	6	47	6	197	47	8	138	8	236	138	2	67	73
New Hampshire	—	—	—	—	—	2	14	2	38	14	7	108	7	244	108	2	95	63
Vermont	6	22	5	10	17	6	44	5	23	36	2	29	2	12	29	1	11	33
Massachusetts	—	—	—	—	—	—	—	—	—	—	9	183	9	204	183	13	234	500
Rhode Island	—	—	—	—	—	—	—	—	—	—	1	24	1	1	24	2	40	81
Connecticut	4	14	4	5	14	4	32	3	9	23	8	130	8	75	130	6	160	218
New England	14	51	13	135	46	18	137	16	267	120	35	612	35	772	612	26	607	968
New York	85	318	80	144	298	51	387	50	119	379	44	731	44	209	731	15	167	521
New Jersey	67	250	67	198	250	56	408	56	179	408	42	636	42	186	636	14	78	507
Pennsylvania	139	489	132	164	466	127	855	115	222	774	66	1,037	60	144	939	11	50	365
Middle Atlantic	291	1,057	279	506	1,014	234	1,650	221	520	1,561	152	2,404	146	539	2,306	40	295	1,393
Ohio	77	281	73	236	269	59	414	53	166	375	44	622	37	142	534	19	141	655
Indiana	39	138	32	48	109	38	246	35	71	230	20	285	15	55	220	10	65	346
Illinois	95	333	84	138	292	78	556	65	180	456	46	685	38	150	585	14	98	501
Michigan	47	165	45	109	158	37	257	37	148	256	35	554	35	209	554	10	65	413
Wisconsin	48	167	46	112	158	23	157	23	86	157	17	240	16	91	225	12	119	430
East North Central	306	1,084	280	643	986	235	1,630	213	651	1,474	162	2,386	141	647	2,118	65	488	2,345
Minnesota	38	132	36	87	125	31	216	28	136	199	16	239	16	121	239	3	23	83
Iowa	42	152	39	97	139	32	180	30	74	169	10	152	9	45	130	8	79	270
Missouri	52	184	50	86	177	28	189	24	62	159	19	285	16	68	238	5	49	173
North Dakota	2	8	2	2	8	6	41	6	14	41	3	51	3	21	51	2	9	65
South Dakota	12	37	10	11	30	7	43	7	24	43	4	59	4	22	59	1	12	25
Nebraska	21	75	16	23	58	9	61	9	18	61	8	121	6	24	94	—	—	—
Kansas	33	117	32	46	114	10	72	9	21	64	20	279	18	63	254	2	14	62
West North Central	200	705	185	352	651	123	802	113	349	736	80	1,186	72	364	1,065	21	186	678

Delaware	3	9	3	4	9	5	29	5	9	29	—	—	—	—	—	—	—	—
Maryland	9	44	7	9	37	7	46	5	14	33	8	101	7	18	90	2	15	74
Dist. of Columbia	—	—	—	—	—	—	—	—	—	—	—	—	—	—	—	—	—	—
Virginia	23	79	22	37	75	19	122	18	49	117	10	135	10	47	135	5	38	186
West Virginia	26	86	23	19	74	15	107	15	28	107	6	112	4	26	80	4	21	117
North Carolina	35	127	33	57	119	23	159	20	52	139	24	320	20	80	268	4	33	148
South Carolina	24	94	23	58	91	22	147	22	76	147	6	116	6	32	116	1	7	37
Georgia	56	192	49	127	170	21	153	21	77	153	14	224	9	41	148	4	39	114
Florida	42	154	39	134	143	23	158	23	111	158	13	188	11	99	163	10	95	335
South Atlantic	218	785	199	445	718	135	921	129	416	883	81	1,196	67	343	1,000	30	248	1,011
Kentucky	33	120	27	33	99	20	141	17	40	122	8	106	7	27	95	4	24	129
Tennessee	26	96	20	31	71	23	161	16	40	109	9	124	7	28	100	2	12	58
Alabama	30	97	28	79	90	26	176	20	100	138	13	93	13	222	93	3	24	106
Mississippi	25	90	22	49	78	13	92	10	31	66	8	215	7	36	192	6	50	191
East South Central	114	403	97	192	338	82	570	63	211	435	38	538	34	313	480	15	110	484
Arkansas	30	105	26	56	88	20	140	16	44	112	7	111	6	30	95	4	54	158
Louisiana	27	99	24	52	88	22	160	19	59	137	10	133	7	31	93	4	47	149
Oklahoma	38	123	31	45	103	23	148	12	24	79	18	271	14	45	213	4	29	135
Texas	109	378	102	206	352	78	542	67	219	463	50	806	42	229	698	5	44	175
West South Central	204	705	183	359	631	143	990	114	346	791	85	1,321	69	335	1,099	17	174	617
Montana	12	40	11	10	36	6	47	4	8	31	4	63	4	18	63	3	18	104
Idaho	18	63	18	27	63	2	11	2	4	11	7	99	5	20	69	2	11	60
Wyoming	13	44	11	16	37	1	7	—	—	—	4	62	4	18	62	1	7	32
Colorado	15	53	15	17	53	10	67	9	16	58	8	116	7	22	110	1	9	45
New Mexico	7	27	5	41	19	10	75	8	25	60	4	62	3	20	44	2	36	54
Arizona	10	37	10	15	37	9	68	7	13	55	1	17	1	6	17	1	10	46
Utah	15	53	10	31	36	11	75	8	53	56	1	17	1	6	17	1	18	29
Nevada	4	13	4	8	13	2	14	2	7	14	1	25	1	14	25	1	7	32
Mountain	94	330	84	165	294	51	364	40	126	285	30	461	26	124	407	12	116	402
Washington	24	80	23	35	76	11	71	10	29	65	12	174	12	59	174	5	48	176
Oregon	23	80	20	30	70	13	93	13	39	93	8	106	7	24	90	2	15	79
California	61	222	56	105	205	44	307	43	139	299	70	1,061	69	333	1,049	16	152	558
Pacific	108	382	99	170	351	68	471	66	207	457	90	1,341	88	416	1,313	23	215	813
UNITED STATES	1,549	5,502	1,419	2,967	5,029	1,089	7,535	975	3,093	6,742	753	11,445	678	3,853	10,400	249	2,439	8,711

SOURCE: U. S. Bureau of the Census, *Land Area and Population of Incorporated Places of 2,500 or more: April 1, 1950*, Series Geo. No. 5, January 1953.

Table D–4. Continued

State and geographic division	Cities of 50,000 to 100,000			Cities of 100,000 to 250,000			Cities of 250,000 to 500,000			Cities of 500,000 to 1,000,000			Cities of 1,000,000 or more			All cities: Total		All cities: Cities reporting area		
	No.	Area	Population	No.	Area	Population	No.	Area	Population	No.	Area	Population	No.	Area	Population	No.	Population	No.	Area	Population
Maine	1	22	78	—	—	—	—	—	—	—	—	—	—	—	—	21	351	21	642	351
New Hampshire	1	32	83	—	—	—	—	—	—	—	—	—	—	—	—	12	268	12	409	268
Vermont	—	—	—	—	—	—	—	—	—	—	—	—	—	—	—	15	128	13	56	115
Massachusetts	10	160	743	6	132	809	—	—	—	1	48	801	—	—	—	39	3,036	39	778	3,036
Rhode Island	3	46	186	1	18	249	—	—	—	—	—	—	—	—	—	7	540	7	105	540
Connecticut	2	51	148	4	78	604	—	—	—	—	—	—	—	—	—	28	1,146	27	378	1,137
New England	17	311	1,238	11	228	1,662	—	—	—	1	48	801	—	—	—	122	5,469	119	2,368	5,447
New York	6	56	468	4	77	611	1	36	332	1	39	580	1	315	7,892	208	11,840	202	1,162	11,812
New Jersey	8	41	507	4	36	505	2	37	738	—	—	—	—	—	—	193	3,551	193	755	3,551
Pennsylvania	9	64	615	4	68	473	—	—	—	1	54	677	1	127	2,072	358	6,583	333	893	6,381
Middle Atlantic	23	161	1,590	12	181	1,589	3	73	1,070	2	93	1,257	2	442	9,964	759	21,974	728	2,810	21,744
Ohio	6	52	365	3	72	529	3	131	955	2	150	1,419	—	—	—	213	5,240	196	1,090	5,101
Indiana	4	56	264	4	99	513	1	55	427	—	—	—	—	—	—	116	2,219	101	449	2,109
Illinois	10	85	682	1	13	112	—	—	—	—	—	—	1	208	3,621	245	6,490	213	872	6,249
Michigan	7	104	516	2	53	340	—	—	—	—	—	—	1	140	1,850	139	4,095	137	828	4,087
Wisconsin	4	46	274	—	—	—	—	—	—	1	50	637	—	—	—	105	1,905	102	504	1,881
East North Central	31	343	2,101	10	237	1,494	4	186	1,382	3	200	2,056	2	348	5,471	818	19,949	749	3,743	19,427
Minnesota	—	—	—	1	62	106	1	52	311	1	54	522	—	—	—	91	1,609	86	535	1,585
Iowa	4	120	296	1	55	178	—	—	—	—	—	—	—	—	—	97	1,228	91	470	1,182
Missouri	2	28	146	—	—	—	1	81	457	1	61	857	—	—	—	108	2,291	99	435	2,207
North Dakota	1	19	88	—	—	—	—	—	—	—	—	—	—	—	—	14	253	14	65	253
South Dakota	1	13	53	—	—	—	—	—	—	—	—	—	—	—	—	25	217	23	82	210
Nebraska	1	24	99	—	—	—	1	41	251	—	—	—	—	—	—	40	607	33	130	563
Kansas	1	12	79	2	44	298	—	—	—	—	—	—	—	—	—	68	907	64	200	871
West North Central	10	216	761	4	161	582	3	174	1,019	2	115	1,379	—	—	—	443	7,112	410	1,917	6,871

Table D–12. Railroad mileage, passenger movement, and freight hauled, 1890-1953

Year[1]	Railroad mileage (1,000) Road owned	Track operated	Passenger miles (billion)	Freight ton miles (billion)
1890	164	200	11.8	76
1891	168	207	12.8	81
1892	172	211	13.4	88
1893	176	222	14.2	94
1894	179	230	14.3	80
1895	181	233	12.2	85
1896	183	239	13.0	95
1897	184	242	12.3	95
1898	186	245	13.4	114
1899	189	250	14.6	124
1900	193	259	16.0	142
1901	197	265	17.4	147
1902	202	283	19.7	157
1903	208	284	20.9	173
1904	214	297	21.9	175
1905	218	307	23.8	186
1906	224	317	25.2	216
1907	230	328	27.7	237
1908	233	334	29.1	218
1909	237	342	29.1	282
1910	240	352	32.3	255
1911	244	363	33.2	254
1912	247	371	33.1	264
1913	250	380	34.7	302
1914	252	387	35.4	289
1915	254	391	32.5	277
1916	254	397	35.2	366
1917	254	400	40.1	398
1918	254	402	43.2	409
1919	253	404	46.8	367
1920	253	407	47.4	414
1921	251	408	37.7	310
1922	250	409	35.8	342
1923	250	413	38.3	416
1924	250	415	36.4	392

Table D–12. Continued

Year[1]	Railroad mileage (1,000) Road owned	Railroad mileage (1,000) Track operated	Passenger miles (billion)	Freight ton miles (billion)
1925	249	418	36.2	417
1926	249	421	35.7	447
1927	249	425	33.8	432
1928	249	428	31.7	436
1929	249	429	31.2	450
1930	249	430	26.9	386
1931	249	430	21.9	311
1932	248	428	17.0	235
1933	246	426	16.4	251
1934	244	422	18.1	270
1935	242	419	18.5	284
1936	240	416	22.5	341
1937	239	415	24.7	363
1938	237	411	21.7	292
1939	235	408	22.7	333
1940	234	406	23.8	375
1941	232	404	29.4	478
1942	229	400	53.7	641
1943	228	399	87.9	730
1944	227	398	95.7	741
1945	227	398	91.8	684
1946	226	398	64.8	595
1947	226	397	46.0	658
1948	225	397	41.2	641
1949	225	397	35.1	529
1950	224	396	31.8	592
1951	223	396	34.6	650
1952	223	395	34.7	618
1953	222	394	32.3	609

[1] Year ending June 30, 1890 to 1915 inclusive; calendar years thereafter.

SOURCE: U. S. Bureau of the Census, *Historical Statistics of the United States, 1789-1945*, and *Statistical Abstract of the United States, 1955.*

Table D–13. Mileage of roads outside of cities, by type, and estimated area of rights of way, 1904-1955

Year	Mileage of rural roads (1,000): Primary state highways	Secondary state highways	County roads under state control	County and local roads[1]	Total	Surfaced roads: Mileage 1,000	% total	Estimated area in rights of way[2] 1,000 acres
1904	—	—	—	—	2,151	154	7.2	9,650
1909	—	—	—	—	2,200	190	8.6	9,860
1914	—	—	—	—	2,446	257	10.5	10,970
1921	203	—	—	2,722	2,925	387	13.2	13,105
1922	227	—	—	2,733	2,960	412	13.9	13,411
1923	252	—	—	2,744	2,996	439	14.7	13,728
1924	261	—	—	2,743	3,004	472	15.7	13,822
1925	275	—	—	2,731	3,006	521	17.4	13,927
1926	288	—	—	2,712	3,000	550	18.3	13,983
1927	293	—	—	2,720	3,013	589	19.6	14,080
1928	306	—	—	2,710	3,016	626	20.8	14,182
1929	314	—	—	2,710	3,024	662	21.9	14,269
1930	324	—	—	2,685	3,009	694	23.1	14,278
1931	329	—	45	2,662	3,036	830	27.3	14,533
1932	358	—	84	2,598	3,040	879	28.9	14,849
1933	346	22	113	2,548	3,029	914	30.2	14,857
1934	325	57	113	2,539	3,034	976	32.1	14,832
1935	332	58	115	2,527	3,032	1,063	35.0	14,894
1936	340	62	115	2,470	2,987	1,156	38.7	14,786
1937	327	75	114	2,444	2,960	1,210	40.8	14,636
1938	327	79	115	2,447	2,968	1,252	42.2	14,688
1939	328	80	114	2,458	2,980	1,291	43.3	14,743

Table D–13. Continued

Year	Mileage of rural roads (1,000)					Surfaced roads		Estimated area in rights of way[2]
	Primary state highways	Secondary state highways	County roads under state control	County and local roads[1]	Total	Mileage 1,000	% total	1,000 acres
1940	329	81	114	2,466	2,990	1,340	44.8	14,797
1941	332	82	114	2,477	3,005	1,383	46.0	14,887
1942	334	84	115	2,471	3,004	1,405	46.8	14,851
1943	333	83	117	2,472	3,005	1,420	47.3	14,898
1944	335	83	117	2,470	3,005	1,429	47.5	14,912
1945	339	84	118	2,471	3,012	1,495	49.6	14,975
1946	342	86	119	2,462	3,009	1,503	49.9	14,980
1947	337	92	120	2,461	3,010	1,543	51.3	14,927
1948	350	85	121	2,451	3,007	1,573	52.3	15,043
1949	358	85	121	2,448	3,012	1,624	53.9	15,196
1950	363	88	122	2,430	3,003	1,699	56.5	15,077
1951	367	92	125	2,423	3,007	1,740	57.9	15,114
1952	371	93	126	2,420	3,010	1,797	59.7	15,172
1953	377	87	127	2,433	3,024	1,865	61.6	15,668
1954	379	89	129	2,446	3,043	1,917	63.0	15,314
1955	387	91	131	2,448	3,057	1,953	63.9	15,456

[1] Including other mileage not on state or county systems.

[2] On basis of average widths as follows: 90 feet for primary state highways, 66 feet for secondary state highways, 50 feet for county roads under state control, and 33 feet for county and local roads. These estimates were provided by letter from Bureau of Public Roads, April 2, 1956.

SOURCE: Bureau of Public Roads, U. S. Department of Commerce, *Highway Statistics, Summary to 1955*, 1957, Table M-200, p. 78.

Table D–14. Number and estimated area of airports and landing fields, and air traffic, 1927-1954

Year	Airports and landing fields in operation			Estimated acreage in airports[1] 1,000 acres	Miles flown (million)			Revenue passenger miles, scheduled airlines (million)
	Public	Limited	Total		Scheduled revenue	Other civil	Total	
1927	—	—	1,036	207	5.9	30.0	35.9	—
1928	—	—	1,364	290	10.5	60.0	70.5	—
1929	—	—	1,550	347	22.7	110.0	132.7	—
1930	—	—	1,782	421	32.6	108.3	140.9	85
1931	—	—	2,093	520	43.1	94.3	137.4	107
1932	—	—	2,117	550	45.9	78.2	124.1	127
1933	—	—	2,188	595	49.3	71.2	120.5	175
1934	—	—	2,297	652	41.5	75.6	117.1	190
1935	—	—	2,368	702	55.9	84.8	140.7	316
1936	—	—	2,342	722	64.3	93.3	157.6	439
1937	—	—	2,299	735	66.8	103.2	170.0	412
1938	—	—	2,374	790	68.6	129.4	198.0	480
1939	—	—	2,280	784	82.9	177.9	260.8	683
1940	—	—	2,331	830	110.1	264.0	374.1	1,052
1941	—	—	2,484	870	134.4	346.3	480.7	1,385
1942	—	—	2,809	970	111.3	293.6	404.9	1,418
1943	—	—	2,769	941	105.4	—	—	1,634
1944	—	—	3,427	1,131	138.7	—	—	2,178
1945	—	—	4,026	1,280	209.0	—	—	3,362
1946	—	—	4,490	1,365	309.9	874.7	1,185.6	5,948
1947	—	—	5,759	1,675	325.1	1,502.4	1,827.5	6,110
1948	—	—	6,414	1,785	338.2	1,469.5	1,807.7	5,981
1949	—	—	6,484	1,720	351.6	1,129.0	1,480.6	6,753
1950	—	—	6,403	1,614	364.3	—	—	8,003
1951	—	—	6,237	1,490	406.1	994.8	1,400.9	10,566
1952	—	—	6,042	1,365	458.6	972.1	1,430.7	12,528
1953	2,903	3,857	6,760	1,440	518.5	1,045.3	1,563.8	14,760
1954	2,783	4,194	6,977	1,326	550.6	—	—	16,769
1955	—	—	—	—	—	—	—	—

[1] Data on area in airports in 1945 and 1954 provided by Civil Aeronautics Administration, based on tabulations made by Production Economics Branch, Agricultural Research Service, U. S. Department of Agriculture. Estimates for other years made by authors.

SOURCE: U. S. Bureau of the Census, *Historical Statistics of the United States, 1789-1945,* and *Statistical Abstract of the United States, 1955.*

Index

NOTE: A major part of the items listed in this Index are grouped under five key headings, representing the five main land-use subject-areas in the book, namely Agriculture, Forestry, Grazing, Recreation, and Urban uses.